# BlackBerry®
## ALL-IN-ONE
## FOR
# DUMMIES®

# BlackBerry® ALL-IN-ONE FOR DUMMIES®

by Tim Calabro, Robert Kao, William Petz, and Dante Sarigumba

Wiley Publishing, Inc.

**BlackBerry® All-in-One For Dummies®**

Published by
**Wiley Publishing, Inc.**
111 River Street
Hoboken, NJ 07030-5774

www.wiley.com

Copyright © 2010 by Wiley Publishing, Inc., Indianapolis, Indiana

Published by Wiley Publishing, Inc., Indianapolis, Indiana

Published simultaneously in Canada

For general information on our other products and services, please contact our Customer Care Department within the U.S. at 877-762-2974, outside the U.S. at 317-572-3993, or fax 317-572-4002.

For technical support, please visit www.wiley.com/techsupport.

Wiley also publishes its books in a variety of electronic formats. Some content that appears in print may not be available in electronic books.

Library of Congress Control Number: 2010935571

ISBN: 978-0-470-53120-4

Manufactured in the United States of America

10 9 8 7 6 5 4 3 2 1

WILEY

# About the Authors

**Tim Calabro** has almost two decades of experience in the field of Information Technology. Mobile communications has been his primary focus for the past several years with specific attention devoted to BlackBerry. While managing the mobile communications support team for a major investment bank, Tim co-developed patent-pending BlackBerry Enterprise Server monitoring software and worked with several large mobile application vendors to design and enhance their products. Tim also holds the distinction of being one of the world's first BlackBerry Certified Handheld Support Specialists and BlackBerry Certified System Administrators.

**Robert Kao** is one well-rounded professional. His ability to translate his technical knowledge and communicate with users of all types led him to co-write *BlackBerry For Dummies, BlackBerry Pearl For Dummies, BlackBerry Storm For Dummies, BlackBerry Bold For Dummies,* and *BlackBerry Curve For Dummies.* He started out as a BlackBerry developer for various financial firms in New York City, that truly global city. Kao is currently the founder of a mobile software start-up. A graduate of Columbia University with a Computer Engineering degree, he currently lives in South Brunswick, New Jersey.

**William Petz** has been a mobility manager at five major investment banks. His responsibilities included global supervision of mobile support, engineering, and provisioning. William was one of the first ever to obtain Research in Motion's BlackBerry Certified Support Specialist and BlackBerry Certified Server Support Specialist certifications. During his career he has co-developed a patent-pending BlackBerry Enterprise Server monitoring application as well as other innovative mobile tools. William is also the founder of Random Events (`RandomEventsNY.com`), an adventure and activity company in New York City.

**Dante Sarigumba** is a long-time user of BlackBerry and a gizmo enthusiast. He is a co-host of the Mobile Computing Authority biweekly podcast. He works for a major investment bank in New York as a software developer and lives in South Brunswick, New Jersey, with his wife, Rosemarie, and two sons, Dean and Drew.

# Dedications

In dedication to my family: A mother and father who have always looked out for me. Siblings that have been supportive throughout my life. Especially a brother that I was always able turn to. I miss you dearly; may you rest in peace.

William Petz

To my father: Your wisdom and advice helped me become the person I am and still guide me to this day. Not a moment passes that you aren't missed. To my son Matthew: You brighten my days and inspire me to new heights. Watching you discover the world reminds me to never lose my desire for learning new things.

Tim Calabro

## Publisher's Acknowledgments

We're proud of this book; please send us your comments at http://dummies.custhelp.com. For other comments, please contact our Customer Care Department within the U.S. at 877-762-2974, outside the U.S. at 317-572-3993, or fax 317-572-4002.

Some of the people who helped bring this book to market include the following:

### Acquisitions and Editorial

**Project Editor:** Rebecca Senninger

**Acquisitions Editor:** Katie Mohr

**Copy Editor:** Heidi Unger

**Technical Editor:** Richard Evers

**Editorial Manager:** Leah Cameron

**Editorial Assistant:** Amanda Graham

**Sr. Editorial Assistant:** Cherie Case

**Cartoons:** Rich Tennant (www.the5thwave.com)

### Composition Services

**Project Coordinator:** Patrick Redmond

**Layout and Graphics:** Samantha K. Cherolis, Kathie Rickard, Christin Swinford

**Proofreaders:** Melissa Cossell, Cynthia Fields, Jessica Kramer

**Indexer:** Steve Rath

### Special Help

Nicole Haims Trevor

---

### Publishing and Editorial for Technology Dummies

**Richard Swadley,** Vice President and Executive Group Publisher

**Andy Cummings,** Vice President and Publisher

**Mary Bednarek,** Executive Acquisitions Director

**Mary C. Corder,** Editorial Director

### Publishing for Consumer Dummies

**Diane Graves Steele,** Vice President and Publisher

### Composition Services

**Debbie Stailey,** Director of Composition Services

# Contents at a Glance

# Table of Contents

# Introduction

*H*i there, and welcome to *BlackBerry All-in-One For Dummies.* If you already have a BlackBerry, this is a great book to have around when you want to discover new features or need something to use as a reference. If you don't have a BlackBerry yet and have some basic questions (such as "What is a BlackBerry?" or "How can a BlackBerry help me be more productive?"), you can benefit by reading this book cover to cover. No matter what your current BlackBerry user status (BUS, for short), this book helps you get the most out of your BlackBerry.

Right off the bat, BlackBerry isn't a fruit you find at the supermarket but rather an always-connected smartphone that has e-mail capabilities and a built-in Internet browser. With your BlackBerry, you're in the privileged position of always being able to receive e-mail and browse the Web.

On top of that, a BlackBerry has all the features you expect from a personal organizer, including a calendar, to-do lists, and memos. Oh, and did we mention that a BlackBerry also has a built-in mobile phone? Talk about multi-tasking! Imagine being stuck on a commuter train: With your BlackBerry by your side, you can compose e-mail while conducting a conference call, all from the comfort of your seat.

That's not all. BlackBerry goes a step further to make it more fun for you to own this device. You can snap a picture with its camera, record a funny video, listen to your music collection, and enjoy watching that video on YouTube.

In this book, you find all the basics, but we also go the extra mile and highlight some of the lesser-known (but still handy) features of the BlackBerry. Your BlackBerry can work hard for you when you need it to and can play hard when you want it to.

## About This Book

*BlackBerry All-in-One For Dummies* is a comprehensive user guide as well as a quick user reference. The book is designed so that you can read it cover to cover if you want, but you don't need to read one chapter after the other. Feel free to jump around while you explore the different functionalities of your BlackBerry.

We cover basic — such as how to make a phone call — and advanced topics — such as developing must-have applications. If you use or want to use a certain function of your BlackBerry, it's likely covered here.

# Who Are You?

We tried to be considerate of your needs, but because we've never met you, our picture of you is as follows. If you find that some of these images are true about you, this might just be the book for you:

✦ You have a BlackBerry and want to find out how to get the most from it.

✦ You don't have a BlackBerry yet and are wondering what one could do for you.

✦ You're looking for a book that doesn't assume that you know all the jargon and tech terms used in the smartphone industry.

✦ You want a reference that shows you, step by step, how to do useful and cool things with a BlackBerry without bogging you down with unnecessary background or theory.

✦ You are tired of hauling your 10-pound laptop with you on trips and are wondering how to turn your BlackBerry into a miniature traveling office.

✦ You no longer want to be tied to your desktop system for the critical activities in your life, such as sending and receiving e-mail, checking your calendar for appointments, getting directions, and surfing the Web.

✦ You like to have some fun, play games, and be entertained from a device but don't want to carry an extra game gadget in your bag.

✦ You're interested in developing applications and what to find the basics.

✦ You just received a BlackBerry from your employer, and you're wondering how it works with the corporate servers.

## Conventions in This Book

A lot of BlackBerry smartphones are in existence, each with their own way of doing things. Trackballs, trackpads, touch screens: We tend to use these terms interchangeably. For the purposes of this book, here's what we intend:

✦ For models with trackballs, highlight the option and press the trackball to select it.

✦ For models with a trackpad, highlight the option and click the trackpad to select it.

✦ For models with a touch screen — Storm and Storm 2— touch-pressing the option selects it.

Whenever you come across instructions — whether we mention one or all three methods — simply substitute the one applicable for your phone.

# What's in This Book

*BlackBerry All-in-One For Dummies* consists of eight books, and each book consists of different chapters related to that book's theme.

## Book I: Smartphone Basics

Book I starts with the basics of your BlackBerry. You know: what it is, what you can do with it, and what elements make it up. We describe how you navigate using the QWERTY keyboard. We also show you how to personalize and express yourself through your BlackBerry. We also give you the must-knows about security and where to go for help when you get into trouble with your BlackBerry. This book wraps up with a chapter outlining the pros and cons of all the existing BlackBerry smartphones currently in existence so if you don't already have one, you can make the right choice.

## Book II: Organization and Productivity

Book II deals with the fact that your BlackBerry is also a full-fledged personal digital assistant (PDA). We show you how to get your BlackBerry to keep your contacts in Contacts as well as how to manage your appointments and meetings in Calendar. You also find out how to use the Clock application to set an alarm, set a timer, and set your device to Bedside Mode. You explore the Password Keeper application to centralize your passwords. We show you how to use the phone part of your smartphone as well as how to control it using your voice by optimizing the hands-free functions.

## Book III: Collaborating, Communicating, and Getting Online

Book III shows you what made BlackBerry what it is today: always-connected e-mail. We also get into another strength of the BlackBerry — Web surfing functionality — but we don't stop there. We point out how you can use other forms of messages, such as text messaging and instant messaging. You also find out about unique forms of messages on the BlackBerry, PIN-to-PIN messages and BlackBerry Messenger.

## Book IV: Desktop Operations

In Book IV, you find details of BlackBerry Desktop Manager and some of the hoops you can put it through with your BlackBerry, including making backups and installing BlackBerry applications from your desktop to your BlackBerry. You also find out how to port data from older devices — BlackBerry or not — to your new BlackBerry. And we didn't forget to cover important stuff, such as data-syncing your appointments and contacts with desktop applications, such as Outlook.

### Book V: Music, Photos, Videos, and TV

Book V starts getting into the fun stuff. Rock your world and use your BlackBerry to play music, watch videos, and take pictures. You also get the scoop on how to record videos and sample ring tones. Plus you get timesaving shortcuts on the Media applications.

### Book VI: BlackBerry Applications

In Book VI, we extend the fun. We show you how to navigate BlackBerry App World to download apps that enhance your lifestyle — whether it's for work (productivity apps), play (social media apps), or to keep in the know (news apps).

### Book VII: Roadmap to Application Development

Think you have a great idea for an application that you can't find in BlackBerry App World or the Internet? In Book VII, we get you on the road to developing your own cool app. We also show you how to distribute your app to the masses. We can't possible cover all the ins and outs of application development — there are entire books devoted to that subject — but this book definitely gets you on the right track,

### Book VIII: Enterprise Communications

If you just received a BlackBerry smartphone from your employer, you may be confused as to how it works. Book VIII outlines all you need to know to keep your smartphone running smoothly with an enterprise system and to stay on the right side of your IT department. Some of the topics we cover are how your smartphone works in a corporate environment, how you can keep your phone secure (your company most likely requires some security measures), and how to integrate your smartphone to your work phone (no one needs to know you're not at your desk, right?).

## Icons Used in This Book

If a paragraph sports this icon, it means we're talking about BlackBerry devices that are provided by your employer.

This icon highlights an important point that you don't want to forget because it just might come up again. We'd never be so cruel as to spring a pop quiz on you, but paying attention to these details can definitely help you.

This book rarely delves into the geeky, technical details, but when it does, this icon warns you. Read on if you want to get under the hood a little, or just skip ahead if you aren't interested in the gory details.

 Here's where you can find not-so-obvious tricks that can make you a BlackBerry power user in no time. Pay special attention to the paragraphs with this icon to get the most out of your BlackBerry.

 Look out! This icon tells you how to avoid trouble before it starts. Be sure to read and follow the accompanying directions.

## Where to Go from Here

Now you can dive in! Give Book I, Chapter 1 a quick look to get an idea of where this book takes you and then feel free to head straight to your chapter of choice.

# Book I

# Smartphone Basics

# Contents at a Glance

# Chapter 1: The BlackBerry Platform in a Nutshell

## In This Chapter

✔ Checking out your BlackBerry behind the scenes

✔ Seeing what your BlackBerry can do

✔ Handling the hardware

There has been a lot of talk in the last few years about smartphones like the BlackBerry (which you are probably carrying), and how they can change the way you work, communicate with others, and the way you are entertained. We're sure that you have an idea of what the BlackBerry smartphone is (it's quite okay if you don't) and you aren't eating it. (We hope you aren't eating it.)

Before we jump right into how the BlackBerry can change your lifestyle, we're just curious, though — what actually convinced you to buy this particular handheld mobile device? Was it BlackBerry Messenger, the always-connected e-mail, the multimedia player to replace your iPod or iPhone, or a really good app that you saw on your friend's BlackBerry? We know; the list could go on and on — and we might never hit the exact reason you got yours. For whatever reason you bought your BlackBerry, congratulations; you made an intelligent choice.

The same smarts that made you buy your BlackBerry are clearly at it again. This time, your intelligence led you to pick up this book, perhaps because your intuition told you that there's more to your BlackBerry than meets the eye.

Your hunch is right. Your BlackBerry *can* help you do more than you thought. For example, your BlackBerry is a whiz at making phone calls and checking e-mails, but it's also a social networking do-it-all smartphone that can update your Facebook account and instant-message with your business partner on another continent. Oh yeah, you can also surf the Web at amazing speed. We're talking *World Wide Web* here, so the sky's the limit. Help is always at your fingertips instead of sitting on your desk at home or at the office. Here are some things that your BlackBerry can help you with:

✦ Need to check out the reviews of that restaurant on the corner?

✦ Need to know — right now — what's showing in your local movie theaters, or what the weather will be like tonight, or what's the best place to shop the sales?

✦ Need to know your current location and get directions to that cozy bed-and-breakfast, or retrieve news headlines, or check stock quotes?

✦ Want to do some online chatting or view some pictures online?

✦ Hankering to network with your old classmates?

You can do all these things (and more) by using your BlackBerry smartphone.

BlackBerry is also a full-fledged *personal digital assistant (PDA)*. Out of the box, it provides you with the organizational tools you need to set up to-do lists, manage your appointments, take care of your address books, and more.

Being armed with a device that's a phone, an Internet connection, a PDA, a GPS (built-in with most BlackBerry models), and a full-on media player all built into one makes you a powerful person. With your BlackBerry (along with this resourceful book), you really can increase your productivity and become better organized. Watch out, world! BlackBerry-wielding power-house coming through!

If you stick with us, you find out all you need to get the most out of your device or maybe even save a troubled relationship. (Well, the last one is a bit of an exaggeration, but we got your attention, right?)

# How It All Works: The Schematic Approach

For those who always ask, "How do they do that?" you don't have to go far; the following sections are just for you.

## The role of the network service provider

Along with wondering how your BlackBerry actually works, you might also be wondering why you didn't get your BlackBerry from RIM (Research In Motion) rather than from a *network service provider* (the company that provides your phone and data service) such as AT&T or T-Mobile. Why did you need to go through a middle person? After all, RIM makes the BlackBerry.

That's an excellent question — and here's the quick-and-dirty answer: RIM needs a delivery system — a communication medium, as it were — for its technology to work. Not in a position to come up with such a delivery system all by its lonesome, RIM partnered (and built alliances across the globe) with what developed into its network service providers — the usual suspects (meaning the big cellphone companies). These middlemen support the wireless network for your BlackBerry so that you can connect to the BlackBerry Internet Service, or BlackBerry Enterprise Server — and get all those wonderful e-mails (and spend so much valuable time surfing the Internet). See Figure 1-1 for an overview of this process.

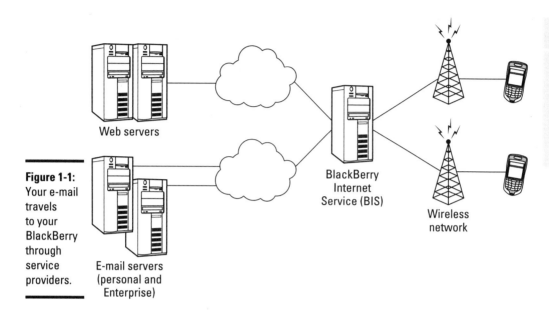

**Figure 1-1:**
Your e-mail travels to your BlackBerry through service providers.

Web servers

E-mail servers (personal and Enterprise)

BlackBerry Internet Service (BIS)

Wireless network

Network service providers don't build alliances for nothing, right? In return, RIM gave them the right to brand their names on the BlackBerry smartphones that they offer for sale. For example, a T-Mobile BlackBerry looks different from a similar model you get from Vodafone.

Which leads to another question: Do BlackBerry functionalities differ from phone model to phone model? Quick answer: In the core BlackBerry applications (such as Message, Tasks, and Address Book), you find no major differences.

Just to keep the score card straight, when we talk about features available from one network service provider that aren't available from others, we point them out.

## Connecting to your computer

Nowadays, a personal computer is a household necessity for most people. People spend so much time using computers and so much information is stored in them that many people find them indispensible. It's no surprise that BlackBerry works hand in hand with your PC, and the USB cable that comes with your BlackBerry does more than just charge your device.

When you want to connect your BlackBerry to your computer, check out Book IV, which helps you use your PC or Mac connection with the help of BlackBerry Desktop Manager and all the utilities that come with it. For instance, in Chapters 2 and 3, you find how to sync your device with the personal information manager (PIM) data that you keep in your PC or Mac.

This includes calendar and task entries, along with phone numbers and e-mail addresses for your contacts. You can also read Chapter 5 for directions about switching from another device (even a non-BlackBerry device) to a new BlackBerry. For example, you find out how to import your contact list into your new BlackBerry. Chapter 4 tells you how to protect your data. Last, Chapter 6 talks about upgrading your BlackBerry operating system with the help of your PC or Mac.

## BlackBerrying the world with your BlackBerry

If you received your BlackBerry from T-Mobile or AT&T, chances are that your BlackBerry will continue to work when you travel to, say, London or Beijing. All you need to worry about is remembering to turn on your BlackBerry (and maybe the extra roaming charges).

Because your BlackBerry is quad band, it works in more than 90 countries. What is *quad band?* Basically, different cellphone networks in different countries operate on different frequencies. For example, the United States operates on 850 and 1900 MHz, Canada operates on 850 and 1900 MHz, and Europe and Asia Pacific operate on 900 and 1800 MHz.

Your quad-band BlackBerry is designed to work on 850 MHz, 900 MHz, 1800 MHz, and 1900 MHz, so you're covered almost wherever you go. Check with your network service provider to see whether your BlackBerry will work at your destination before you hop on a plane, just to be sure.

Nothing stands still in this world, and this saying is proven by the fact that GSM (Global System for Mobile Communications) has spawned *High-Speed Downlink Packet Access (HSDPA)* technologies that have been growing because they work on the GSM phone infrastructure. This HSDPA is now available in the United States through most major network service providers. HSDPA competes in the marketplace against CDMA's (Code Division Multiple Access) EV-DO (Evolution-Data Optimized).

What's it to you? CDMA and GSM aren't compatible. Your phone works on only one technology. When you travel outside North America, you face the burning question: CDMA or GSM? (Read: Will my BlackBerry work on this country's network or won't it?)

If you currently work with GSM, you should be okay to travel outside the United States. Most non–North American countries are on GSM networks. If you're a CDMA kind of person, not to worry; you should be fine to travel, as many CDMA BlackBerry smartphones are also equipped to handle international GSM networks. When in doubt, talk to your network service provider.

## Knowing your BlackBerry history

Your BlackBerry is truly a wondrous thing, boasting many features beyond the ordinary mobile phone. And its "sudden" popularity didn't happen overnight. Like any other good product, BlackBerry has come a long way from its (relatively humble) beginnings.

In the days when the Palm Pilot ruled the PDA world, Research In Motion (RIM, the maker of the BlackBerry) was busy in its lab, ignoring the then-popular graffiti input method, and designing a device with a QWERTY keyboard — the kind of keyboard that people were already used to from working on their computers. RIM didn't stop there, however. It added an always-connected e-mail capability, making this device a must-have among government officials as well as finance and health professionals.

To meet the needs of government officials and industry professionals, RIM made reliability, security, and durability the priorities when manufacturing its devices. Today, the BlackBerry comes from the same line of RIM family products, inheriting all the good genes while boosting usability and adding more functions to its core BlackBerry applications. As a result, BlackBerry is popular among not only *prosumers* (professional customers) but also consumers. Starting with BlackBerry Pearl, RIM has been targeting the mainstream consumer market. Clearly, with BlackBerry, RIM is winning the hearts of consumers while maintaining its hold on the enterprise market.

## *Oh, the Things You Can Do!*

In the BlackBerry world, always-connected e-mail used to be the primary factor that made BlackBerry very attractive and is likely first in the long list of reasons you got yours. And, if you need to go global, you can use your BlackBerry in more than 100 countries. Just hop off your flight, turn on your BlackBerry, and *voilà:* You can receive and send e-mails if you're in Hong Kong, London, or Paris. Your significant other can get in touch with you wherever you are — just to say hi or to remind you that you promised Aunt Edna a bottle of Chanel No. 5.

One caveat here — you need to make sure that your network service provider has the technology to go global. See the preceding section of this chapter for more info. Generally speaking, you can receive and send e-mails just like you do when you're at home.

Although e-mail is BlackBerry's strength, that's not the only thing it can do. The following sections go beyond e-mail to point out some of the device's other major benefits.

## Social networking

Want to update your Facebook fans on your whereabouts? Tweet the latest news on a stock? Upload a quick snapshot (taken with your BlackBerry) to Flickr? You can do all of that with your BlackBerry, and we tell you how in Book III.

Do you have buddies who also have BlackBerry smartphones? You'll never be out of touch because you have BlackBerry Messenger to keep you connected. (See Book III, Chapter 4 for instructions on how to use BlackBerry Messenger to chat with friends and colleagues.)

If you were socially connected online before you got your BlackBerry, then with your BlackBerry, you'll be socially connected always. (Is that such a good thing? Well, that's up to you!)

## All-in-one multimedia center

Previously, many people hesitated to buy a BlackBerry due to the lack of multimedia functions. They wanted a camera and audio and full video playback. BlackBerry has changed all that and has more features than you may expect. Not only does BlackBerry have a high-resolution camera — but it also has a memory slot for a microSD chip (see Book I, Chapter 2) on which you can store your photos. What does that mean? Well, it means your BlackBerry can function as the following:

✦ A music player

✦ A video player

✦ A portable flash drive

✦ Your personal photo collection

## Internet at your fingertips

Yup, with the new BlackBerry Web browser, you can surf the Net nearly as smoothly as you do on a desktop computer. Even better, you can continue chatting with your friends through instant messenger, just as if you never left your office. You'll get an alert when your stock is tanking (if you have the eTrade app installed). True, that isn't fun, but you want this information as fast as possible.

Intrigued? Read how BlackBerry can take full advantage of the Web in Book III, Chapter 1.

## Me and my great personal assistant

You might be saying, "But I'm really a busy person, and I don't have time to browse the Web. What I *do* need is an assistant to help me better organize my day-to-day tasks." If you can afford one, by all means go ahead and hire a personal assistant. The next best thing is a personal *digital* assistant (PDA). Just like people come in many flavors, so do many PDAs.

Whip out that BlackBerry of yours and take a closer look. That's right; your BlackBerry is also a full-fledged PDA, helping you do all this and much more:

✦ Remember contact information for all your acquaintances. (See Book II, Chapter 1.)

✦ Manage your appointments. (See Book II, Chapter 2 for help with using BlackBerry Calendar.)

✦ Keep a to-do list. (See Book II, Chapter 3.)

## A computer in the palm of your hand

Remarkable communication device? Check.

Full-fledged PDA? Check.

Full-featured media player? Check.

These capabilities are just the tip of the iceberg. Don't underestimate the device because of its size: Your BlackBerry is also a powerful computer.

Need convincing? Here goes. Out of the box, with no fiddling, it comes with a great set of organizational and productivity tools in the form of programs. Software developers besides RIM are taking advantage of this growing market — which means that hundreds of applications are out there for you. For example, you can download graphic-intensive games or a mortgage calculator right from BlackBerry App World wirelessly (more on that in Book VI, Chapter 1).

## Look Dad, no hands!

Your BlackBerry is equipped with an earphone that doubles as a mic for hands-free talking. This accessory is your doctor's prescription for preventing the stiff neck that comes from wedging your BlackBerry against your ear with your shoulder. At the minimum, it helps free your hands so that you can eat Chinese takeout. Some places require you by law to use an earphone while driving and talking on a phone.

We recommend that you avoid using your cellphone while driving, hands-free or not.

But RIM didn't stop with just your standard wired earphones. BlackBerry also supports cool wireless earphones based on Bluetooth technology. How could a bizarrely colored tooth help you here? *Bluetooth* is the name for a (very) short-distance wireless technology that connects devices. See Book II, Chapter 4 for how to connect your BlackBerry to a Bluetooth headset.

# Putting a Sentry on Duty

The virtual world isn't exempt from general human nastiness; in fact, every day a battle is fought between those who are trying to attack a system and those who are trying to protect it. A computer connected to the Internet faces the risk of being cracked by a hacker or infected by a virus. Viruses try to replicate themselves and generally bug you.

Fortunately, security is a BlackBerry strong point. Viruses often come as e-mail attachments. However, BlackBerry supports very few file types out of the box (mostly images and documents). You won't face threats from e-mails with these attachments.

In addition, if your BlackBerry was given to you by your employer, any data transitions to or from the BlackBerry are *encrypted* (coded) to prevent snooping.

RIM also has a signature process for application developers, which forces them to identify themselves and their programs if they're developing any applications for the BlackBerry that need to integrate with either BlackBerry core applications or the OS.

Remember the I Love You and Anna Kournikova viruses? These are virtual evils transmitted through e-mail, scripts, or sets of instructions in the e-mail body or attachment that can be executed either by the host e-mail program or, in the case of an attachment, by the program associated with the attached file. Fortunately, the BlackBerry smartphone's Messages application (through which you view, send, and receive e-mail) doesn't support scripting languages, which can execute harmful programs. BlackBerry's viewer for such attachments doesn't support scripting either, so you won't face threats from e-mails having these attachments.

The security measures RIM implemented on the BlackBerry platform have gained the trust of the U.S. government as well as many of the Forbes Top 500 enterprises in the financial and health industries.

# Chapter 2: Navigating Your BlackBerry

## In This Chapter

✔ BlackBerry Anatomy 101

✔ Using touch-screen navigation

✔ Understanding general navigation guidelines

✔ Using common shortcut keys

*P*rogress being what it is, each new model of BlackBerry seems better equipped than the last. Generally speaking, the most significant feature changes take place under the hood.

Your BlackBerry smartphone comes in one of three forms:

✦ A large screen with a QWERTY keyboard and trackpad

✦ A slightly smaller screen with a SureType keyboard and trackpad (Pearl)

✦ A very large screen with touch-screen navigation (Storm)

What is a trackpad? What can you do with it? When did BlackBerry go touch screen? How can you navigate your BlackBerry better with these methods? Those are some of the questions that we answer in this chapter. Bear with us, and you will be master of your BlackBerry device in no time.

## Anatomy 101: The Body and Features of Your BlackBerry

In this and the following sections, we show you all the keys and features on a BlackBerry Bold, Pearl Flip, Curve 8900, and BlackBerry Storm. The four models are shown in Figures 2-1, 2-2, 2-3, and 2-4 respectively.

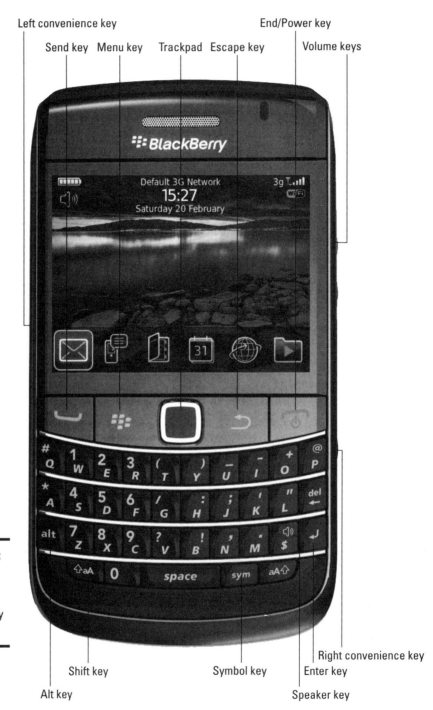

Left convenience key

End/Power key

Send key    Menu key    Trackpad    Escape key

Volume keys

**Figure 2-1:**
Main
features
on a
BlackBerry
Bold 9700.

Shift key      Symbol key      Enter key

Right convenience key

Alt key      Speaker key

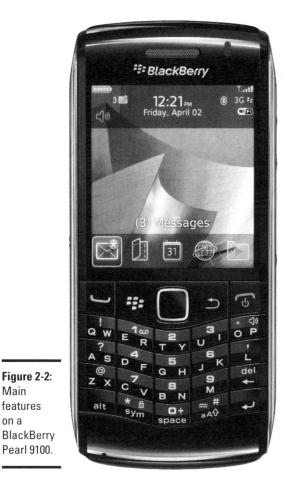

**Figure 2-2:**
Main
features
on a
BlackBerry
Pearl 9100.

Media Keys

Last Track    Play/Pause    Next Track

**Figure 2-3:**
BlackBerry
Curve 8500
series,
similar to a
BlackBerry
Bold 9700.

**Figure 2-4:**
Main
features
on a
BlackBerry
Storm.

These are the major features of a BlackBerry:

✦ **Display screen:** The graphical user interface (GUI) on your BlackBerry.

✦ **QWERTY keyboard:** The input for your BlackBerry — very straight-forward. The keyboard that you use with your computer has the same QWERTY layout.

✦ **SureType keyboard:** The input for your BlackBerry Pearl. We cover how to type with the SureType keyboard in the later section, "SureType keyboard."

✦ **Touch screen:** The graphical user interface (GUI) on your BlackBerry Storm. It lets you point to things using your finger instead of using the trackpad. As for typing, you will be typing on a virtual keyboard instead of on a QWERTY keyboard.

✦ **Virtual QWERTY keyboard:** Tilt your Storm to the side, and its virtual QWERTY keyboard appears on the screen for you to type with.

✦ **Virtual SureType keyboard:** The Storm SureType keyboard appears while you're typing in the Storm's upright (vertical) position. Each virtual SureType key contains two letters. We cover how to type with the SureType keyboard in the later section, "SureType keyboard."

✦ **Escape key:** This key is used to cancel a selection or return to a previous page within an application. If you hold it down, it returns you to the Home screen from any program.

✦ **Menu key:** This key displays the full menu of the application you are using.

If you owned an older BlackBerry (one with a trackwheel), the Menu key essentially replaces the trackwheel click but with added functions, like quick access to the short menu. With the new BlackBerry, you can also click the trackpad to confirm selection choices.

✦ **Trackpad:** You navigate the display screen of your BlackBerry Bold 9700, Pearl 9100, and Curve 8500 series with the trackpad. It allows you four directional movements. When you press the trackpad, the short menu of the application you are using appears.

✦ **Convenience keys:** Depending on the BlackBerry model you have, you have one or two convenience keys. By default, the convenience keys are preprogrammed to open an application. In Book I, Chapter 3, we show you how to reprogram the convenience keys so that they display the programs you use the most.

✦ **MicroSD slot:** Although it is hidden inside your BlackBerry and can be revealed only by removing the battery, the microSD slot is a crucial element to your BlackBerry media experience.

✦ **Send key:** Press this key to go straight to the Phone application, regardless of which application you are currently using. When you are already in the Phone application, the Send key starts dialing the number you entered.

✦ **End key:** This key ends your call. If you're not on a call, press this key to go to the Home screen from wherever you are. Pressing and holding the End key can turn your BlackBerry on or off.

✦ **Mute key:** Mutes the call, which means that you can hear the person on the other end of the line, but he can't hear you.

Two types of context menus can appear on your BlackBerry, as follows:

✦ **Full menu:** Press the Menu key to get a full menu that lists all the options and features you can perform.

✦ **Short menu:** Press the trackpad to get an abbreviated list of the full menu. (See Figure 2-5.)

**Figure 2-5:** Example of a short menu.

## Display screen

When you first turn on your BlackBerry, the screen displays the *Home screen,* which is your introduction to your BlackBerry's graphical user interface (GUI). The different icons represent the different applications found in your BlackBerry. See Figure 2-6 for an example of what your Home screen might look like.

If you tilt your BlackBerry Storm sideways, the screen adjusts its orientation.

Depending on the theme you're using, your applications may be listed in text form rather than as icons. Your GUI look depends on how you've customized the fonts and theme. For more on personalizing your BlackBerry, see Book I, Chapter 3.

**Figure 2-6:**
Your
BlackBerry
smartphone
might
come with
a different
default
theme.

## Menu key

The Menu key brings up the full menu for the application you are using.
When you are on the Home screen, pressing the Menu key displays a list of
applications installed on your BlackBerry smartphone. If you want to change
the order of the applications in the list, see Book I, Chapter 3.

When you are on the Home screen, the behavior of the Menu key depends
on the BlackBerry theme. The behavior just described is based on the
default theme. See Book I, Chapter 3 for more on changing themes.

## QWERTY keyboard

For those non-BlackBerry Pearl owners, this section is for you.

Unlike some PDA manufacturers (they know who they are) Research In Motion
(RIM) includes the same QWERTY keyboard you know and love from your personal computer on your BlackBerry smartphone. We think that is a great decision because it means that you don't have to master some new way of writing
"graffiti or whatever" to get data into your BlackBerry. All you have to do is type
on a familiar keyboard — and you most likely already know how to do that.

Whether you use your pinky or your index finger, how you type on your
BlackBerry is up to you. However, most people find that typing with two
thumbs is the most efficient method.

## SureType keyboard and the multitap option

If you have the BlackBerry Pearl or BlackBerry Storm, this section is for
you. The Pearl doesn't have a full QWERTY keyboard; rather, it works with

a QWERTY-based keyboard known as the *SureType* keyboard. As for the BlackBerry Storm, it has both virtual QWERTY and virtual SureType.

With the BlackBerry Storm, when you tilt your Storm sideways, the SureType keyboard appears. However, you can change the default upright keyboard setting so that you are using QWERTY by default no matter how you hold your BlackBerry Storm. You can change this setting by following these steps:

1. **From the BlackBerry Home screen, scroll to the Options icon and select it.**

2. **Navigate to the Screen/Keyboard setting and select it.**

   The Screen/Keyboard screen appears with various customizable fields.

3. **In the Default Keyboard setting, select QWERTY.**

4. **Press the Menu key and select Save to save your changes.**

The idea of SureType is that many keys share letters (refer to Figure 2-2 to see how this looks) and that the SureType technology is smart enough to find what key combinations come up with the words you want. Basically, with SureType, you can now type with only one thumb, and the more you type with SureType, it learns the words that you frequently use.

Here are tips to speed the learning curve when using SureType technology:

+ **Always finish typing a word before correcting it.** This way, SureType learns what you want to type next time.

+ **If SureType gets the word you're typing right on the first try, simply use the Space key to move on instead of clicking the trackpad or pressing Enter.**

+ **Take advantage of Custom Wordlist, which is a list of words that you choose.** You can find more on this later in the "Custom Wordlist" section.

+ **Type! Type! Type!** Because SureType learns how you type, the more you use it, the smarter it becomes in adapting to your style.

### SureType versus multitap

On your Pearl or Storm's virtual SureType, you can type in another mode: *multitap.* The regular way — at least we think of it as the regular way — is the multitap approach. The best way to explain multitap is by example: Say you want to type an **h** character on the SureType. You search out the *h* on the SureType keyboard but then notice to your dismay that the *h* shares a key with the letter *g*. What's a person to do? Do you really want to go through life writing *GHello* for *Hello?*

Actually, there's a perfectly easy solution to your problem. To get the letter *g,* you tap the GH key once. To get the letter *h* — the second letter in the key's pair — you tap the key twice — hence, the term *multitap.*

To switch from SureType or multitap, press the Menu key and select Multitap.

What about SureType? When you are in SureType mode, it tries to help you do the communication thing by figuring out what word you are typing. For example, if you want to type the word **hi**, you start by pressing the GH key and then the UI key. Doing so prompts SureType to display a list of words it thinks might be what you're aiming for. If the first listed word is what you want, simply press the Space key. The word is automatically selected, and you can continue typing. If what you really wanted to type appears a little later in the list, simply scroll to it by using the trackpad and press the trackpad to select it. Over time, SureType learns the words you're most likely to use and sticks those at the top of the list. Yup, that's right; it gets smarter the more you use it.

Regardless of which mode you are in, if you want to type numbers, you simply press the Alt key with the number you see on the SureType screen.

### SureType Custom Wordlist

SureType keeps all the words it has learned in a safe place, a *Wordlist.* You can review and add to your SureType Wordlist using the Custom Wordlist option. (Using this option to add words or proper names to the list means that SureType doesn't have to learn them when you are in the act of typing.)

To see or add words by using the Custom Wordlist option, follow these steps:

1. **From the Home screen, press the Menu key and then select the Options (wrench) icon.**

2. **Select Custom Wordlist.**

   The Custom Wordlist opens. This is where you can see all the words that SureType has learned. (If you purchased your BlackBerry recently, there might be only a few words or even no words in this list, depending on how often you've used SureType mode to type.)

3. **From within Custom Wordlist, press the Menu key and then select New.**

   A dialog box prompts you to type a new word.

4. **Type the new word.**

5. **To save your changes, press the Menu key and then select Save.**

Getting people's names right is tough with SureType, but thankfully, you can make sure that SureType automatically learns all the names in Address Book as follows:

1. **From the Home screen, press the Menu key and then select the Options icon.**

2. **Select Language.**

   The Language option screen appears, where the handy Input Option icon makes its home.

3. **Scroll to the Input Option icon and select it.**

   The Fast Options screen appears with the following options:

   - *Frequency Learning:* The word used most frequently appears first in the SureType Wordlist while you type.

   - *Auto Word Learning:* SureType learns as you type.

   - *Use Address Book as Data Source:* SureType learns all the names in your Address Book.

4. **Make sure that the Use Address Book as Data Source option is turned on.**

   If it isn't, scroll to this field, press the trackpad, and then select On from the drop-down list.

5. **To save your changes, press the Menu key and then select Save.**

You can hide the virtual keyboard by pressing the Menu key and touch-pressing Hide Keyboard.

Whether you use your pinky or your index finger, how you type on your BlackBerry is up to you. However, most people find that typing with two thumbs is the most efficient way to type on a BlackBerry Storm. And like the SureType keyboard with its Custom Dictionary, the QWERTY keyboard helps you as you type so you can get the word out (literally) with fewer keystrokes.

## Escape key

Simple yet useful, the Escape key allows you to return to a previous screen or cancel a selection. The Escape key is the arrow key to the right of the trackpad. On the Storm, it's located to the right of the Menu key.

## Trackpad

You can perform two functions with the trackpad: scrolling and pressing. When you scroll with your trackpad, you can navigate the display screen in four directions. In a text-filled screen such as the body of an e-mail, you can usually navigate through the text in four directions.

Depending on where you are on the BlackBerry's screen, different situations determine what happens when you press the trackpad, also called the *track-pad click.* Here are the details:

✦ **Display a drop-down list.** When you are in a choice field, pressing the trackpad displays a drop-down list of choices for that field.

✦ **Confirm a choice.** The trackpad can also function as a confirmation key. For example, when you need to select a choice in a drop-down list, you can press the trackpad to confirm the highlighted choice.

✦ **Display a short menu.** When you are in a text-filled screen (an e-mail body or Web page, for instance), pressing the trackpad displays a short menu, which is just an abbreviated version of the full menu. You get the full menu by pressing the Menu key.

## The microSD slot

Your BlackBerry comes with some internal memory, 1GB if you have the BlackBerry Bold or Storm. If you are a music or video fan, you know that 1GB is not going to keep you entertained for a long commute. But no need to worry. The folks at Research In Motion have incorporated a microSD slot into your BlackBerry so that you can add extended memory and store all the media files you want in your BlackBerry.

You can purchase a microSD card separately for a relatively low price these days. At the time of this writing, a 4GB microSD card costs about $20.

# General Navigation Guidelines

In this section, we go over general shortcuts and navigation guidelines.

On a Web page or an e-mail full of text, you can perform the following tasks:

✦ **Move to the top of the page.** Press the T key.

✦ **Move to the bottom of the page.** Press the B key.

✦ **Move to the top of the next page.** Press the Space key.

✦ **Select a line.** Press and hold the Shift key and scroll the trackpad horizontally.

✦ **Select multiple lines.** Press and hold the Shift key and scroll the track-pad vertically.

✦ **Copy selected text.** Press and hold the Shift key and press the trackpad.

✦ **Cut selected text.** Press and hold the Shift key and press the Delete key.

✦ **Paste text.** Press and hold the Shift key and press the trackpad.

✦ **Insert an accented character.** Hold down a letter key and scroll the trackpad.

✦ **Insert a symbol.** Press the Sym key and press the letter below the symbol.

✦ **Activate caps lock.** Press the Alt and right Shift keys.

✦ **Activate num lock.** Press the Alt and left Shift keys.

Here's how to navigate the BlackBerry Pearl:

✦ **Move to the top of the page.** Press the ER key.

✦ **Move to the bottom of the page.** Press the CV key.

✦ **Move to the next page.** Press the M or Space key.

✦ **Move to the previous page.** Press the UI key.

✦ **Move to the next line.** Press the BN key.

✦ **Move to the previous line.** Press the TY key.

✦ **Activate num lock.** Press Shift and Alt.

✦ **Switch between multitap and SureType mode.** When typing in a text field, press and hold the * key.

Here's how to navigate the BlackBerry Storm:

✦ **Move to the top of the page.** Finger swipe down.

✦ **Move to the bottom of the page.** Finger swipe up.

✦ **Move to the next page.** Finger swipe left.

✦ **Move to the previous page.** Finger swipe right.

## Switching applications

When you are navigating in an application, an option called Switch Application appears when you press the Menu key. Switch Application, which is similar to Alt+Tab in Windows, lets you multitask between applications. (See Figure 2-7.)

Another way to switch applications is to press the Alt and Escape keys. (The Alt key is located to the left of the Z key, and the Escape key is the arrow key to the right of the trackpad.)

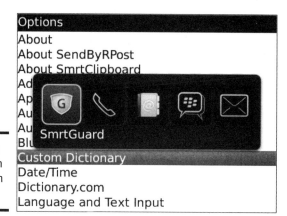

**Figure 2-7:**
The Switch
Application
menu.

If you always use a particular application, such as the Tasks application, you can program the convenience key so that you can get to your favorite application even more quickly than by using the Switch Application function. See Book I, Chapter 3.

## Selecting options

Throughout this book, you see examples of options that you can select. The easiest way to change the value in a field is to first use the trackpad to scroll to the field. Then press the trackpad to display a drop-down list of choices (see Figure 2-8), and finally, press the trackpad again on your choice.

For the Storm, simply point at the drop-down list, press the screen, and then select from the list by pressing the screen again to select.

**Figure 2-8:**
An example
of an option
field's drop-
down list.

---

## Other Pearl shortcuts

The Pearl has a few other shortcuts you may find useful:

✔ **Switch between current profile and vibrating profile.** From the Home screen, press and hold the # key. If you don't know what a profile is, see Book I, Chapter 3.

✔ **Key-lock your Pearl.** From the Home screen, press and hold the * key.

---

## General Keyboard Shortcuts

If you have a BlackBerry Pearl or Pearl Flip, this section does not apply to you. All the shortcuts described here are for QWERTY-based BlackBerry models only.

If you have a BlackBerry Storm, the Home screen shortcuts apply, but you must first select the keyboard display while you are on the Home screen. To do this, simply press the Menu key and select Show Keyboard. When the keyboard is showing, tilt the Storm sideways. You can follow the Home screen shortcuts that follow. To hide the keyboard, press the Menu key and select Hide Keyboard.

In many instances in this book, when we ask you to go to a BlackBerry application (Profile, for example), you have to first scroll to it from the Home screen and then click the trackpad. You might be thinking "Hey, there must be a shortcut for this," and you're right. This and the following sections cover such general keyboard shortcuts, all in the name of making your life easier. (Shortcuts that are more application-specific are covered in the chapter dealing with the particular application.)

Before you get all excited about shortcuts, you need to take care of one bit of housekeeping. (This doesn't apply to the Storm.) To use some of these general keyboard shortcuts, you first have to make sure that the Dial from Home Screen setting — buried deep within the Phone application — is turned off.

Inquiring minds want to know, so we'll tell you. The Dial from Home Screen option is designed for users who make frequent BlackBerry phone calls. It allows them to place phone calls directly from the Home screen without having to manually enter the Phone application. If you are not a frequent phone user and want to access all applications with a press of a button, get ready to ditch Dial from Home Screen.

Here's how to turn off the Dial from Home Screen setting:

*1.* **From the BlackBerry Home screen, select the Phone application.**

*2.* **Press the Menu key and then select the Options icon.**

A screen that lists a range of options appears.

*3.* **Select General Options.**

The General Options screen appears.

*4.* **Highlight the Dial from Home Screen field and then select No.**

Doing so shuts down the Dial from Home Screen option, enabling you to use Home screen shortcuts.

*5.* **To confirm your change, select Save from the menu that appears.**

If you're a frequent phone user on your BlackBerry, as opposed to an e-mail or Internet user, you might not want to turn off the Dial from Home Screen feature.

## Using Home screen shortcuts

After you disable the Dial from Home Screen feature, you are free to use any Home screen shortcut. (The name for these shortcuts is actually a pretty good fit because you can use these shortcuts only while you are on the Home screen.)

Okay, here goes. To call up an application listed in Table 2-1, press the corresponding key listed in the Home Screen Shortcuts column.

| Table 2-1 | Home Screen Shortcuts |
|---|---|
| *Application* | *Shortcut Key* |
| Messages | M |
| Saved Messages | V |
| Compose | C |
| Search | S |
| Contacts | A |
| Tasks | T |
| Profile | F |
| Browser | B |

*(continued)*

**Table 2-1** *(continued)*

| Application | Shortcut Key |
| --- | --- |
| Calendar | L |
| Calculator | U |
| MemoPad | D |
| Keyboard Lock | K |
| Phone | P |

## *Other (non-Home screen) shortcuts*

The following shortcuts can be used at any time, regardless of which screen you're currently in — or whether you have Dial from Home Screen enabled, for that matter:

✦ **Soft Device Reset (also known as the *3-Button Salute*):** Pressing Alt+Shift+Del forces a manual soft reset, which is just what you need when your BlackBerry has crashed or when you install an application and it needs a manual reset. A hard reset can be done by pulling out the battery from the back of the BlackBerry. What's the difference between a soft reset and a hard reset? Without getting into the technical jargon, from a BlackBerry user's perspective, a hard reset takes longer and is usually the last resort to solve any issues before contacting the help desk.

✦ **HelpME:** In the BlackBerry world, SOS is actually spelled Alt+Shift+H. Use it when you're on the phone with technical support. (It gives support personnel info such as your BlackBerry PIN, memory space, and version number so that they have information about your BlackBerry when they try to troubleshoot your problems.)

*Note:* Your BlackBerry PIN isn't a security password; rather, it is a unique number that identifies your BlackBerry, sort of like a serial number. But unlike a serial number, you can message another BlackBerry by using PIN-to-PIN messages. For more on PIN-to-PIN messages, see Book III, Chapter 3.

# Chapter 3: Configuring and Using Your BlackBerry

## In This Chapter

✔ **Putting your stamp on your BlackBerry device**

✔ **Saving battery power**

✔ **Watching your BlackBerry's back**

✔ **Blocking spam e-mail and unwanted SMS messages**

*R*egardless of how long you've had your BlackBerry — one week, one month, one year, or five years — you'll want to have it around for as long as you possibly can. (Or, at least until you have the bucks for that way-cool new model that's surely coming down the pipe.) And, for the duration that you *do* have your device, you'll want to trick it out so that your BlackBerry doesn't feel and sound exactly like the millions of other BlackBerry devices out there. (C'mon, admit it — your BlackBerry is definitely a fashion statement, so you better feel comfortable with what that statement is saying.)

In addition to customizing your BlackBerry so that it expresses the inner you, you want to make sure that you keep your BlackBerry in tip-top shape by watching out for things such as your BlackBerry battery life and information security. Luckily for you, this chapter puts any and all such worries to rest by filling you in on all you need to know to keep your BlackBerry a finely tuned (and yet quirkily personal) little PDA.

## Making Your BlackBerry Yours

BlackBerry devices are increasingly popular, so much so that over 28 million BlackBerry smartphones are out there serving the needs of people like you. Because of this fact, we're certain that finding ways to distinguish your BlackBerry from your colleagues' is high on your list of priorities.

Your wish is our command. Follow the tips and techniques outlined in the following sections and you, too, can have your very own personalized BlackBerry.

## Branding your BlackBerry

Like any number of other electronic gadgets that you possibly own, your BlackBerry comes to you off the shelf fitted with a collection of white-bread factory settings. This section helps you put your name on your BlackBerry, both figuratively and literally. You can start by branding your name on your BlackBerry. Follow these steps:

*1.* **Press the Menu key, and select the Options icon.**

*2.* **Scroll through the list of options and select Owner.**

You see fields to enter your owner information.

*3.* **Enter your name in the Name field and your contact information in the Information field.**

The idea here is to phrase a message (like the one shown in Figure 3-1) that would make sense to any possible Good Samaritan who might find your lost BlackBerry and want to get it back to you.

If you lock or don't use your BlackBerry for a while, the standby screen comes on, displaying the owner information that you entered. Read how to lock your BlackBerry, either manually or by using an auto setting, in the section "Keeping Your BlackBerry Safe" later in this chapter.

*4.* **Confirm your changes by pressing the Menu key and then select Save.**

**Figure 3-1:** List your owner info here.

## Choose a language, any language

Branding your BlackBerry with your own John Hancock is a good start, but setting the language to your native tongue so that you don't need to hire a translator to use your BlackBerry is equally important — and equally easy. By default, it will be English.

Here's how you choose a language:

**1. From your BlackBerry Home screen, select the Options (wrench) icon.**

Depending on your BlackBerry, the Options icon may be in the Settings folder.

**2. Scroll through the list of options and select the Language setting.**

Here you can choose the language and input method of your choice.

**3. First select the Language field, and then select your native tongue.**

Depending on your network provider, as well as what region (North America, Europe, and so on) you're in, the language choices you have can vary. Most handhelds sold in North America default to English or English (United States).

If your network provider supports it, you can install more languages onto your BlackBerry smartphone by using Application Loader in BlackBerry Desktop Manager. For more information on Application Loader, see Book IV, Chapter 1.

**4. Press the Menu key and select Save to confirm your changes.**

Isn't it great when you can actually read what's on the screen?

## Typing with ease using AutoText

Even the most devoted BlackBerry user has to admit that typing on a full keyboard is easier than thumb-typing on a BlackBerry. In an attempt to even the score a bit, your BlackBerry comes equipped with an AutoText feature, which is a kind of shorthand that can cut down on how much you have to type. Even if you have the BlackBerry Pearl with SureType technology, you can still benefit by using the AutoText feature. (For more on SureType technology, see Book I, Chapter 2.)

AutoText basically works with a pool of abbreviations that you set up. You then just type an abbreviation to get the word you associated with that abbreviation. For example, after setting up *b/c* as an AutoText word for *because,* any time you type **b/c** and press space, you automatically get *because* onscreen.

Your BlackBerry comes with a few default AutoText entries. Here are some useful ones:

✦ **mypin:** Displays your BlackBerry PIN.

✦ **mynum:** Displays your BlackBerry phone number.

✦ **myver:** Displays your BlackBerry model number and OS version.

The whole AutoText thing works best if you set up your own personal code, mapping your abbreviations to their meanings. (This is why we're discussing AutoText as part of our personalization discussion.) To set up your own code, do the following:

*1.* **From your BlackBerry Home screen, select the Options (wrench) icon.**

*2.* **Select the AutoText option.**

Here, you can choose to see (or search for) existing AutoText words or create new ones.

*3.* **Press the Menu key, and then select New.**

The AutoText screen appears, as shown in Figure 3-2.

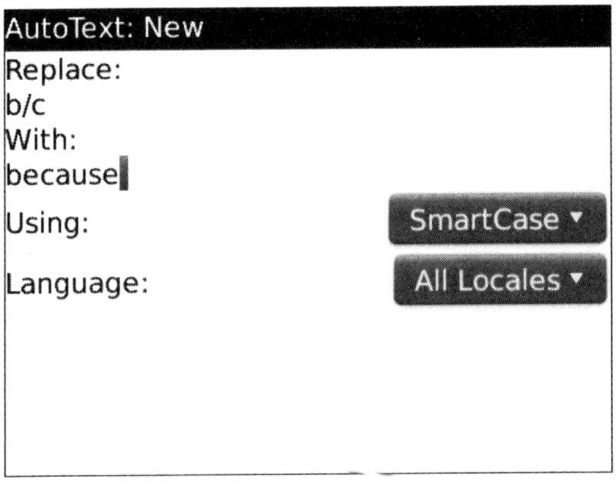

**Figure 3-2:** Create AutoText here.

> AutoText: New
> Replace:
> b/c
> With:
> because|
> Using:              SmartCase ▼
> Language:           All Locales ▼

*4.* **In the Replace field, enter the characters that you want to replace (in this example,** b/c**). In the With field, type what you want to replace your characters with (in this example,** because**).**

*5.* **In the Using field, choose between the SmartCase and Specified Case options.**

- *SmartCase:* Capitalizes the first letter when the context calls for that, such as the first word in a sentence.

- *Specified Case:* Replaces your AutoText with the exact text found in the With field.

For example, say you have the AutoText *bbg* set up for the term *black berryGoodies.com* and you want it to appear as is, in terms of letter cases. (The first *b* is not capitalized.) If you were to choose SmartCase for this particular AutoText, it would be capitalized as the first word in a sentence, which is not what you want. On the other hand, if you use Specified Case, your AutoText always appears as *blackberryGoodies.com* no matter where it is in the sentence.

**6. Scroll to the Language field and then select All Locales.**

Our preference for any self-created AutoText is All Locales. What this means is that regardless of the language input method (for example, English UK, English US, or French), any self-created AutoText is available for you to use. So, in the case of the AutoText bbg (blackberryGoodies. com), whether you are typing in French or Chinese, you can use this AutoText. On the other hand, if you select only the French input method for bbg as the Language field, you would be able to use this only if your input method is set to French in the Language option.

You can choose the input method in the Language options. We go over choosing a language input method next.

**7. Confirm your changes by pressing the Menu key and select Save.**

If you specify a language input method other than All Locales, your input method setting in the Language option must match the Language field in AutoText to use your newly created AutoText. Follow these steps:

**1. From your BlackBerry Home screen, select the Options (wrench) icon.**

**2. Scroll through the list of options and select the Language setting.**

Here you can choose the language and input method.

**3. Scroll to the Input Method field and then select the input method you want from the list.**

For your new AutoText setting to work (assuming that you didn't choose All Locales as the language for your AutoText), this option must match the input method set in your Language option.

**4. Confirm your changes by pressing the Menu key and select Save.**

## Getting your dates and times lined up

Having the correct date, time, and time zone is important when it comes to your BlackBerry for what we hope are obvious reasons. Many of the fine features that make up the BlackBerry core experience, as it were, depend on the date, time, and time zone being accurate.

Need an example? How about your BlackBerry calendar events? Imagine, if you will, that you have set up a make-or-break meeting for 9 a.m. (in your time zone) with a client in Paris, France, which is in who-knows-what time zone. You definitely want to be on time for that appointment, but you probably won't be if you're planning on having your BlackBerry remind you and you haven't set up the appropriate date, time, and time zone. Follow these steps to do that:

1. **From your BlackBerry Home screen, select the Options (wrench) icon.**

2. **Select the Date/Time setting.**

   The Date/Time screen appears.

3. **For the Time Zone field, select a time zone.**

   The Date/Time screen confirms the time zone that you chose.

   If you travel to a different time zone, you need to adjust this field because it does not adjust automatically.

4. **For Auto Update Time Zone, select Prompt.**

   When you travel across different time zones, your BlackBerry can detect the time zone.

5. **For the Use Network Time field, select No.**

   By selecting Yes, your date and time source is set to your service provider's server time. (See Figure 3-3.) There are times when the service provider does not have the time zone information available; you would then have to adjust it on your own.

   If you always set the time a few minutes earlier than the actual time (so you can be on time for those important meetings), set this field to No.

6. **For the Time and Date fields, select the proper date, hour, and minutes.**

   Here, you adjust the date and time to current hours and minutes. These fields only show if Step 4 is set to No.

7. **Press the Menu key and then select Save from the menu that appears.**

   Doing so saves your date and time settings in perpetuity — a really long time, in other words.

**Figure 3-3:**
Set the time source of your BlackBerry to your network provider's clock.

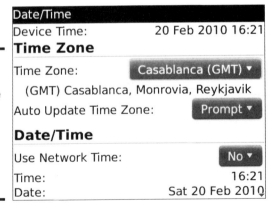

## Customizing your screen's look and feel

Right up there with making sure that your date and time settings are accurate is getting the display font, font size, and screen contrast to your liking. Now we know that some of you don't give a hoot if your fonts are Batang or Bookman as long as you can read the text, but we also know that some of you won't stop configuring the fonts until you get them absolutely right. For all you tweakers out there, here's how you play around with your BlackBerry's fonts:

1. **From the BlackBerry Home screen, scroll to the Options icon and press the trackball, trackpad, or touch screen.**

2. **Scroll to the Screen/Keyboard setting and then press the trackball, trackpad, or touch screen.**

   The Screen/Keyboard screen appears with various customizable fields, as shown in Figure 3-4.

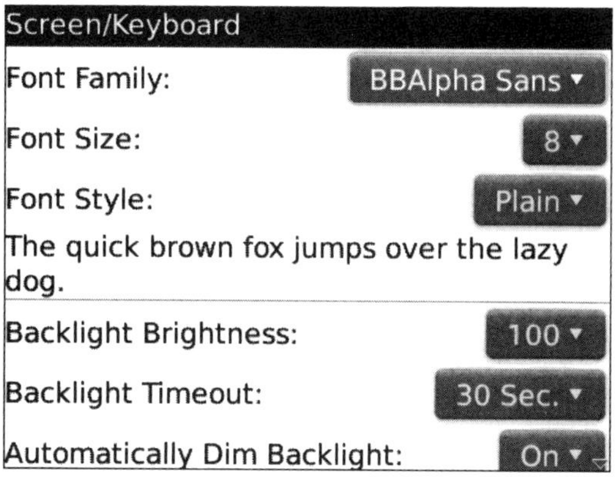

**Figure 3-4:**
The Screen/
Keyboard
screen,
waiting for
personal-
ization.

| Screen/Keyboard | |
| --- | --- |
| Font Family: | BBAlpha Sans ▾ |
| Font Size: | 8 ▾ |
| Font Style: | Plain ▾ |
| The quick brown fox jumps over the lazy dog. | |
| Backlight Brightness: | 100 ▾ |
| Backlight Timeout: | 30 Sec. ▾ |
| Automatically Dim Backlight: | On ▾ |

3. **Select the Font Family field and then select a font from the drop-down list.**

   You can choose from three to ten fonts, depending on your provider.

4. **Select the Font Size field, and then select a font size.**

   One thing to keep in mind is that the smaller the font size, the more you can see onscreen; however, a smallish font is harder on the eyes.

   *Note:* As you scroll up and down the list of fonts and font sizes, notice that the text *The quick brown fox jumps over the lazy dog* in the background takes on the look of the selected font and size so that you can preview what the particular text looks like. (In case you were wondering, this sentence uses every letter in the alphabet.)

5. **Press the Menu key and choose Save to confirm your changes.**

You can also adjust Font Style to Bold, Italic, or Plain using the same steps you used to adjust the Font Size.

With fonts out of the way, it's time to change the brightness of your screen as well as a few other viewing options, including programming the convenience key(s) to exactly what is convenient for you:

1. **From the BlackBerry Home screen, scroll to the Options icon and press the trackball, trackpad, or touch screen.**

2. **Scroll to the Screen/Keyboard setting and press the trackball, trackpad, or touch screen.**

   The Screen/Keyboard screen appears with its various customizable fields. (Refer to Figure 3-4.)

3. **Select the Backlight Brightness field and select the desired brightness from the drop-down list.**

   You can choose from 0 to 100, where 0 is the darkest and 100 is the brightest.

4. **Select the Backlight Timeout field and choose the amount of time before the backlight times out.**

   You can choose from ten seconds up to two minutes. The lower this setting is, the less time you'll have backlighting after you press each key. However, a low setting helps you conserve battery life.

5. **Select the Left Side Convenience Key Opens field, and then choose which application you want your left-side convenience key to open when you press it. You can do the same for the right side convenience key.**

   Check Chapter 2 in this book to find out more about convenience keys.

6. **Select the trackpad Horizontal Sensitivity field and choose how sensitive you want the trackpad to be for horizontal movements.**

   You can choose from 20 to 100, where 20 is the least sensitive and 100 is the most sensitive. *Note:* This is not applicable for the BlackBerry Storm.

7. **Select the Trackpad Vertical Sensitivity field and choose how sensitive you want the trackpad to be for vertical movements.**

   Again, 20 is the least sensitive, and 100 is the most sensitive. Keep in mind that if your trackball is too sensitive, it will be hard to control. *Note:* This is not applicable for the BlackBerry Storm.

8. **Press the Menu key and select Save to confirm your changes.**

## Choosing themes for your BlackBerry

Your BlackBerry is preloaded with different *themes,* or predefined sets of looks (wallpaper, fonts, and menu layouts) for your BlackBerry.

Regardless of what BlackBerry model you have, follow these steps to change your theme:

1. **From the BlackBerry Home screen, select the Options icon.**

2. **Select the Theme setting.**

   You see a list of available themes.

3. **Scroll to and select the theme you want.**

   You see a preview of the theme you've selected. See Figure 3-5.

4. **Press the Menu key and choose Activate.**

   It may take a few seconds for the new theme to be applied. You should notice the changes in the Home screen once they are applied.

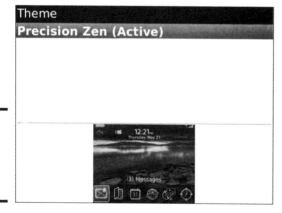

**Figure 3-5:**
Preview of
currently
selected
theme.

 You can download other themes. When you get new themes, they will appear in Step 2 above when you are selecting available themes. Just remember that you have to use your BlackBerry, not your PC, to access the following URLs:

✦ http://mobile.blackberry.com

✦ http://blackberrywallpaper.com

## Wallpaper for your BlackBerry

Like your desktop PC, you can customize the BlackBerry Home screen with personalized wallpaper. You set an image to be your BlackBerry Home screen background by using the BlackBerry Media application. Follow these steps:

1. **From the BlackBerry Home screen, choose the Media icon.**

   The Media application opens, where you see different categories: Music, Video, Ringtones, Pictures, and Voice Notes.

2. **Scroll to and select the Pictures category.**

   You're presented with these options:

   - *Sample Pictures:* Shows pictures that came with your BlackBerry.
   - *All Pictures:* Shows all pictures on your BlackBerry.
   - *Picture Folders:* Lets you select a specific folder.

3. **Scroll to and select one of the options.**

   This lists all the pictures in the chosen option.

4. **Select the picture you want to use for your Home screen background.**

   The selected picture appears in full-screen view.

5. **Press the Menu key and then select Set as Home Screen Image.**

   The picture is now your new Home screen wallpaper.

6. **Press and hold the Escape key to return to the Home screen and see the result.**

You can download free wallpapers from the following Web sites (as long as you use your BlackBerry, not your PC, to access the URLs):

- http://mobile.blackberry.com
- www.blackberrywallpapers.com
- www.blackberrygoodies.com/bb/wallpapers

## Letting freedom ring

The whole appeal of the BlackBerry phenomenon is the idea that this little electronic device can make your life easier. One of the ways it accomplishes this is by acting as your personal reminder service — letting you know when an appointment is coming up, a phone call is coming in, an e-mail has arrived, and so on. Your BlackBerry is set to bark at you if it knows something it thinks you should know, too. Figure 3-6 lists the kinds of things your BlackBerry considers bark-worthy, ranging from browser alerts to task deadlines.

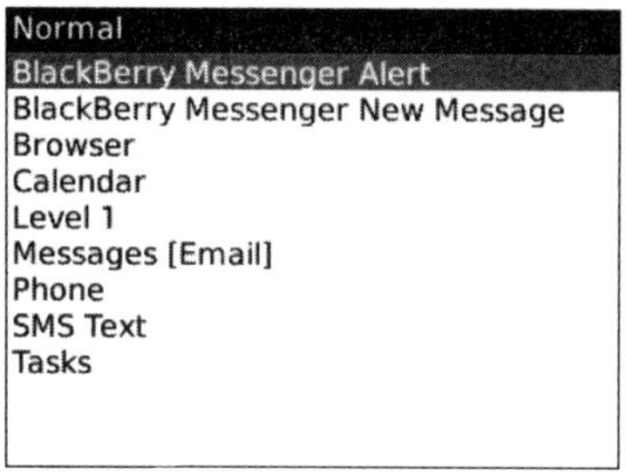

**Figure 3-6:**
Set
attention-
needy
applications
here.

```
Normal
BlackBerry Messenger Alert
BlackBerry Messenger New Message
Browser
Calendar
Level 1
Messages [Email]
Phone
SMS Text
Tasks
```

Different people react differently to different sounds. Some BlackBerry barks might be greatly appreciated by certain segments of the population, whereas other segments might react to the same sound by pitching their BlackBerry under the nearest bus. The folks at Research In Motion (RIM) are well aware of this and have devised a great way for you to customize how you want your BlackBerry to bark at you — they call it your *sound profile.*

You can jump right into things by using a predefined profile, or you can create your own profile. The upcoming sections look at both approaches.

Whether you create your own profile or customize a predefined profile, each profile is divided into several categories that represent the application for which you can define alerts.

In the Sound application, the Sound items are organized into the following categories:

✦ **Phone:** Alerts you if you have an incoming call or voice mail.

✦ **Messages:** Alerts you if you have an incoming e-mail, SMS, MMS, or BlackBerry PIN message. Additionally, you can set different alerts for each individual e-mail account.

✦ **Instant Messages:** Alerts you if you have any BlackBerry Messenger Alerts; if you have third-party instant messaging installed (such as Google Talk), you can set the alerts here as well.

✦ **Reminders:** Alerts you if you have set up calendar reminders, tasks reminders, or e-mail follow-up flags.

✦ **Other:** This category is a notification setting for third-party applications, such as Facebook, as well as the Browser application.

You can personalize all the listed applications according to how you want to be alerted. Because how you customize them is similar, we use one application, Messages, as an example in the text that follows.

After this, we go over creating a profile from scratch. You may be wondering why you need to create a profile if you can personalize the predefined ones. If your needs are different from the predefined settings, creating a profile is the way to go.

### Using and tweaking the preset profile settings

If you're like us and get more than 200 e-mails daily, you probably don't want your BlackBerry sounding off 200 times a day. You can set up your BlackBerry so that it notifies you only if an e-mail has been marked urgent, requiring your immediate attention. You can do this by setting the notification for your Messages application to None for both In Holster and Out of Holster. Then, in the Level 1 option (refer to Figure 3-6), you can set your desired notification for both In Holster and Out of Holster. That way, you conveniently filter any unnecessary e-mail notifications, leaving just the urgent stuff to sound off to you.

As an example, we will take the Normal preset profile and tweak how the Message application will alert you.

Follow these steps to customize alerts for your BlackBerry:

1. **From the BlackBerry Home screen, select the Sounds application.**

   A pop-up screen appears, listing different profiles (Silent, Vibrate, Normal, Loud, Medium, Phone Calls Only, All Alerts Off).

2. **Select Edit Profiles, which appears at the end of the list.**

   A screen appears, listing different profiles.

3. **Highlight the Normal profile in the list, press the Menu key, and then select Edit.**

   The Normal screen appears, listing the applications with alert capabilities mentioned in the preceding section (refer to Figure 3-6).

4. **Expand the Messages heading by selecting it and then select any of the e-mail accounts you have.**

   A screen appears with options to set the ring tone, LED, and vibration.

5. **For Ring Tone, you can set the following options:**

   - *Ring Tone:* The ring tone you want.

   - *Volume:* How loud you want the ring tone, from Silent to 10 (the loudest).

   - *Count:* The number of times the ring tone repeats, from 1 to 3.

   - *Play Sound:* Whether the ring tone will play while your BlackBerry is in or out of the holster or whether it will always play.

6. **For LED, set it On or Off.**

7. **For vibration, set it to On, Off, or Custom.**

   If you choose Custom, you have the following options:

   - *Length:* How long each vibration lasts: Short, Medium, or Long.
   - *Count:* The number of times the vibration occurs; you can choose 1, 2, 3, 5, or 10.
   - *Vibrate:* Whether the vibration will occur while your BlackBerry is in or out of holster or will always vibrate.

8. **Press the Menu key and then select Save.**

## Creating your own profile

You need to know which applications on your BlackBerry have alert capabilities because then you can personalize each "Hey, you!" to your liking. You can have your BlackBerry so personalized that you can tell whether you have a phone call or an incoming message just by how your BlackBerry sounds.

If you're already familiar with the different applications and are clear how you want each one to alert you, go on and create your own profile. As we mention earlier, you can achieve the same result by personalizing the predefined profiles that come with your BlackBerry. However, if you like to keep the predefined profiles the way they are, create a new profile by following these steps:

1. **From the BlackBerry Home screen, select the Sound application.**

   A pop-up screen appears, listing different profiles.

2. **Select Edit Profiles, which appears at the end of the list.**

   A screen appears, with an Add Custom Profile line and a list of profiles.

3. **Select Add Custom Profile.**

   A New Custom Profile screen appears, prompting you to name your profile.

4. **In the Name field, enter a name for your profile.**

   For this example, just type **MyOwnProfile**.

5. **Configure your new profile.**

   To customize each of the categories of applications, refer to the preceding section, "Using and tweaking the preset profile settings," Steps 3–7.

6. **Press the Menu key and then select Save.**

   Your newly created profile appears in the Profile screen.

7. **Select My Profile.**

   You can start to use your newly created profile.

You can switch between your current profile and the Quiet profile by pressing and holding the # key.

Regardless of whether the ring tone is for an incoming call or an incoming e-mail, you can download more ring tones to personalize your BlackBerry. Additionally, you can use any MP3 file in your Media application as your personalized ring tone. Follow these steps:

1. **From the Home screen, press the Menu key and then select the Media application.**

2. **In Media, select the Music category.**

   Doing so brings up various music classifications such as Artist, Album, and Genres.

3. **Highlight the music file you want to use for your ring tone.**

4. **Press the Menu key and then select Set as Phone Tune.**

   This sets the music file as your new phone tune.

5. **Press and hold the Escape key to return to the Home screen.**

# Setting Bedside Mode

Bedside mode turns your BlackBerry into a clock and prevents it from interrupting you with calls or messages while you sleep. It dims the clock screen, disables the LED, and can even be configured to shut the wireless antenna off. When you exit Bedside mode, all of your settings return to normal.

Bedside mode has several options, which you can configure with these steps:

1. **From your BlackBerry Home screen, select the Clock icon.**

2. **Press the Menu key and select Options.**

3. **Navigate down to the Bedside Mode section (shown in Figure 3-7).**

   You can adjust these Bedside mode options:

   - *Disable LED:* If you choose Yes, your BlackBerry LED will not flash when you receive new messages. If you choose No, your BlackBerry LED will flash as usual.

   - *Disable Radio:* Choose to disable the BlackBerry wireless antenna. If you choose Yes, your BlackBerry's wireless antenna will turn off when your device enters Bedside mode and then turn back on when it leaves Bedside Mode. This prevents your BlackBerry from receiving any incoming calls or messages. If you choose No, your wireless antenna will remain on when in Bedside mode.

- *Dim Screen:* Defines whether the clock display dims while your BlackBerry is in Bedside mode.

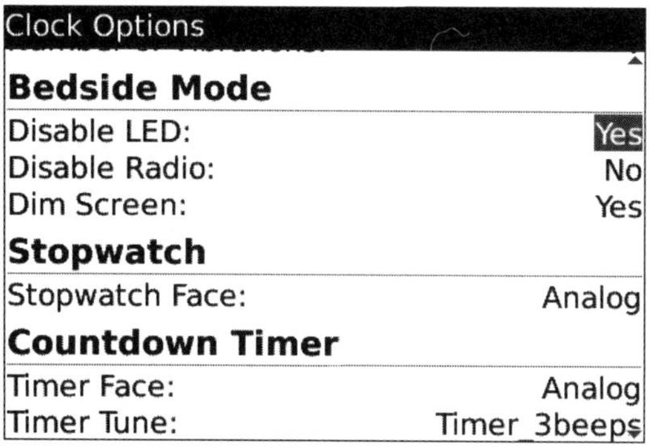

**Figure 3-7:**
Bedside
mode
options.

```
Clock Options                              ▲
Bedside Mode
Disable LED:                             Yes
Disable Radio:                            No
Dim Screen:                              Yes
Stopwatch
Stopwatch Face:                       Analog
Countdown Timer
Timer Face:                           Analog
Timer Tune:                    Timer_3beeps
```

4. **Navigate to the option you would like to change and make the changes.**

5. **When you are done setting Bedside mode options, press the Menu Key and select Save.**

You can also configure your BlackBerry to automatically enter Bedside mode while charging — especially useful if you charge your BlackBerry every night while you sleep. Follow these steps:

1. **From your BlackBerry Home screen, select the Clock icon.**

2. **Press the Menu key and select Options.**

3. **Navigate to the When Charging option in the Clock section and select Enter Bedside Mode.**

4. **When you are done setting Bedside mode options, press the Menu Key and select Save.**

# Keeping Your Battery at Optimal Use

When it comes to mobile devices, battery life is key. Although BlackBerry smartphones typically sport better battery life than other devices, even they are prone to a dead battery eventually. So how can you do everything you want with your BlackBerry while squeezing the most possible juice out of the battery? We tell you how you can get a little extra mileage out of your BlackBerry's charge in the following sections.

## Using Auto On/Off

Auto On/Off is one of the easiest ways to save energy on your BlackBerry smartphone. Auto On/Off does exactly what you would expect; it turns your BlackBerry off and on at defined times. Because you're probably not using your BlackBerry while sleeping, you can set it to turn off when you are definitely asleep and have it turn on a few minutes before your alarm goes off.

If you're using your BlackBerry as your alarm clock, you don't have to worry that your alarm won't ring. Your BlackBerry rings at the time you set, ignoring any Auto On/Off or Standby mode settings.

You can configure Auto On/Off by following these steps:

1. **From the BlackBerry Home screen, choose the Options icon.**

2. **Select Auto On/Off.**

3. **Set the Weekday, Weekend, and Turn On/Turn Off options you would like.**

4. **Press the Menu key and choose Save.**

## Turning off unused BlackBerry connections

Your BlackBerry has several different connectivity options, including cellular antenna, Bluetooth, and Wi-Fi. These options are great when you need them, but there is no reason to have them turned on when you're not using them. These connectivity options drain your battery faster. Turning them off will help extend your battery life.

Turn off all connections when you're in areas where you have no connection, like the subway or on an airplane. If your BlackBerry doesn't have reception, it starts to actively look for an available network, which drains your battery faster, which is why turning all connections off conserves some of your battery power.

Follow these steps to turn off connections that you don't need:

1. **On the BlackBerry Home screen, select the Manage Connections (antenna) icon.**

2. **Navigate to the connection that you want to turn off and deselect the check box.**

   To turn back on the connection, you just check the check box.

3. **Press the Escape key to return to the BlackBerry Home screen.**

If you turn off all connections, you won't receive e-mails, phone calls, or messages until you turn your mobile connection back on.

## Optimizing your Backlight settings

The backlight is one of the biggest battery drains on your BlackBerry. Although the standard settings are good for most users, if you really want to extend battery life, consider lowering your backlight brightness and timeout. Follow these steps to adjust your backlight settings:

1. **From your BlackBerry Home screen, select the Options icon.**

2. **Choose Screen/Keyboard.**

3. **Select Backlight Brightness and choose a value lower than 100 (the brightest).**

   The lower the setting, the less battery life will be used, but also the dimmer the screen, which may make it harder to see in some cases.

4. **Select Backlight Timeout and choose a value lower than 2 min.**

   The lower the setting, the less the battery life that will be used, but the backlight will also time out faster when a key is not pressed — which can get annoying.

## Closing unused BlackBerry applications

The more BlackBerry applications you have running at once, the more battery power your battery consumes. Applications that run in the background, such as Facebook, Twitter, Viigo, and instant messaging applications, significantly decrease your battery even when you're not actively using them. Besides saving battery life, closing applications that you're not using makes your BlackBerry respond faster.

The best way to prevent these applications from draining your battery is to turn them off when you're not using them. To do so, press the Menu key and choose Exit or Shut Down instead of pressing the Escape key to send the application to the background.

Some applications need to run in the background; these applications won't show up as open applications. Each of these background processes takes up a small amount of memory and battery life. In the event that your device is having an issue, or the battery is draining quickly, you may want to try starting your BlackBerry smartphone in Safe mode, which loads only the Research In Motion default applications and no third-party applications. You can then troubleshoot and remove the application that could be causing the problem.

---

# Taking care of your BlackBerry

If you're like us, you want to get the most out of your BlackBerry until it finally transmits its last e-mail or you're ready to trade up to the latest model. These tips will help ensure that your BlackBerry stays at its operational best.

✔ **Keep your BlackBerry dry.** If your BlackBerry gets wet, remove the battery and dry your device was thoroughly as possible before reinserting it.

If your BlackBerry ever becomes completely soaked, for example, you dropped it in a puddle, all may not be lost. Try placing your BlackBerry in a bowl of rice overnight. The rice will draw out most of the moisture, and you may be pleasantly surprised by your BlackBerry booting up when you reinsert the battery.

✔ **Don't use or store your BlackBerry in dusty, dirty areas.** If dust gets inside your BlackBerry, it could malfunction or stop working completely.

✔ **Don't store your BlackBerry in abnormally hot or cold temperatures.** Abnormal temperature differences can permanently damage the internal circuitry of your BlackBerry.

✔ **Avoid dropping or shaking your BlackBerry smartphone.** Either can cause physical damage.

✔ **Avoid using harsh chemicals, cleaning solvents, or strong detergents to clean your BlackBerry smartphone.** You should use only a soft, dry, clean cloth to wipe the surface of your BlackBerry.

✔ **Do not paint your BlackBerry smartphone.** Paint could get into your BlackBerry through a small crack or other opening and may result in a malfunction.

✔ **Reset your BlackBerry from time to time by removing the battery for a few seconds and then reinserting it.** Resetting your BlackBerry periodically will clear anything that may be running in memory and help ensure optimum performance.

---

Follow these steps to enter into Safe mode:

*1.* **Remove your BlackBerry battery for about 10 seconds, and then reinsert it.**

*2.* **Wait for the LED light to turn off.**

*3.* **Press and hold the Escape key until your device finishes booting up.**

A dialog box appears, asking if you want to stay in Safe mode.

*4.* **Select OK.**

At this point, you can uninstall the application that you think might be causing problems. You can do this by going to the BlackBerry Options (wrench icon) and select Application options, and then uninstall the trouble-causing application.

5. **When you are finished with Safe mode, remove your BlackBerry battery for about 10 seconds, and then reinsert it again.**

   Make sure you do *not* hold the Escape key while your BlackBerry smartphone is booting up. This will allow it to boot normally.

# Keeping Your BlackBerry Safe

The folks at Research In Motion take security seriously, and so should you. Always set up a password on your BlackBerry. If your BlackBerry hasn't prompted you to set up a password, you should immediately do so. Here's how it's done:

1. **From the BlackBerry Home screen, select the Options (wrench) icon.**

2. **Select the Password option.**

3. **Select the Password field and choose Enabled.**

   All this does for now is enable the Password feature. You won't be prompted to type a password until you save the changes you just made.

4. **Choose the Set Password button.**

   At this time, you're prompted to enter a new password, as shown in Figure 3-8.

   Note that if you have set a password before, the button will be called Change Password.

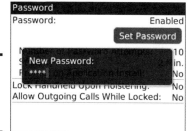

**Figure 3-8:** It's time to enter a new password.

5. **Type a password, and then type it again for verification.**

   From this point on, whenever you lock your BlackBerry and want to use it again, you have to type the password. How do you lock your BlackBerry? Good question. Keep reading.

Remember that when you set your password on a BlackBerry Pearl, you must make sure that you know what letters your password uses and not just which keys you pressed. You need the same password if you link your BlackBerry with BlackBerry Desktop Manager for synchronization. For more on BlackBerry Desktop Manager, see Book IV.

Setting up your password is a good first step, but just having a password won't help much if you don't take the further step of locking your BlackBerry when you're not using it. (You don't want people at the office, or those sitting at the next table at the coffee shop, checking out your e-mails or phone history when you take a bathroom break, do you?) So, how do you lock your BlackBerry? Let us count the ways . . . we came up with three.

Your BlackBerry can lock automatically after timeout (also known as security timeout) by following these steps:

1. **From the BlackBerry Home screen, select the Options icon.**

2. **Select the Password option.**

   The Password screen appears.

3. **Select the Security Timeout field and then choose the desired timeout option.**

   The preset times range from 1 minute to 1 hour.

4. **Press the Menu key and select Save.**

If you're more the hands-on kind of person, you can go the manual lockout route by scrolling to the Lock icon on your BlackBerry Home screen and pressing the trackball, trackpad, or touch screen.

Also, if you have a full QWERTY keyboard device, pressing K while you're at the Home screen does the same thing. Make sure to turn off the Dial from Home Screen option first. See Book I, Chapter 2 for more info on using Home screen shortcuts.

No matter what route you take to lock your BlackBerry, you use your (newly created) password to unlock it when you get back from wherever you've been.

## Blocking That Spam

With your existing BlackBerry operating system (OS), you can block certain e-mails, SMS numbers, or BlackBerry PINs from getting to your inbox. It's like having your own spam blocker on your BlackBerry!

To set up your personal spam blocker, follow these steps:

*1.* **From the BlackBerry Home screen, select the Options icon.**

*2.* **Select Security Options.**

*3.* **Select the Firewall option.**

The Firewall screen opens.

*4.* **Highlight the Status field and select Enable.**

The spam blocker is enabled.

*5.* **Under Block Incoming Messages, select the option(s) you would like to block:**

- *SMS:* Select this check box if you want to block SMS (text) messages.

- *PIN:* Select this check box if you want to block BlackBerry PIN messages.

- *BlackBerry Internet Service:* Select this check box if you want to block e-mail messages (for example, the e-mail account that you set up from Google or Yahoo! mail).

- *Enterprise Email:* Select this check box if you are within a large corporate e-mail network and want to block the enterprise e-mail.

*6.* **In the Except Messages From area, select the desired options.**

- *Contacts:* Select this check box if you want to block everything except the e-mails and phone numbers in your Contacts.

- *Specific Addresses:* Select this check box if you want to block everything specified by you. (You can set up the list below.)

*7.* **If you chose Specific Address, press the Menu key and select Configure Exception.**

The Firewall exception screen opens.

*8.* **Press the Menu key and select the desired options.**

- *Add Email:* You can specify the e-mails you want to block by selecting this check box.

- *Add PIN:* You can specify the BlackBerry PIN you want to block by selecting this check box.

- *Add Phone Number:* You can specify the SMS number you want to block by selecting this check box.

# Getting help from online resources

We know what you're thinking; all of the information provided in this chapter is fantastic, but what if you have configuration questions that we haven't covered? Never fear; you can use these links to find the answers you seek.

From your BlackBerry

✔ `http://mobile.blackberry.com`: RIM's mobile Web site has a Help & Support section with links for Getting Started, BlackBerry Answers, BlackBerry Support forums, and App World Mobile Support. It also has a search field right at the top so you can find the fastest answers possible.

From your PC

✔ `www.blackberry.com`: RIM's official Web site offers a variety of helpful documents. It also has a search feature so you can find the closest matches to your BlackBerry questions.

✔ `http://forums.crackberry.com`: Crackberry forums provide a wealth of information contributed by new and veteran BlackBerry users alike. If you can't find your question already answered, you can pose the question and have the knowledge you seek in no time.

✔ `www.blackberryforums.com`: BlackBerry Forums are similar to the forums at Crackberry.com. You can view questions, answers, and other content from fellow BlackBerry users and BlackBerry experts.

# Chapter 4: Connectivity Options

## In This Chapter

✔ Wireless carrier connection modes explained

✔ Connecting with Wi-Fi

✔ Tethering your BlackBerry to your laptop

*I*t all starts here. You can have the newest, biggest, fastest BlackBerry smartphone on the market, but it's practically useless if it doesn't have a wireless data connection. A wireless data connection is required for you to send and receive e-mail, SMS text and PIN-to-PIN messages, browse the Web, and utilize features in most of the third-party applications available today. If you have ever shopped for a cellphone, you have probably heard terms like GSM, 3G, CDMA, and EV-DO. There are so many acronyms out there that it's easy to get overwhelmed.

Fear not! In this chapter, we discuss the various wireless carrier connection modes, what they mean, and how fast they are. We also delve into the world of Wi-Fi connectivity. Finally, we show you how to tether your BlackBerry to your PC as a modem using a USB cable or Bluetooth.

## Choosing a Wireless Carrier Connection Mode

If you purchase a data plan, you do *not* have a choice on the type of service. The connection modes available on your BlackBerry vary based on the model you have and the wireless carrier that you purchase your monthly data plan through. Regardless of which carrier you have, your connection mode is classified under one of two technologies: Global System for Mobile Communications (GSM) or Code Division Multiple Access (CDMA).

Tables 4-1 and 4-2 give you an at-a-glance view of each connection mode and its features. We talk about each of these in more detail in the following sections.

To see a full list of models and what connection mode they support, check out Book I, Chapter 5.

| Table 4-1 | GSM-Based Connection Modes | |
|---|---|---|
| *Connection Mode* | *Available Features* | *Download Speed* |
| GSM | Phone calls and SMS text messages | Data is unavailable. |
| gprs | Phone calls and SMS text messages (in an area that supports it, but connected to it) | Data is unavailable. |
| GPRS | Phone calls and SMS text messages | 35 Kbps |
| EDGE | Phone calls, SMS text messages, PIN-to-PIN messages, e-mail, Web browsing, and third-party application and data downloads | 135 Kbps |
| 3g | Phone calls and SMS text messages (in an area that supports it, but connected to it) | Data is unavailable. |
| 3G (HSDPA) | Phone calls, SMS text messages, PIN-to-PIN messages, e-mail, Web browsing, and third-party application and data downloads | 1.7 Mbps |

| Table 4-2 | CDMA-Based Connection Modes | |
|---|---|---|
| *Connection Mode* | *Available Features* | *Download Speed* |
| 1x (Note the lower-case *x*.) | Phone calls and SMS text messages | Data is unavailable. |
| 1X | Phone calls, SMS text messages, PIN-to-PIN messages, e-mail, Web browsing, and third-party application and data downloads | 50 Kbps |
| 1xev | Phone calls and SMS text messages | Data is unavailable. |

| Connection Mode | Available Features | Download Speed |
|---|---|---|
| EV-DO Rev 0 (1XEV) | Phone calls, SMS text messages, PIN-to-PIN messages, e-mail, Web browsing, and third-party application and data downloads | 150 Kbps |
| EV-DO Rev A (1XEV) | Phone calls, SMS text messages, PIN-to-PIN messages, e-mail, Web browsing, and third-party application and data downloads | 701 Kbps |

# Global System for Mobile Communications (GSM)

GSM is the technology that GPRS, EDGE, and HSPA (3G), are based on. It operates on the 850, 900, 1800, and 1900 MHz frequencies, and it's the most widely used cellular technology globally. Many carriers, including AT&T, T-Mobile, and Vodafone, utilize GSM technology. If you travel, it's important to know what frequency your BlackBerry smartphone uses so you know where in the world your BlackBerry smartphone will work.

If your BlackBerry shows GSM in the top-right corner, you don't have a data connection and can't send or receive e-mails and PIN-to-PIN messages, browse the Web, or utilize third-party applications that require a data connection. However, you can place and receive phone calls as well as send and receive SMS text messages.

### General Packet Radio Service (GPRS)

GPRS is an add-on to GSM technology that allows a BlackBerry to have data and access the Internet. It is possible to lose GPRS and have it fall back to GSM so that you can only make calls and send SMS messages. A GPRS connection supports a download speed of approximately 35 Kbps.

If your BlackBerry smartphone shows GPRS in the top-right corner, you can use all the features you know and love (phone calls, SMS, e-mail, Web browsing, and applications). However, because this is the oldest and slowest connection mode, certain tasks like loading large Web pages or downloading applications will take longer than they do on faster data connections. Your device also can't process data and voice at the same time, which means you can talk on the phone or send and receive e-mail, but you can't do both at the same time.

### Enhanced Data Rates for GSM Evolution (EDGE)

EDGE is a revision to GPRS technology and offers improved download speeds of about 135 Kbps. This improved download speed allows for faster loading of larger Web pages and faster data access for third-party applications. Like GPRS, EDGE doesn't support data and voice at the same time, which means that you can't send and receive e-mail or log into an application while speaking on the phone.

If your BlackBerry smartphone shows EDGE in the top-right corner, you can use all the features of your BlackBerry with reasonably fast load times for Web pages, application downloads, and application data. This is the fastest connection mode that models like the BlackBerry 8820, BlackBerry Curve, and BlackBerry Pearl Flip support.

### High-Speed Packet Access (HSPA), also known as 3G

HSPA is the latest enhancement to GSM technology. It consists of two mobile communications protocols, High-Speed Upload Packet Access (HSUPA), which is dedicated to sending information to the Internet, and High-Speed Downlink Packet Access (HSDPA), which is to download information. Although the theoretic maximum download speed of HSPA is up to 14 Mbps, the current version of HSPA only supports up to 1.7 Mbps. This is still almost ten times faster than EDGE, which is a significant improvement.

If your BlackBerry smartphone shows 3G in the top-right corner, you are using HSPA and can enjoy the fastest load times available for loading Web pages, downloading applications, and loading application data. Best of all, your device can handle data and voice at the same time, which means that you can talk on the phone while sending and receiving e-mail or browsing the Web. The BlackBerry Bold is currently the only BlackBerry smartphone that supports 3G.

## Code Division Multiple Access (CDMA, 1x)

CDMA is a technology that carriers like Verizon Wireless and Sprint use to transfer voice and data to your BlackBerry smartphone. CDMA allows multiple senders to use the same channel to communicate by assigning a unique code to each person transmitting. This allows for much more data to be used over the same channel as other technologies. CDMA allows for higher speeds than most of the other connections; however, it is not commonly used outside of North America.

CDMA is not global. If you plan on taking your phone outside of North America, then you should use a GSM service. Wireless carriers like Sprint and Verizon have dual devices that will use GSM when you travel internationally. It is best to ask your local carrier if your phone will work where you plan to travel, to ensure you choose the right one.

If your BlackBerry shows 1x (lowercase x) in the top-right corner, it means you don't have a data connection and can't send or receive e-mails and PIN-to-PIN messages, browse the Web, or utilize third-party applications that require a data connection. However, you can still place and receive phone calls as well as send and receive SMS text messages.

### Connection mode 1X (uppercase X)

If your device shows 1X (uppercase X) in the top-right corner, it means that you can use all the features of your BlackBerry (phone calls, SMS, e-mail, Web browsing, and applications). However with only 50 Kbps download speeds, certain tasks like loading large Web pages or downloading applications will take longer than they do on faster data connections.

### Evolution-Data Optimized (EV-DO)

EV-DO is based on CDMA technology and provides higher speeds for downloading and uploading data. It comes in these two versions:

✦ **Rev 0:** EV-DO Rev 0 boasts a slightly higher download speed of approximately 150 Kbps compared to 50 Kbps using 1X. This lets you download files such as music, pictures, and Web pages quicker.

✦ **Rev A:** With speeds of 701 Kbps exceeding Rev 0 by over 550 Kbps, downloading large files can be done even quicker.

If your BlackBerry shows 1XEV in the top-right corner, you can take advantage of the fastest Web page load times, application download speeds, and third-party application data speeds available for CDMA devices. Older models, including BlackBerry 8330 and BlackBerry Pearl 8130, use EV-DO Rev 0, whereas newer models, including BlackBerry Storm and BlackBerry Tour, come with EV-DO Rev A.

## Exploring Wi-Fi Connectivity Options

Wi-Fi is a way to connect to the Internet without using the carrier's network, such as GSM and EV-DO. The difference is that GSM and EV-DO signals come from big antennas placed all over the country covering miles of land, and Wi-Fi is from a box that you install in your home or store that connects to your home or business Internet connection. These Wi-Fi boxes cover only a few hundred feet.

Wi-Fi is a useful feature available on the BlackBerry Bold (9000), BlackBerry Curve (8900), BlackBerry Pearl Flip (8220), and several older models, including the 8820, Curve 8320, and Pearl 8120.With this feature, you can access data at Wi-Fi speeds, which provides faster downloads, Web browsing, and a better experience for streaming media content (think YouTube). Best of all,

Wi-Fi makes it possible to use your BlackBerry in places where there is no wireless carrier signal.

You need an active data plan with your wireless carrier to receive data over Wi-Fi on your BlackBerry. When Wi-Fi is active, your BlackBerry chooses the fastest path available for data. Wi-Fi cannot be used if your BlackBerry does not have a data plan with your carrier. Carriers require this so that you still have to pay a monthly service charge to use their service and cannot use their device freely on Wi-Fi all the time.

Follow these steps to connect to a Wi-Fi hot spot, your company's wireless router, or your own wireless router:

1. **From your BlackBerry Home screen, choose Settings or Setup⇨Set Up Wi-Fi**

   The Wi-Fi Connections screen appears.

2. **Choose Scan For Networks.**

   Your BlackBerry scans for available Wi-Fi connections and returns a list of what was found. See Figure 4-1.

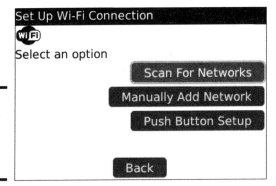

**Figure 4-1:** Scan for or manually add Wi-Fi networks.

3. **Choose the Wi-Fi network connection you would like to add.**

4. **(Optional) If the Wi-Fi network you choose has security enabled, enter the key, and then choose Next.**

5. **Enter a name for this Wi-Fi connection, and then choose Finish.**

   Your newly added connection is listed in the Wi-Fi Connections screen, as shown in Figure 4-2.

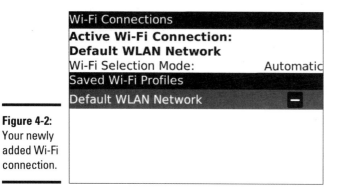

**Figure 4-2:**
Your newly
added Wi-Fi
connection.

6. **Press the Escape key twice to return to the BlackBerry Home screen.**

7. **From the Home screen, choose the Manage Connections (the antenna) icon.**

8. **On the menu that comes up, make sure Wi-Fi is selected.**

9. **Press the Escape key to return to the BlackBerry Home screen.**

Your BlackBerry now automatically switches between Wi-Fi and your wireless carrier if you leave your office or home.

Your BlackBerry can also make and receive phone calls over Wi-Fi using either Unlicensed Mobile Access (UMA) or Voice over IP (VoIP), which we discuss in the next sections. Although both of these protocols work over Wi-Fi, they differ in how they work and why they are used.

## Unlicensed Mobile Access (UMA)

Unlicensed Mobile Access (UMA) makes it possible to seamlessly connect GSM networks (available through providers such as T-Mobile and AT&T in the U.S.) with Wi-Fi by treating a Wi-Fi network like a cell tower. In a UMA-enabled BlackBerry, calls are transferred without interruption from a cellular network to a Wi-Fi network. If you walk into range of a Wi-Fi network that has a connection profile on your BlackBerry, you are switched automatically. Walk out of Wi-Fi range, and you're back on GSM. Wireless carriers use UMA to offer a more attractive calling plan with higher profit for them and lower costs for you. They can do this because cell calls sent over a cellular data network cost about five cents a minute for that transport, but over a Wi-Fi network that the carrier doesn't own, they cost only a penny a minute. The cost is even less than a penny when it's over a "Bring Your Own Wi-Fi" network, such as the one in your home or office.

UMA is not available for every carrier, and costs may vary from free to a few dollars per month. The benefits of using UMA are no/low cost per minute that do not subtract from your minute plans. Depending on your cellular coverage, with UMA over Wi-Fi your call quality is always clear and in most cases sounds like talking from a land line. UMA also is great for those areas like a basement or buildings that don't have voice coverage, but can access a Wi-Fi spot.

### Voice over Internet Protocol (VoIP)

Voice over Internet Protocol (VoIP) allows telephone calls to be made over computer networks. It converts analog voice signals into digital data packets and supports real-time, two-way transmission of conversations using Internet Protocol (IP). Many larger companies use VoIP in their desk phone systems. Wi-Fi comes into play when these companies are looking for ways to extend desk phone functionality to mobile devices like BlackBerry smartphones. Research In Motion provides a solution called Mobile Voice System, or MVS, which extends VoIP to BlackBerry devices. Check out Book VIII, Chapter 4 for more information on MVS.

## Using Your BlackBerry as a Modem

Have you ever been somewhere with your laptop and wanted to access the Internet, but there was no Wi-Fi signal to be found? You could carry a WAN card provided by a wireless carrier, but why carry around extra equipment and pay more for an additional monthly data plan when you already carry your BlackBerry? You even have two options for connecting your BlackBerry as a modem: through a USB cable or via Bluetooth technology.

Some wireless carriers provide their own software to use your BlackBerry smartphone as a modem, like Verizon's VZ Access Manager or AT&T's Connection Manager, which simplifies the process of using your BlackBerry as a modem. Ask your wireless carrier's customer service whether any connection software is available and, if so, how to download it.

Using your BlackBerry as a modem significantly increases the amount of data that it uses. This can become quite costly if you do not have an unlimited data plan, so proceed with caution if your monthly plan has limits on the amount of data you can use.

### BlackBerry as a USB modem

Using your BlackBerry as a USB modem is easy; follow these steps:

*1.* **Make sure you have the latest BlackBerry Desktop Manager installed and configured.**

See Book IV, Chapter 1 for more information on installing and configuring BlackBerry Desktop Manager.

*2.* **Open Desktop Manager, connect your BlackBerry to the USB cable, and enter your security password if prompted.**

*3.* **From the main screen of Desktop Manager, click IP Modem.**

The IP Modem icon will appear only if your BlackBerry smartphone is permissioned from the carrier to allow for tethering of your BlackBerry. *Tethering* is what carriers call the ability to use your BlackBerry smartphone as a wireless modem that will connect your laptop to the Internet.

*4.* **Check your Connection Settings.**

On the left panel of the IP Modem screen in Desktop Manager, your carrier should be displayed next to Current Connection Profile. If the carrier is incorrect, select Configure to manually change it to the same carrier that your BlackBerry smartphone has. (See Figure 4-3.)

**Figure 4-3:**
Desktop
Manager
IP Modem
screen.

*5.* **Select Connect.**

After establishing a connection, Desktop Manager displays a connection window that has information about the duration that you have been connected and the amount of data that you have used. (See Figure 4-4.) At this point, you are able to browse the Internet on your laptop.

**Figure 4-4:**
Desktop
Manager IP
Modem with
an active
connection.

## BlackBerry as a Bluetooth modem

Using your BlackBerry as a Bluetooth modem is simple; just stick to the
instructions that follow, and you'll be surfing wirelessly in no time.

*1.* **Make sure you have the latest BlackBerry Desktop Manager installed
and configured.**

This is important, as Desktop Manager installs the required modem drivers as part of its setup process, which allows you to find the modem in
Step 3. See Book IV, Chapter 1 for more information on installing and
configuring BlackBerry Desktop Manager.

*2.* **On your PC, choose Start⇨Control Panel⇨Phone and Modem.**

*3.* **Click the Modems tab, click the Standard Modem over Bluetooth link
that is listed, and then click Properties.**

The Standard Modem over Bluetooth link Properties dialog box appears.
(See Figure 4-5.)

*4.* **Click the Advanced tab and enter the initialization command for your
wireless carrier.**

The initialization command varies based on your carrier. The easiest
way to find the correct command is to perform a Google search for
"Initialization command for *carrier name*." For example, if you want to
find the initialization command for T-Mobile, you can perform a Google
search for *initialization command for T-Mobile*.

*5.* **Click OK to close the Standard modem over Bluetooth link screen, and
then click OK on the Phone and Modem screen to close that as well.**

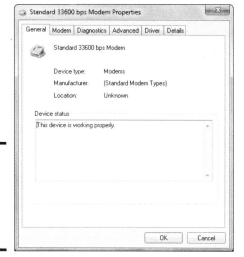

**Figure 4-5:**
The
Standard
Modem
Properties
screen.

6. **Choose Start⇨Control Panel.**

7. **Click Network and Sharing Center.**

8. **Choose Set Up a New Connection or Network.**

9. **Choose Set Up a Dial-Up Connection, and then click Next.**

10. **Choose the Standard Modem over Bluetooth link, and then click Next.**

11. **Enter *99# for the dial-up phone number. (See Figure 4-6.)**

**Figure 4-6:**
Enter the
dial-up
phone
number,
username,
password,
and
connection
name.

*12.* **Enter the username and password required for your wireless carrier.**

The username and password vary based on your carrier. Some carriers, including T-Mobile, don't even require a username and password for dial-up access. The best way to find the correct username and password is to perform a Google search for "APN credentials for *carrier name*." For example, if you want to find the username and password for AT&T, you can perform a Google search for *APN credentials for AT&T*.

*13.* **Enter a name for your connection (we use BlackBerry Bluetooth Modem), and then click Connect.**

Your computer tests the newly created connection to verify that it works. If it fails, you may need to retry Steps 4 and 12 with a different combination of APN and/or username.

*14.* **Click Finish to exit the wizard once it is done.**

*15.* **Open the Network and Sharing Center and click Connect to a Network.**

*16.* **Choose BlackBerry Bluetooth Modem and click Connect.**

# Chapter 5: Checking Out the Latest BlackBerry Models

## In This Chapter

✔ Choosing your BlackBerry flavor

✔ Getting a summary of BlackBerry model features

✔ Considering the pros and cons of each BlackBerry model

**G**ood news! There are ten different BlackBerry models to choose from. Better news! Each model comes in one of three flavors. In this chapter, we describe each flavor so that you can make the right choice for your business and personal needs. Then we detail the features of each model. We also cover some things you need to know about carriers and coverage. Finally, we give you a look at the details you need to know about each BlackBerry model so that you can confirm that you're making the right choice before you make a purchase.

## Choosing Your Favorite BlackBerry Flavor

With ten different BlackBerry models available, making a choice may seem overwhelming. Don't worry! We've grouped your options into three flavors: SureType, touch screen, and QWERTY.

Each BlackBerry model has the following features built in:

✦ **Dialing, texting, and e-mail:** Okay, we know we sound like Captain Obvious, but it's important to know that you can use your smartphone to make a simple phone call, as well as send text messages and e-mail just like you'd expect.

✦ **Expandable memory:** All BlackBerry models include an expandable memory card slot. This means that you can insert a microSD memory card, which gives you lots more room to store video, pictures, and more.

✦ **Built-in camera:** All BlackBerry models have a built-in camera with a flash.

✦ **Social media syncing and Internet access (as long as you pay for it):** No matter which BlackBerry you purchase, you can use it to access the Web and update your status on sites like Facebook, MySpace, Twitter, and Foursquare. That's right; as long as you pay your provider for the privilege, you can update your status anywhere, anytime, 24/7. You can also upload pictures from your BlackBerry directly to the social networking site of your choice.

Various models of BlackBerry are available exclusively through specific cellphone carriers. While we don't go into the specifics of which phone models are offered by which carrier, you need to know that carriers handle voice and data coverage in different ways. There are two types of carriers — Global System for Mobile Devices (GSM) and Code Division Multiple Access (CDMA). See the sidebar, "Understanding carrier options," as well as Book I, Chapter 4 for more information about the differences between GSM and CDMA models.

## Being sure with SureType

A *SureType* BlackBerry has a traditional phone keypad, with two letters per key. One perk of a SureType option is that these phones are close to the size of a traditional cellphone and will fit in your pocket nicely. SureType phones are good options if you're looking for a compact device, don't care about high-resolution media, and don't want to spend much money. If this sounds like you, the BlackBerry Pearl Flip 8220 or 8230 should be the first models you take a look at.

## Getting in touch with a touch-screen phone

The first benefit of a touch-screen BlackBerry is the fact that you don't have to do a whole lot of typing. In addition, BlackBerry smartphones with touch screens include a nice, large screen for viewing media. Touch-screen BlackBerry smartphones include the BlackBerry Storm and Storm 2. They tend to be mid- to high-range in price, and because of the newness of the touch-screen technology, there are some limitations you should look into. We tell you more about those limitations in the "Touch-screen BlackBerry smartphones" section later in the chapter.

## Getting QWERTY

If you want the best of both worlds, you might enjoy a *QWERTY* device, which means it has a full keyboard. QWERTY devices combine the convenience of fast typing with a respectable screen size. The majority of BlackBerry models are available in this flavor, which means that you have a variety to choose from. Some QWERTY phones are inexpensive (even free), but others can be quite pricy. One size does not fit all.

## Understanding carrier options

When considering a type of carrier, you have a couple of options. The first is Global System for Mobile Communications (GSM). The other option is Code Division Multiple Access (CDMA).

GSM carriers provide better global coverage, and faster download speeds in areas where 3G networks are available. This type of carrier is a good choice if you plan to travel internationally or if you plan to use your BlackBerry to surf the Web and use other media features.

CDMA carriers provide better voice quality on calls and have coverage in some of the few areas that GSM does not. This is a good option if you want to use your BlackBerry primarily locally and for making phone calls. CDMA-based devices tend to have poorer battery life than GSM based devices.

Read Book I, Chapter 4 for more information on the differences between the carrier network types.

# Summarizing BlackBerry Bonus Features

Each BlackBerry flavor provides a specific method for accessing the data screen, dialing, typing e-mails, accessing the Web, and taking photos. (Refer to the preceding "Choosing Your Favorite BlackBerry Flavor" if you haven't already read that section.) In addition, each model has bonus features that go beyond the basics. Start by deciding which features are "must haves" to narrow your choices. Here are some features that are available on various BlackBerry models:

✦ **Wi-Fi capability:** No wireless carrier signal? No problem! With Wi-Fi capability, your BlackBerry can connect to the Internet using a wireless network at your home, office, or even many public places like Starbucks or McDonald's. If Wi-Fi is a must for you, skip the BlackBerry Tour, Pearl Flip 8230, and BlackBerry Storm.

✦ **GSM 3G capability:** The BlackBerry Bold 9000 and BlackBerry Bold 9700 have GSM 3G capability, which means that they can download data faster and browse the Web more quickly than all other BlackBerry models. These models also allow you to talk on the phone and send an e-mail (or surf the Web) at the same time.

✦ **Global Positioning System (GPS):** A BlackBerry with GPS capability can determine its location and use it to provide location-based services, including turn-by-turn directions and local points of interest. Want this feature? If so, the BlackBerry Curve 8520 and BlackBerry Pearl Flip 8220 aren't right for you since they don't have it. All other BlackBerry models have GPS built-in.

✦ **Additional UMTS wireless antenna:** The BlackBerry Bold 9000, BlackBerry Bold 9700, BlackBerry Tour, BlackBerry Storm, and BlackBerry Storm 2 come with an additional wireless antenna that supports the 2100 MHz UMTS (Universal Mobile Telecommunications System) signal used in Japan and South Korea. These BlackBerry models can access 3G data speeds in the U.S. and also work anywhere in the world (as long as there's a wireless signal available, of course).

Table 5-1 gives you an overview of which features are available on the BlackBerry models.

**Table 5-1        BlackBerry Models and Features**

| Model | Keyboard | Basic Features | Wi-Fi | GSM 3G | GPS | UMTS Antenna | Carrier |
|---|---|---|---|---|---|---|---|
| Bold 9000 | QWERTY | Camera, expandable memory, Web access | Yes | Yes | Yes | Yes | GSM |
| Bold 9700 | QWERTY | Camera, expandable memory, Web access | Yes | Yes | Yes | Yes | GSM |
| Curve 8520 | QWERTY | Camera, expandable memory, Web access | Yes | No | No | No | GSM |
| Curve 8530 | QWERTY | Camera, expandable memory, Web access | No | No | Yes | No | CDMA |
| Curve 8900 | QWERTY | Camera, expandable memory, Web access | Yes | No | Yes | No | GSM |
| Pearl Flip 8220 | SureType | Camera, expandable memory, Web access | Yes | No | No | No | GSM |
| Pearl Flip 8230 | SureType | Camera, expandable memory, Web access | No | No | Yes | No | CDMA |
| Storm | Touch screen | Camera, expandable memory, Web access | No | No | Yes | Yes | CDMA |
| Storm 2 | Touch screen | Camera, expandable memory, Web access | Yes | No | Yes | Yes | CDMA |
| Tour | QWERTY | Camera, expandable memory, Web access | No | No | Yes | Yes | CDMA |

# *Getting the Down and Dirty about Your Favorite BlackBerry Models*

In this section, we give you more details on what each model has to offer. We group the discussion by flavor — SureType, touch screen, and QWERTY. For an overview of these flavors, refer to the earlier section "Choosing Your Favorite BlackBerry Flavor." For a bird's-eye-view of all the extra features each model has to offer, refer to the preceding section "Summarizing BlackBerry Bonus Features."

## *SureType BlackBerry smartphones*

A SureType keyboard is similar to a traditional cellphone. Each key has two letters on it. SureType technology learns and suggests the words you type most often using a combination of keys. This type of keyboard takes some time to get used to, but once you adjust to it and it learns your typing habits, you can type on it almost as quickly as a full QWERTY keyboard. Read Book I, Chapter 2 for more information on SureType keyboards.

There are two models of SureType BlackBerry smartphones, the Pearl Flip 8220 and the Pearl Flip 8230. Each model is a flip phone — and it's also a BlackBerry. All in one! The greatest benefit of the Pearl Flip is that it's small enough to fit in your pocket. That's not what you usually think of when you think of BlackBerry, is it?

Here's some more information about each model:

✦ **Pearl Flip 8220:** The SureType keyboard has big, soft buttons that are easy to use. Although the display is smaller than on other models, it does a decent job of displaying your pictures, videos, and e-mail. The Pearl Flip 8220 also has a smaller display on the outside that shows the time and a preview of incoming text messages, e-mails, and instant messages.

  • *Pros:* The Pearl Flip 8220 is great if you want the global dependability of GSM technology, spend a lot of time on the phone, and want the functionality of a BlackBerry without holding a bulky device up to your ear. It also has the added bonus of Wi-Fi capability and tends to be on the lower end of the cost spectrum. Figure 5-1 shows the BlackBerry Pearl Flip 8220 along with its CDMA-based counterpart, the BlackBerry Pearl Flip (8330).

  • *Cons:* There are a few reasons this may not be the BlackBerry for you. It has the smallest screen of all BlackBerry models, so it's not the best BlackBerry model if you want to view photos or videos. This model also lacks all the bells and whistles that you're probably looking for in a BlackBerry. It has no 3G functions, no GPS, and no UMTS antenna, which means it won't work in Japan or South Korea.

Send
Mute

Volume keys
End/Power

Front Display

**Figure 5-1:**
The Black
Pearl Flip
8220 and
Pearl Flip
8230 look
the same.

Alt

Escape

Right convenience key

Trackball

Symbol

Enter

Menu

Left convenience key

✦ **BlackBerry Pearl Flip 8230:** The BlackBerry Pearl Flip 8230 is almost
identical to the Pearl Flip 8220.

• *Pros:* The Pearl Flip 8230 has clear voice call quality because it uses a
CDMA carrier. The phone is compact. It has a comfortable SureType
keyboard and smaller high-resolution display that does a good job
of displaying your pictures, videos, and e-mails. This model includes
GPS capability, which the Pearl Flip 8220 lacks.

• *Cons:* The Pearl 8230 has all the same cons as the 8220 — except for
its GPS functions.

## Touch-screen BlackBerry smartphones

Touch-screen BlackBerry smartphones use *SurePress* technology. A SurePress screen feels like a keyboard because it clicks when you press it! This touch-screen interface gives you a unique navigation experience and gives you application options that aren't available on other models. It also lets you choose between a full QWERTY keyboard, which is the same layout used on your PC, or a SureType keyboard, where keys share letters.

Some applications available for older BlackBerry models don't work properly on touch-screen BlackBerry smartphones.

Both BlackBerry touch-screen models are created by CDMA carriers, so you will enjoy great sound quality. At this time, there are no GSM-based touch-screen BlackBerry models available.

Here's some more information about each model:

✦ **BlackBerry Storm:** On its large, bright display, pictures, videos, Web sites, and more look great. This bad boy has 1GB of built-in memory. The touch-screen interface offers you the choice of a SureType keyboard or QWERTY keyboard — it just depends on how you hold the phone.

   • *Pros:* This model also has an additional UMTS antenna, so you can use it anywhere in the world — as long as you can get a wireless data signal.

   • *Cons:* The Storm's larger display makes it bigger and bulkier than the other BlackBerry models. You may find it's a little too bulky to comfortably type on or stick in your pocket. Another con is that it has a shorter battery life than other models. The touch screen doesn't support *multitouch* capability (touching multiple parts of the screen at the same time), which can slow you up when typing because you have to completely release the screen after each letter. Another consideration is that touch-screen versions of apps tend to take up more memory. Surprisingly, the BlackBerry Storm also has no Wi-Fi capability. This model is also at the middle to high end of the BlackBerry cost spectrum, so be prepared to pay a little extra if you want the flashiness of touch screen.

Our overall recommendation is that the Storm is a good option if you're a media power user who likes to browse the Web, listen to music, and watch videos more than you like to send e-mails and text messages. Figure 5-2 gives you an idea of what the Storm looks like.

Left convenience key

Lock    Mute

**Figure 5-2:**
The
BlackBerry
Storm and
BlackBerry
Storm 2 look
the same.

Send    Escape    Right convenience key

Menu    End/Power

✦ **BlackBerry Storm 2:** Offers the same features as the Storm, including a great media experience. It also improves on a lot that the first model doesn't have quite right.

- *Pros:* The Storm 2 has Wi-Fi capability. It also uses a new *multitouch technology,* which allows you to touch multiple parts of the screen at the same time. It has twice as much built-in memory as the original Storm, so you can install more apps. With the same size display screen as the Storm, the Storm 2 is slightly thinner and lighter — it feels a lot less clunky in your pocket.

- *Cons:* Even though the BlackBerry Storm 2 is multitouch capable, the technology is so new that few applications support it. Also, although the Storm 2 is trimmer and slimmer than the Storm, it's still a little bulky compared to the other BlackBerry models. Figure 5-2 shows the BlackBerry Storm 2.

## QWERTY BlackBerry smartphones

The majority of the latest BlackBerry models come with a full QWERTY keyboard, which is the same keyboard layout that you use on your PC. The QWERTY keyboard is essential if you e-mail, text, and IM a lot. These devices are the most comfortable to type on, but the full keyboard makes them larger than a SureType flip phone. You also don't get the larger screen size of a touch-screen phone because the keyboard takes up more space.

### GSM BlackBerry smartphones with QWERTY keyboards

The following BlackBerry models couple the global dependability of GSM technology with the typing-based friendliness of a full QWERTY keyboard:

✦ **BlackBerry Bold 9700:** In the GSM and full QWERTY arena, the BlackBerry Bold 9700 is the Rolls-Royce of smartphones.

- *Pros:* It has it all — trackpad navigation and a compact and lightweight design with a comfortable keyboard — great for typing or talking. It also has 3G capability and a UMTS antenna, so it works everywhere in the world. We've used 'em all, and all we can say about this model is, "Wow!" It feels great in your hand, whether you're typing or using the phone, and fits perfectly in your pocket. The high-resolution screen also makes videos, pictures, and even e-mails look fantastic.

The *trackpad* is a motion-sensitive square, located where the track-ball is on other models. You swipe your thumb on the trackpad to move around and press it to make a selection. If you've ever used a trackball BlackBerry model (or even one of those old-school laptops with the mouse nub), you probably know that they attract dirt and can easily get stuck. The trackpad eliminates these annoyances.

- *Cons:* The BlackBerry Bold 9700 is not without flaws. It has a smaller screen than the BlackBerry Bold 9000 and much less installed memory (256MB, as opposed to 1GB in the Bold 9000). Its higher-resolution screen and expandable memory card slot help make up for these shortcomings, though. Because of all the bells and whistles that come with this model, it's at the top of the cost spectrum.

Our expert opinion is that if you're looking for a high-end BlackBerry, and money is no object, you're done shopping. Figure 5-3 gives you a glimpse of the BlackBerry Bold 9700.

**Figure 5-3:**
The
BlackBerry
Bold 9700.

✦ **BlackBerry Bold 9000:** This beast of a phone lives up to its namesake.
It's a big, bold, powerful smartphone.

• *Pros:* The Bold 9000 has a large, high-resolution screen — the largest
of any QWERTY BlackBerry model. It has 3G capability and an additional UMTS wireless antenna, which means it's a necessity if you're
a global jet setter. It also comes with 1GB of built-in memory, so you
have plenty of room for your favorite apps. It has a large, soft-keyed
keyboard, making it very comfortable to type on.

• *Cons:* The large screen and keyboard are nice, but they also make
the BlackBerry Bold 9000 the heaviest and bulkiest of the QWERTY
BlackBerry smartphones. This beast may be too big — and heavy —
for you, and it definitely won't fit in your pocket. The larger screen
also tends to drain the battery, so be prepared to charge it frequently. All of the features packed into this model also place it at the
higher end of the cost spectrum, so be prepared to shell out a few
extra dollars if this looks like the model for you.

Our overall recommendation is that if you're a world traveler who loves
to e-mail and text as much as you love to browse the Web and view
streaming media, this is an ideal phone for you. (It's an extra bonus if
you're over six feet tall, because at that height, the phone won't look
so ginormous.) Figure 5-4 helps familiarize you with the anatomy of the
BlackBerry Bold.

Right convenience key

Mute          Volume keys

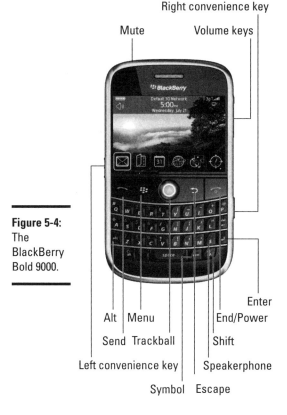

**Figure 5-4:**
The
BlackBerry
Bold 9000.

Enter

Alt | Menu          End/Power

Send  Trackball       Shift

Left convenience key          Speakerphone

Symbol   Escape

✦ **BlackBerry Curve 8900:** Another phone that lives up to its namesake, the Curve 8900 is sleek and effective.

- *Pros:* The BlackBerry Curve 8900 does an excellent job of displaying high-quality video content and pictures. The compact keyboard is comfortable to type on. This model is also small and light, and fits nicely in your pocket. It's also at the low to mid-range of the cost spectrum, so you can get a decent BlackBerry that won't leave you broke.

- *Cons:* The BlackBerry Curve 8900 doesn't have 3G capability, so you'll see slower download and browsing speeds. It also lacks the additional UMTS antenna, so it doesn't work in Japan or South Korea.

If you're looking for a device that can comfortably function as a phone, in addition to handling all of your e-mail, text, and media needs, take a look at the BlackBerry Curve 8900. Figure 5-5 gives you a glimpse of the BlackBerry Curve 8900.

Left convenience key    Volume keys

Lock    Mute    Right convenience key

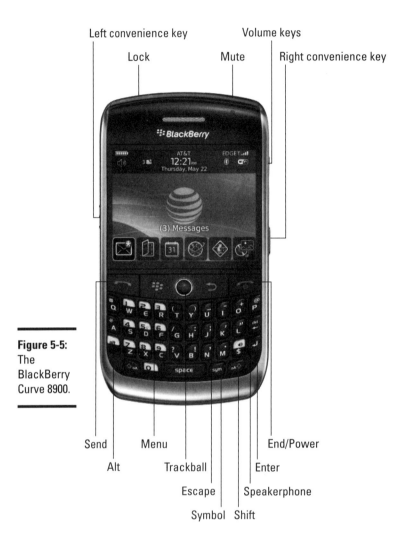

**Figure 5-5:**
The
BlackBerry
Curve 8900.

Send    End/Power

Alt    Enter

Trackball

Escape    Speakerphone

Menu

Symbol  Shift

+ **BlackBerry Curve 8520:** The BlackBerry Curve 8520 has some unique innovations that make it sleek and efficient.

- *Pros:* The Curve 8520 uses the trackpad navigation method instead of a trackball. The Curve 8520 has media buttons along the top edge, which let you pause, play, and skip songs without having to unlock your BlackBerry. This model is also the perfect size and weight for your hand and your pocket.

- *Cons:* The BlackBerry Curve 8520 has the lowest screen resolution of all the QWERTY BlackBerry models. Your pictures and videos still look pretty good, but they don't look as great as they do on the other QWERTY models. This model also lacks 3G as well as GPS capability, so you want to skip this device if either of these is on your must-have list. Finally, the BlackBerry Curve 8520 does not have an additional UMTS antenna, which means it doesn't work in Japan or South Korea.

Our expert opinion is that if you're looking for a starter phone, you could do worse than the Curve 8520. It comes with a lower price tag than all the other BlackBerry devices — it frequently comes free with a new contract. If you're looking for a cost-efficient, entry-level BlackBerry with a cool new navigation method and easy music navigation, look no further. But if you want every bell and every whistle (and you're willing to pay the piper), skip this model. Figure 5-6 gives you a glimpse of the BlackBerry Curve 8520.

**Figure 5-6:**
The
BlackBerry
Curve 8520.

### CDMA BlackBerry smartphones with QWERTY keyboards

The following BlackBerry models couple the superior call quality of CDMA technology with the typing-friendly layout of a full QWERTY keyboard:

✦ **BlackBerry Curve 8530:** The perfect-sized smartphone for texting, making calls, or just hanging out in your pocket.

  • *Pros:* Similar to the Curve 8520, this model of Curve also uses track-pad navigation and built-in media buttons along the top edge. The Curve 8530 is also one of only two CDMA BlackBerry smartphones with Wi-Fi capability — the other one is the BlackBerry Storm 2. It's also at the lower end of the BlackBerry cost spectrum. In some cases, it's even free with a carrier contract.

  • *Cons:* Of course, there are a few things you should be aware of if you're considering the BlackBerry Curve 8530. This model has a lower screen resolution than most other BlackBerry smartphones. Like the BlackBerry Tour, you can't send e-mails while talking due to no GSM 3G capability, and it's slower than GSM-based 3G devices like the BlackBerry Bold 9000 and BlackBerry Bold 9700. Also, the BlackBerry Curve 8530 doesn't have the additional UMTS antenna required for it to work in Japan or South Korea.

In our expert opinion, if you're looking for a cost-effective, entry-level BlackBerry with a cool new navigation method and easy music navigation, look at the Curve 8530. The only major difference between this phone and the Curve 8520, which we discuss earlier, is that the 8520 is a GSM phone and the 8530 has the superior call quality of CDMA. If you want a smartphone with more features, keep shopping. Figure 5-7 gives you a glimpse of the BlackBerry Curve 8530.

✦ **BlackBerry Tour:** The Tour has amazing voice call quality (probably better than a regular land line).

  • *Pros:* The Tour's data connection is much faster than the BlackBerry Curve 8900 and BlackBerry Curve 8520. The keyboard is a good size for typing, but it's not overwhelming or bulky. This model also includes an additional UMTS antenna, which means you can use it anywhere in the world that a wireless carrier connection is available. This model also offers a decent array of features at a mid-level cost that won't break the bank.

  • *Cons:* There are a few things you should note if you're considering the BlackBerry Tour. First, it doesn't have Wi-Fi capability. Also, its data connection speed is slower than GSM-based 3G devices like the BlackBerry Bold 9000 and BlackBerry Bold 9700. You can't use the phone while e-mailing, Web browsing, or texting. The Tour is known to have a shorter battery life.

**Figure 5-7:**
The
BlackBerry
Curve 8530.

Overall, the BlackBerry Tour is ideal if you're a world traveler who
wants the full QWERTY keyboard experience along with the superior call
quality of CDMA. However, if Wi-Fi is a must, skip this device. Figure 5-8
shows the BlackBerry Tour.

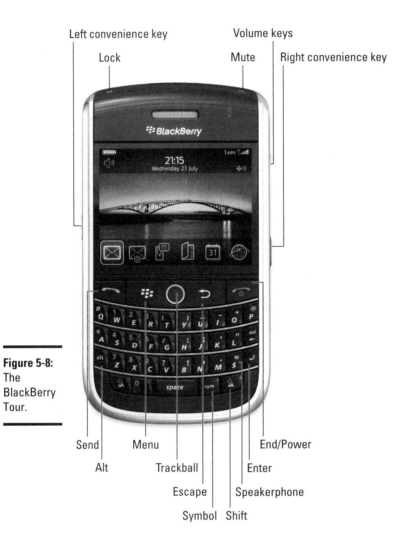

Left convenience key

Lock

Volume keys

Mute

Right convenience key

**Figure 5-8:**
The
BlackBerry
Tour.

Send

Alt

Menu

Trackball

Escape

Symbol Shift

End/Power

Enter

Speakerphone

# Chapter 6: Checking Out the Device Software 5.0 Features

## In This Chapter

✓ Exploring the new features in BlackBerry Handheld Firmware 5.0

✓ Discovering enhancements to existing Firmware features

*C*utting-edge technology is cutting edge specifically because it's always changing and evolving. Research In Motion (RIM) — the company that makes BlackBerry smartphones, firmware, and BlackBerry Enterprise Server software — is no exception to this rule. Its Device Software continues to evolve. (*Firmware* is the operating system and set of core applications that make your BlackBerry smartphone work and perform at its best.) Over the years, the firmware has incorporated wireless synchronization, spell checking, and HTML e-mail support.

In this chapter, we explore the new features and existing feature enhancements in BlackBerry Device Software 5.0.

## New Features in Device Software 5.0

BlackBerry Device Software 5.0 includes a bunch of new features that are designed to give you a better experience than ever before. Here is a rundown of the new features — as well as what they can do for you:

✦ **Watch the progress of your BlackBerry startup.** A new device startup screen (see Figure 6-1) shows the BlackBerry logo with a progress bar while the BlackBerry is starting up. This feature is great because you can gauge how long the system will take to boot up. It's certainly better than staring at the endlessly spinning hourglass.

**Figure 6-1:** The new BlackBerry startup screen added in Firmware 5.0.

✦ **Set your defaults with ease.** From the BlackBerry Home screen, you can set the default downloads folder, wallpaper, and theme for your BlackBerry. To change your defaults, follow these steps:

1. *From the BlackBerry Home screen, press the Menu key.*

2. *Select Options on the menu that appears. (See the left side of Figure 6-2.)*

Home Screen Preferences opens. (See the right side of Figure 6-2.)

**Figure 6-2:**
Device
Software 5.0
adds an
option to
customize
your Home
screen.

✦ **Change the layout of icons.** You can reset the icon layout of your chosen Home screen theme without affecting the other themes on your BlackBerry. Follow these steps:

1. *Go to the BlackBerry Home screen and press the Menu key.*

2. *Select Options.*

3. *Scroll down and select Reset Icon Arrangement. (See the left side of Figure 6-3.)*

4. *Choose Yes to confirm that you would like to reset your Home screen icon arrangement. (See the right side of Figure 6-3.)*

**Figure 6-3:**
Select
Reset Icon
Arrange-
ment, then
select Yes to
reset Home
screen icons.

✦ **Quickly switch between applications.** Simply hold the Menu key to bring up the Application Switcher. (See Figure 6-4.)

Scroll left or right to the app you want to switch to then press to select it.

**Figure 6-4:**
Press and
hold the
Menu key to
bring up the
Application
Switcher.

✦ **Add a personal e-mail account.** A new Java-based interface makes
   adding *BIS* (short for BlackBerry Internet Service) e-mail accounts —
   better known as personal e-mail accounts — a breeze.

   Read Book III, Chapter 2 for more information on adding personal e-mail
   accounts.

# Reviewing Device Software 5.0 Enhancements

RIM has made significant improvements to some of the most popular and
important features that you have already grown to love in earlier versions of
BlackBerry firmware. In the following sections, we discuss these improvements.

## Making BlackBerry e-mail even better

BlackBerry Device Software 5.0 introduces several improvements to BlackBerry
e-mail. These improvements make it easier for you to do the following:

✦ **Figure out the date an e-mail was sent or received.** The date header
   always shows at the top of the messages list, even when you scroll down a
   long list of messages sent and received on the same day. This change makes
   it much easier to keep track of when messages were sent or received.

✦ **Automatically resize photos before you attach them to an e-mail.** RIM
   has created a new photo-sending option that allows you to resize images
   based on a predefined list of standard desktop computer resolutions.

Follow these steps to choose a resolution when sending a photo:

*1.* **Compose a new e-mail message.**

   Read Book III, Chapter 2 for more information on composing and sending e-mails.

*2.* **Press the Menu key and select Attach File.**

*3.* **Browse to and select the picture you would like to attach.**

   The Select an Image Size screen automatically appears.

**4.** **Select the desired option. (See Figure 6-5.)**

- *Original:* Keeps the resolution you used to take the picture.
- *Large:* Resizes the picture to 1024 x 768 pixels.
- *Medium:* Resizes the picture to 800 x 600 pixels.
- *Small:* Resizes the picture to 640 x 480 pixels.

**5.** **Press the Menu key and select Send.**

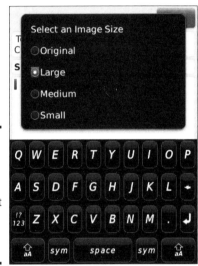

**Figure 6-5:**
BlackBerry
Firmware
5.0 lets you
choose what
resolution
to send
photos in.

# Improving the BlackBerry Calendar

RIM has introduced some long-awaited enhancements to the calendar. If your BlackBerry smartphone is running Device Software 5.0 and has been activated on a *BES* (BlackBerry Enterprise Server), you're in luck.

You're on a BES if you can send and receive corporate e-mail, calendar appointments, and other information using your BlackBerry.

Here's a list of innovations:

✦ **View attachments to calendar appointments.** You can easily view calendar appointment attachments right from your BlackBerry Calendar.

✦ **Forward appointments.** You can forward appointments directly from your BlackBerry calendar application to anyone you want via e-mail.

To view a BlackBerry calendar attachment, follow these steps:

1. **On the BlackBerry Home screen, select the Calendar icon.**

2. **Open a calendar appointment that has an attachment.**

3. **Highlight the attachment, press the Menu key, and select Open Attachment.**

To forward a calendar appointment from your BlackBerry, follow these steps:

1. **On the BlackBerry Home screen, select the Calendar icon.**

2. **Open the calendar appointment you would like to forward.**

3. **Press the Menu key and choose Forward.**

4. **Add the recipient(s) you would like to send the appointment to.**

5. **Press the Menu key and choose Send.**

Read Book II, Chapter 2 for more on using your BlackBerry calendar.

## Noticing improvements to BlackBerry Browser

BlackBerry Browser has been enhanced in several ways:

✦ **Experience the Web as it really is.** With improved support for Ajax and JavaScript pages, BlackBerry smartphones with Device Software 5.0 can display Web content that did not show properly in older BlackBerry firmware versions. This is a real improvement to your Web experience!

✦ **Enjoy streaming media.** BlackBerry Browser displays videos without opening the video player. The result? Think full-screen streaming videos from YouTube. Nice!

✦ **Love true Web interoperability with your apps.** Without getting too technical on you, new BlackBerry apps can include a browser search bar. These apps can therefore search the Web without going to BlackBerry Browser.

Read Book III, Chapter 1 for more on using your BlackBerry Browser.

Happy browsing!

## Improving BlackBerry Maps

BlackBerry Maps has a few special enhancements. Cheers to never being lost again.

✦ **Get speedy access to your destination.** Maps render twice as fast as they used to. Panning and zooming are noticeably faster as well.

✦ **Use improved address search tools.** The improved address search handles requests more intelligently than the previous incarnations did.

✦ **Customize your search.** The improved search screen lists options based on favorites, recent searches, searches from contacts, or searches by address.

✦ **Access layers of map information.** See what you want to see (and only what you want to see) with the new layers feature. This feature lets you toggle between points of interest, local addresses, favorites, and routes (among other options) so that you can customize your BlackBerry Maps view.

✦ **Add photo geotags.** *Photo geotagging* (attaching location information to a photo) can be toggled as a layer. When enabled, photos on the device appear as thumbnails on the local map being displayed.

Customizing your BlackBerry Maps view is easy if you use these steps:

*1.* **On your BlackBerry Home screen, choose the Maps icon.**

*2.* **Press the Menu key and choose Layers.**

*3.* **Highlight the layer you would like to appear on the map.**

*4.* **Press the trackball, trackpad, or touch screen to enable the layer item.**

*5.* **Repeat Steps 3 and 4 for each layer item you would like to show.**

*6.* **Press the Escape key to return to the map. It will refresh with the layers you have added.**

Read Book VI, Chapter 6 for more information on map applications.

## Using the new, improved BlackBerry Messenger

BlackBerry Device Software 5.0 introduces an all-new version of BlackBerry Messenger with a redesigned user interface and several new features:

✦ **Create an avatar.** You can take a picture using your BlackBerry camera, or use a photo saved on your device. Either way, you can use the image as your avatar. The picture appears next to your name on your buddies' contact lists as well as on your BlackBerry Messenger screen.

✦ **Create Home screen icons for your frequent BlackBerry Messenger contacts.** You can add BlackBerry Messenger contacts directly to your BlackBerry Home screen for easy access!

To set up your avatar, just follow these steps:

1. **Open BlackBerry Messenger.**

2. **Press the Menu key and choose My Profile.**

3. **Press the Menu key and choose Change Display Picture.**

4. **Browse to and select the picture you would like to use as your avatar.**

   The picture will appear in full screen with a box over a section of it.

   You will probably need to crop the picture.

5. **Scroll to the area of the picture you would like to use. Then press the Menu key and select Crop and Save.**

   Don't worry; this doesn't alter the original picture. It just creates a separate cropped version within BlackBerry Messenger.

6. **Press the Menu key and choose Save.**

Follow these steps to add a BlackBerry Messenger contact as an icon on your Home screen:

1. **Open BlackBerry Messenger.**

2. **Highlight a contact and press the Menu key.**

3. **Select Show on Home Screen.**

   In you Instant Messaging folder, your contact is added as an icon with the contact's BlackBerry Messenger display name. If the contact has an avatar, the icon will show it. Otherwise, it will show a generic BlackBerry Messenger contact icon.

For more information on messaging using BlackBerry Messenger, check out Book III, Chapter 3.

# Book II

# Organization and Productivity

| 14 Jan 2010 | | | Week 2 | | | | 14:21 |
|---|---|---|---|---|---|---|---|
| Jan 2010 | Mon 11 | Tue 12 | Wed 13 | Thu 14 | Fri 15 | Sat 16 | Sun 17 |
| 09:00 | | | | | | | |
| 10:00 | | | | | | | |
| 11:00 | | | | | | | |
| 12:00 | | | | | | | |
| 13:00 | | | | | | | |
| 14:00 | | | | | | | |
| 15:00 | | | | | | | |
| 16:00 | | | | | | | |
| 17:00 | | | | | | | |

Daily Rocks - Day's Priority

09:00 - 10:00

# Contents at a Glance

# Chapter 1: Managing and Locating Your Contacts

## In This Chapter

✔ **Exploring BlackBerry Contacts**

✔ **Adding, viewing, editing, and deleting contacts**

✔ **Transferring contacts from cellphones to your BlackBerry**

✔ **Finding a contact in Contacts**

✔ **Organizing Contacts**

✔ **Sharing BlackBerry contacts**

✔ **Synchronizing Facebook contacts**

The idea of storing contacts was around long before the BlackBerry was conceived. The Contacts feature (formerly called Address Book) on the BlackBerry serves the same function as any list of contacts: It is an organizational tool that gives you a place to record information about people. This tool gives you a central place from which you can retrieve information so that you can reach your contacts by phone, cellphone, e-mail, snail-mail, or the speedy messaging of PIN, SMS, MMS, or BlackBerry Messenger. Depending on the type of work you do, Contacts is likely an essential tool, and your BlackBerry is there at the ready.

You can benefit from using BlackBerry Contacts if you answer yes to any of the following questions:

✦ Do you travel?

✦ Do you meet clients frequently?

✦ Do you spend a lot of time on the phone?

✦ Do you ask people for their phone numbers or e-mail addresses more than once?

✦ Do you carry around the old-fashioned paper day planner with a section allocated for recording contacts? Or do you write phone numbers of acquaintances on the back of business cards?

✦ Is your wallet or purse full of these "important" business cards, but you can never seem to find the ones you need when you need them?

Regardless of how you keep your contact information, it's time to get down to business and organize your contacts by using your BlackBerry. In this chapter, we show you how to make your BlackBerry a handy, timesaving tool for managing your contact information. Specifically, you find out how to add, change, and delete contacts as well as how to locate them later. You'll also be amazed at how well Contacts is integrated with all the other BlackBerry features you've come to know and love — phoning contacts, adding invites to your meetings, adding contacts to BlackBerry Messenger, and composing e-mails.

If you're one of those, um, stubborn folks who insist that you don't need Contacts — "I'm doing just fine without it, thank you very much!" — think of it this way: You've been using a virtual contact list all the time — the one buried inside your cellphone. And that address book often isn't even a very good one! When you're ready to join the rest of us in this millennium, read this chapter to see how to transfer all that good contact info from an old phone into your new BlackBerry-based Contacts.

Current BlackBerry models use one of three navigation methods — trackball, trackpad, or touch screen. One thing they all have in common is that you press the navigation tool to make a selection. Throughout this book, whenever we provide steps that simply say "press," we mean to press the trackball, trackpad, or touch screen, depending on which BlackBerry model you have.

## Accessing Your Contacts

The good people at Research In Motion make it easy for you to find Contacts. Start by taking a look at the BlackBerry Home screen. The Contacts icon looks like an old-fashioned address book. (Remember those?) If you have a hard time locating it, Figure 1-1 shows what it looks like with a Pearl Flip and a Bold. (For more information about the BlackBerry models and their differences, check out Book I, Chapter 2.) To describe Contacts in this chapter and for most of the illustrations throughout this book, we use the BlackBerry Bold model.

**Figure 1-1:**
The
Contacts
icon as
depicted on
BlackBerry
Pearl Flip
(left) and
BlackBerry
Bold (right).

Contacts

Opening Contacts couldn't be simpler. On a BlackBerry Storm, you just touch-press the Contacts icon. On a BlackBerry Bold, Curve, or Pearl, follow these steps:

**1.** **Use the trackball or trackpad to highlight the Contacts icon.**

**2.** **Press the trackball or trackpad.**

Your BlackBerry Contacts feature is accessible from a number of applications, including Phone, Messages, and Calendar. For example, say you're in Calendar and you want to invite people to one of your meetings or appointments. Look no further. Contacts is on the menu, ready to lend a helping hand.

You can also get to Contacts by pressing A while at the Home screen. Go to Book I, Chapter 2 for more on Home screen shortcuts.

## Working with Contacts

Getting a new gizmo is always exciting because you just know that your newest toy is chock-full of features you're dying to try out. Imagine having a new BlackBerry, for example. The first thing you'll want to do is try to call or e-mail someone, right? But wait a sec. You don't have any contact information yet, which means you're going to have to type in someone's e-mail address each time you send an e-mail. What a hassle.

It's time to get with the plan. Most of us humans — social creatures that we are — maintain a list of contacts somewhere, whether in an e-mail program such as Outlook or Lotus Notes, in an old cellphone, or maybe on a piece of paper kept tucked away in a wallet. We're pretty sure that you have some kind of list somewhere. The trick is getting that list into your BlackBerry device so that you can access the info more efficiently. The good news for you is that the "getting contact info into your BlackBerry device" trick isn't a hard one to master. Stick with us, and you'll have it down pat by the end of this chapter.

Often, the simplest way to get contact information into your BlackBerry is to enter it manually. However, if you've invested a lot of time and energy in maintaining some type of contacts application on your desktop computer, you might want to hot-sync that data into your BlackBerry. (For more on synchronizing data, check out Book IV. It gives you details on how to synchronize some of your desktop application data, including contacts, e-mails, appointments, and memos.) You can also transfer your old cellphone contacts to your BlackBerry. Intrigued? Then check out the "Transferring Contacts from Cellphones" section, later in this chapter. Do you have an address book in another smartphone, such as a Palm Treo or a Windows Mobile? No problem. Most of these devices allow you to sync to a desktop computer application such as Outlook. But this is not a book about other smartphones, so refer to your other device's instructions on how to sync it to your desktop. After the data is in the desktop, you can check out Book IV for details on synchronizing it to your BlackBerry.

## Creating a contact

Imagine you've just run into Jane Doe, an old high school friend you haven't seen in years. Jane is about to give you her number, but you don't have a pen or pencil handy to write down her information. Are you then forced to chant her phone number to yourself until you can scare up a writing implement? Not if you have your handy BlackBerry smartphone on you.

With BlackBerry in hand, follow these steps to create a new contact:

1. **On the BlackBerry Home screen, select the Contacts application.**

   Contacts opens. You can also access Contacts from different applications. For example, see Book III, Chapter 2 to find out how to access Contacts from Messages.

2. **Select the Add Contact option.**

   The New Contact screen appears, as shown in Figure 1-2.

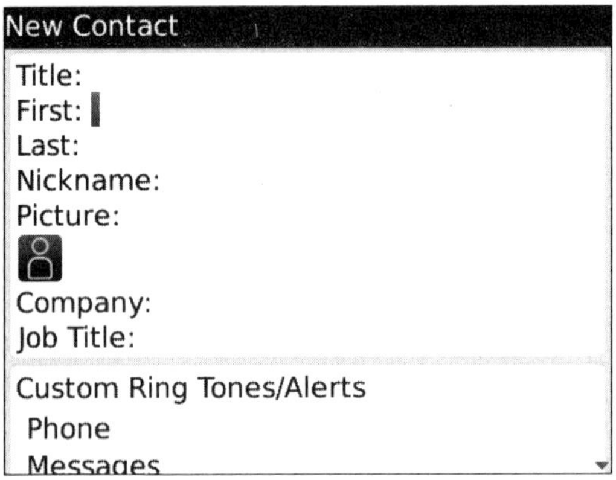

**Figure 1-2:**
Create a
new contact
here.

3. **Scroll through the various fields, stopping and entering the contact information that you feel is appropriate.**

   Use your BlackBerry keyboard to enter this information. For an overview of the different keyboard types, see Book I, Chapter 2. Note, however, that all BlackBerry keyboards look (and work) alike. When entering an e-mail address, press the Space key to insert an at symbol (@) or a period (.).

   We don't think you can overdo it when entering a person's contact information. You should strive to enter as much info as you possibly can. Maybe the benefit won't be obvious now, but in the future when your

memory fails you or your boss needs a critical piece of info that you just happen to have at the ready, you'll thank us for this advice.

To create another new blank e-mail field for the same contact, press the Menu key and then select Add Email Address. You can have up to three e-mail addresses per contact.

If a contact has an extension for his or her phone number, no problem. When calling such a contact from your BlackBerry, you can instruct BlackBerry to dial the extension after the initial phone number. When entering the phone number into the New Contact or Edit Contact screen, type the primary phone number, press the Menu key, select Add Pause from the menu that appears, and then add the extension number.

Book II
Chapter 1

Managing and
Locating Your
Contacts

4. **When you finish entering the contact information, press the Menu key and then select Save.**

   At this point, Jane Doe is added to the list, as shown in Figure 1-3.

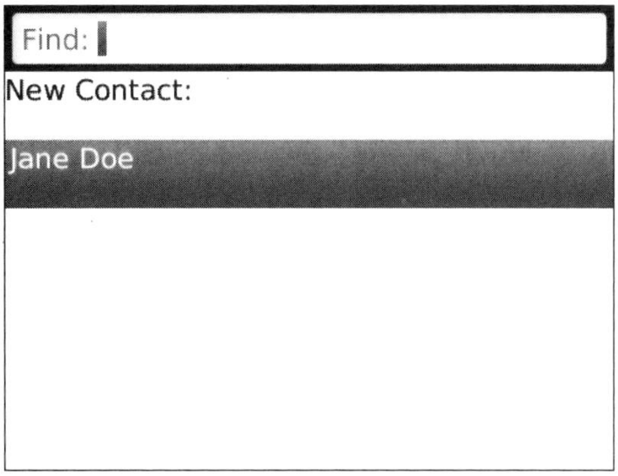

**Figure 1-3:**
The
Contacts
screen after
adding Jane
Doe.

| Find: |
| --- |
| New Contact: |
| Jane Doe |

The menu is always available through the Menu key, but just for convenience and if you have a model that has a trackball or a trackpad, we prefer to use the trackball or the trackpad when a menu is available through it. Also, when a menu is available through a press of a trackball or trackpad, it's always a shortened menu. The application is smart enough to figure out which menu items to display based on where you are.

### Taking notes

The Notes field on the New Contact screen (you might need to scroll down a bit to see it) is useful for adding a unique description about your contact. For example, use the field to jog or refresh your memory with tidbits such as *Knows somebody at ABC Corporation* or *Can provide introduction to a Broadway agent.* Or perhaps your note is something personal, such as *Likes golf; has 2 children: boy, 7 & girl, 3; husband's name is Ray.* It's up to you. Again, the more useful the information, the better it will serve you.

### Adding your own fields

Perhaps you'd like to add contact information that doesn't fit into any of the available fields. Although you can't really create additional fields from scratch, you can commandeer one of the User fields for your own purposes.

The User fields are located at the bottom of the screen; you have to scroll down to see them. Basically, you can use these fields any way you want (which is great), and you can even change the field's name. (Face it, *User field* is not that helpful as a descriptive title.) For example, you can rename User fields to capture suffixes (such as MD, PhD, and so on). Or how about profession, birth date, hobbies, school, or nickname? When it comes down to it, you decide what information is important to you.

Keep in mind, though, that changing the field name for this particular contact changes it for all your contacts.

To rename a User field, follow these steps:

1. **Scroll to the bottom of the screen to navigate to one of the User fields.**

2. **Press the Menu key and then select Change Field Name.**

   *Note:* The Change Field Name selection on the menu appears only if the cursor is in a User field.

3. **Use the keyboard to enter the new User field name.**

4. **Press the Enter key to save.**

   You're all set.

### Adding a picture to a contact

Most modern phones can display a picture of the caller. And BlackBerry is no stranger to this neat feature. For this to work, you first need the obvious: a picture of your friend or acquaintance. After you have the picture in a digital format that your device supports, you can get it into your BlackBerry through e-mail or copy it to the microSD card. Don't worry — we won't leave you helpless if you don't know how to get media to that microSD card; Book V, Chapter 3 is your gateway to media satisfaction.

Your BlackBerry supports the following standard picture formats:

✦ **BMP:** Bitmap file format (.bmp).

✦ **JPEG:** Developed by the Joint Photographic Experts Group committee. A JPEG (.jpg) file is typically compressed to a tenth of its size with little perceptible loss of image quality.

✦ **PNG:** A Portable Network Graphics (`.png`) file is a bitmapped image that employs a lossless data transmission.

✦ **TIFF:** Tagged Image File Format (`.tif`), a popular file format.

✦ **WBMP:** Wireless Bitmap (`.wbmp`) file format, which is optimized for mobile devices.

After a picture is inside the BlackBerry, adding it to one of your contacts is a snap. Follow these steps:

*1.* **On a contact's Edit screen, press the Menu key and then select Add Picture. (See Figure 1-4.)**

**Figure 1-4:**
Add a
picture
here.

*2.* **Navigate to the drive and folder that contain the picture.**

You can use multiple locations for storing media files such as pictures. Book V, Chapter 3 gives you the scoop.

*3.* **Select the picture you want.**

The picture you choose is displayed in full on the screen.

*4.* **Position the rectangle on the face.**

Contacts uses a tiny image, just enough to show the face of a person. The rectangle you see here indicates how the application crops the image.

*5.* **Press the Menu key and then select Crop and Save.**

You're all set. Just save this contact to keep your changes.

*6.* **Press the Menu key and select Save.**

### Assigning a tone

Oh no, you've been awakened with your BlackBerry ringing on your night-stand. Wouldn't it be nice if you could tell who's calling? Ring tones help you decide whether to ignore the call or get up and answer it. We hope you can easily switch your brain back to Sleep mode if you decide to ignore the call.

Follow these steps to assign a ring tone to one of your contacts:

1. **On the Contact screen, select Custom Ring Tone/Alerts⇨Phone; on the Bold, Curve, or Pearl, press the Menu key and select Add Custom Ring Tone.**

2. **Press the trackball or touch screen.**

   A list of ring tones from which you can choose is displayed. You can also select Browse to navigate to the drive and folder containing the ring tone. You can use multiple locations for storing media files such as ring tones. Book V, Chapter 3 gives you the scoop.

3. **Select the ring tone you want.**

   The ring tone appears in the Custom Phone Tune field. To save these changes, save this contact.

4. **Press the Menu key and select Save.**

## Adding contacts from other BlackBerry applications

When you receive an e-mail message or get a call from someone, you have contact information in Messages or Phone. (RIM makes this easy for you because Messages and Phone can recognize phone numbers or e-mail addresses and then highlight that information for a quick cut and paste.) Maybe that info isn't complete, but you definitely have at least an e-mail address or a phone number. Now, if you're pretty sure that you'll be corre-sponding with this person and he or she isn't yet listed in Contacts, it's just logical that you'd want to add the information.

If you have a sharp eye, you might have noticed that Phone lists only out-going numbers. That's half of what you need. Oddly enough, you can access the history of incoming phone calls inside Messages. To do so, in Messages, press the Menu key and select View Folder⇨Phone Call Logs.

Like e-mail, a call log entry stays in the list as long as you have free space in your BlackBerry. This could mean months or years, depending on your model and usage. When BlackBerry runs out of space, it starts reclaiming space a chunk at a time by deleting read e-mails and phone call logs, starting from the oldest.

When you have an e-mail or a call log open, just scroll to an e-mail address or a phone number and press the trackball while that piece of information is highlighted. An Add to Contacts option pops up on the menu. (This particular menu item is located at the bottom of the list, so you might have to scroll to see it.) Select Add to Contacts, and a new New Contact screen appears, pre-filled with that particular piece of information. Now just enter the rest of the information you know about the person, and you can save it to Contacts. This is just one more sign of BlackBerry's ongoing attempt to make your life easier.

## Viewing a contact

Okay, you just entered your friend Jane's name into your BlackBerry, but you have this nagging thought that you typed the wrong phone number. You want to quickly view Jane's information. Here's how you do it:

*1.* **On the BlackBerry Home screen, select the Contacts application.**

Contacts opens.

*2.* **Select the contact name you want.**

Selecting the contact name is the same as opening the menu and choosing View — just quicker.

View mode displays only information that's been filled in, as shown in Figure 1-5. (It doesn't bother showing fields in which you haven't entered anything.)

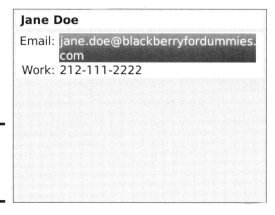

**Figure 1-5:**
View
mode for a
contact.

## Editing a contact

Change is an inevitable part of life. Given that fact, your contact information is sure to change as well. If you want to keep the information you diligently put in Contacts current, you're going to have to do some updating now and then.

Book II
Chapter 1

Managing and
Locating Your
Contacts

To update a contact, follow these steps:

1. **On the BlackBerry Home screen, select Contacts.**

   Contacts opens.

2. **Scroll to and highlight a contact name, press the Menu key, and then select Edit.**

   The Edit Contact screen for the contact name you selected makes an appearance.

   In Contacts (or any BlackBerry application, for that matter), displaying a menu involves a simple press of the Menu key. You see the Edit option on the menu right below View.

3. **Scroll through the various fields of the Edit Contact screen, editing the contact information as you see fit.**

   If you want to replace only a few words or letters located in the middle of a field (instead of replacing all the text), scroll while pressing and holding the Alt key (located to the left of the Z key) to position your cursor precisely on the text you want to change. Then make your desired changes. This tip is not applicable to Storm or Storm 2.

4. **Press the Menu key and then select Save.**

   The edit you made for this contact is saved.

When you're editing information and you want to totally replace the entry with a new one, it's much faster to first clear the contents, especially if you have a lot of old data. When you are in an editable field (as opposed to a selectable field), just press the Menu key and then select Clear Field. This feature is available in all text-entry fields and for most BlackBerry applications.

## Deleting a contact

When it's time to eradicate somebody's contact information in Contacts (whether it's a case of duplication or a bit of bad blood — yes, we admit to have occasionally stricken somebody from our Contacts list in a fit of pique), the BlackBerry OS makes it easy to delete a contact.

Here's how to delete a contact:

1. **On the BlackBerry Home screen, select Contacts.**

   Contacts opens.

2. **Scroll to and highlight a contact name you want to delete, press the Menu key, and then select Delete.**

   (If you don't initially see the Delete option, scroll down to the bottom part of the screen.) A confirmation screen appears, as shown in Figure 1-6.

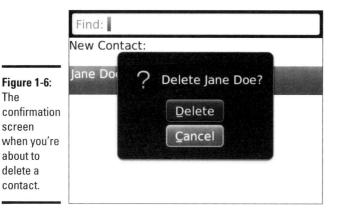

**Figure 1-6:**
The confirmation screen when you're about to delete a contact.

3. **On the confirmation screen, select Delete.**

   The contact you selected is deleted and disappears from your contact list.

Sometimes dealing with the confirmation screen can be a pain if you want to delete several contacts in a row. If you are 100 percent sure that you want to ditch a number of contacts, you can suspend the Confirmation feature by setting the Confirm Delete option to No on the Contacts Options screen. See the "Setting preferences" section, later in this chapter, for more on Contacts Options.

It also pays to spend a little bit of time adding your own contact record(s). We recommend adding at least one record for your business contact info and one for your personal contact info. This saves you time having to type your own contact information every time you want to give it to someone. You can share your contact record by sending it as an attachment to an e-mail. (See the later section, "Sharing a Contact.".

# Transferring Contacts from Cellphones

Suppose you have an old cellphone that has served you well for many years. Most likely you've accumulated contacts on that phone by painstakingly typing them each time in the past. You'd really like to have the same contacts on your BlackBerry as you have on your cellphone (or vice versa), but you just cannot bear the thought of typing them all again. You think to yourself, "There has to be a way." Good news: You're right; there is.

## Copying contacts from a cellphone SIM card

We've copied contacts from Nokia, Ericsson, Motorola, and Samsung cellphones — and in theory, our method should also work on other phones.

The trick is to use the SIM card as an external storage device — sort of like a floppy disk. What exactly is a SIM card? The acronym stands for *subscriber identity module*. The non-head-scratching definition describes an electronic chip that is capable of storing information such as your phone numbers and contacts. Basically, you store your old phone's contact list on the SIM card, insert that SIM card into your BlackBerry, and then upload the contacts from your SIM card. (For all the gory details, check out the steps later in this section.)

This sounds easy to do, but it might be tricky depending on the type of phone you have. A caveat: You can do this only on a GSM phone. (GSM is short for *Global System for Mobile Communications*, probably the most popular standard for mobile phones in the world.) To tell whether you have a GSM phone, first check whether you have a SIM card. To do that, take out your cellphone battery. Behind it, you should see a SIM card, looking (we hope) like the one you see in Figure 1-7. If you don't have a SIM card, your phone is not a GSM phone.

**Figure 1-7:**
Transfer phone info from your cellphone with a SIM card.

If your cellphone is a GSM phone and you're determined to forge ahead and transfer your cellphone contact info to your BlackBerry, we're here to help. Obviously, we can't give instructions for all the types of phones out there. For the purpose of showing you what's what, we use a Nokia 6300 phone. (Why the Nokia 6300? That's the one the dartboard told us to use.) Please check the manual of your phone for the equivalent steps.

No more digressions! If you want to know the steps for copying contacts from a Nokia phone to a BlackBerry by using a SIM card, here they are:

*1.* **Take out the SIM card buried inside your BlackBerry and put it in the Nokia phone.**

Most people are uncomfortable doing this, but taking out a SIM card and putting it back in is no big deal. The SIM card is usually behind the battery, so you have to slide or take off the back cover of the device to get to it. On the BlackBerry, the back cover has a groove where you can put your thumb and push the cover out. A Nokia phone has a simple locking mechanism @— a tiny bump on the back of the phone — which you slide down to unlock the back cover. When the cover is off, remove the battery and you can see the SIM card. Slide the SIM card's plastic enclosure;

it should pop open. You can then remove the existing SIM card from the enclosure and replace it with the BlackBerry SIM card.

Because this step requires you to remove the batteries from both your BlackBerry and your Nokia phone, both devices are going to power off (obviously).

Then you resuscitate both, starting with the Nokia. After you put the battery back in the Nokia phone, it resets, and you should see the display come up.

If your phone doesn't recognize your BlackBerry SIM card, perhaps the phone is locked. Phone providers do this all the time: They lock the phones to their network, making it unusable in other networks. If this is the case, call your phone provider and ask for instructions on how to unlock your phone.

2. **On the Nokia phone, select Names by pressing the top of the rightmost top button.**

   Note that the display above the rightmost button shows Names. Names is the equivalent of Contacts. The phone displays a list of contacts.

3. **Select Options by pressing the leftmost top button.**

   The Names menu appears, as shown in Figure 1-8. Note the Mark All menu item.

**Figure 1-8:**
The Nokia
6300 Names
menu.

There is no trackball for scrolling and selecting, but you can use the middle button, which has a silver edge around it. You can scroll up by pressing the top silver edge or scroll down by pressing the lower silver edge.

**4. Select Mark All.**

You can make the selection by pressing the middle button while Mark All is highlighted.

**5. Select Options by pressing the leftmost top button.**

The Names menu appears.

**6. Select Copy Marked.**

Another menu appears, displaying two options: From Phone to SIM Card and From SIM Card to Phone.

**7. Select From Phone to SIM Card.**

**8. Select All and then select Keep Original from the menu that follows.**

A confirmation screen appears.

**9. Select OK to confirm the copy.**

Your phone starts copying the contents of the Phone Book to the SIM card. While it's making the copy, the screen displays a bar that moves back and forth. If you have many contacts on your phone, this process can take some time, so be patient. When the contacts are loaded into the SIM card, the screen displays the number of contacts that were copied. You can proceed with the next step.

**10. Take out the SIM card from the Nokia phone and put it back into your BlackBerry.**

Reinserting the SIM card and battery resets your BlackBerry.

**11. On the BlackBerry Home screen, select Contacts.**

Contacts opens.

**12. In Contacts, press the Menu key and then select SIM Phone Book.**

It might take some time to load the contacts from your SIM card; how long depends on how many contacts you've saved to the card. (You'll see a progress bar on the screen.) After the contacts are loaded, they are listed on the screen, and you can start browsing or copying them to your Contacts list.

The SIM Phone Book menu item is located toward the bottom of the menu, as shown in Figure 1-9. Depending on your model, you might need to scroll down to see it.

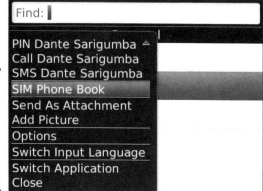

**Figure 1-9:**
The
Contacts
menu,
showing the
SIM Phone
Book option.

*13.* **To copy a contact to Contacts, just highlight the contact, press the Menu key, and then select Add to Contacts.**

Your contact has now found its way to the BlackBerry Contacts. You can repeat this step for all the contacts you want to copy. Although this is a tedious process, it's still a lot better than trying to type each one on your BlackBerry.

SIM cards do have a limited capacity. If all your contacts don't fit on the SIM card on your first try, here are two ways you can add leftover contacts to your BlackBerry:

✦ **One by one:** If only one or two contacts are not included, you might be better off just typing them into your BlackBerry.

✦ **In batches:** If you have many more contacts left, you probably should try doing multiple saves. The trouble here is that you have to figure out what contact info was saved on your first try — which, of course, lets you know what you still have to copy. We recommend loading such contacts in batches. To do that, first load a batch of names onto your BlackBerry and then delete the names from the SIM card. (That gives you a clean slate.) Repeat Steps 2 and 4 from the preceding list, and when you get to Step 5, select One by One from the menu rather than All. This allows you to select the remaining contacts to copy to the SIM card one at a time, instead of copying the whole shebang one more time. When the remaining contacts are in the SIM card, you can copy the second batch of names and load them to your BlackBerry by using the preceding steps. Repeat as needed.

### Copying a contact to a SIM card

You can reverse the info-import process, too. That is, you can copy from your BlackBerry Contacts list to a cellphone. "Can this be done?" you implore. To which we reply, "Why, certainly, with a little help from BlackBerry experts." For, truth be told, this feature of Contacts is probably one of the most difficult tricks to figure out unless somebody shows you how it's done.

Follow the steps below to copy BlackBerry contact info to the BlackBerry SIM card:

*1.* **Start by viewing the contact information.**

Follow the steps in the "Viewing a contact" section, earlier in this chapter.

*2.* **On the View screen, scroll to a Phone Number field, press the Menu key, and then select Copy to SIM Phone Book.**

The Copy to SIM Phone Book feature (see Figure 1-10) shows up only when you position the cursor in a Phone Number field.

**Yosma Sarigumba**
*YOS*

Copy to SIM Phone Book
Help
Copy
Edit
Delete
Activity Log
Call Mobile
PIN Yosma Sarigumba
SMS Yosma Sarigumba

**Figure 1-10:**
Copy
BlackBerry
info to a
cellphone.

*3.* **On the Phone Book Entry screen, press the Menu key and then select Save.**

This operation is a snap, and the screen immediately returns to the View Contact screen shown as a result of Step 1.

## Copying Contacts from Desktop Applications

Most of us have desktop applications that we use to maintain our network — you know, Microsoft Outlook, IBM Lotus Notes, or Novell GroupWise. A word to the wise: You do not want to maintain two Contacts lists, one in your BlackBerry and one on your desktop computer. That's a recipe for disaster.

Luckily for you, RIM makes it easy to get your various contacts — BlackBerry, desktop, laptop, whatever — in sync. Your BlackBerry comes with BlackBerry Desktop Manager (BDM), a collection of programs, one of which is Intellisync, a program that helps you sync between your device and the PC software. It also allows you to set up and configure the behavior of the program, including how the fields in the desktop version of Contacts map to the Contacts fields in your BlackBerry. But you're not going to read about it here. For that, check out Book IV, which has complete details on how you can use the Intellisync feature of BlackBerry Desktop Manager to synchronize with the applications in your device — including, of course, your Contacts list.

## Looking for Someone?

Somehow — usually through a combination of typing skills and the shuttling of data between various electronic devices — you've created a nice, long list of contacts in Contacts. Nice enough, we suppose, but useless unless you can find the phone number of Rufus T. Firefly at the drop of a hat.

That's where the Find screen comes in. In fact, the first thing you see in Contacts when you open it is the Find screen, as shown in Figure 1-11.

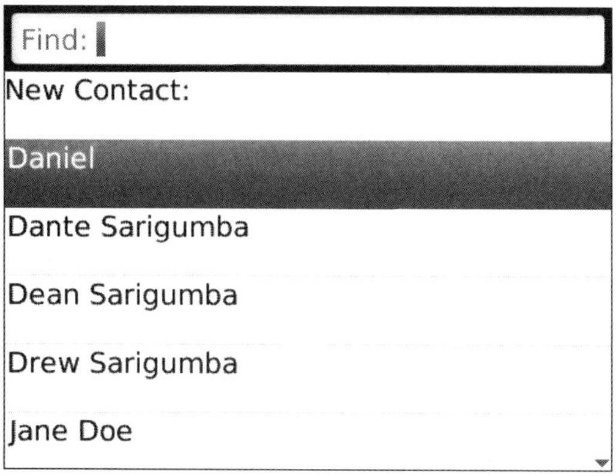

**Figure 1-11:**
Your search
starts here.

| Find: |  |
|---|---|
| New Contact: | |
| Daniel | |
| Dante Sarigumba | |
| Dean Sarigumba | |
| Drew Sarigumba | |
| Jane Doe | |

You can conveniently search through your contacts by following these steps:

*1.* **In the Find field, enter the starting letters of the name you want to search for.**

Your search criterion is the name of the person. You could enter the last name or first name or both. The list is usually sorted by first name and then last name. As you type the letters, notice that the list starts shrinking based on the matches to the letters that you enter. Figure 1-12 illustrates how this works.

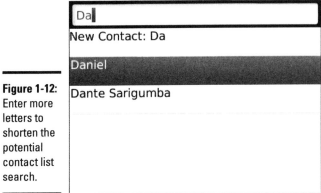

**Figure 1-12:** Enter more letters to shorten the potential contact list search.

2. **Scroll and highlight the name from the list of matches.**

   If you have multiple matches, scroll through the list to find the person's name.

   If you have a long list in Contacts and you want to scroll down a page at a time, just hold down the Alt key (it's located to the left of the Z key) and scroll. You get where you need to go a lot faster.

3. **Press the Menu key and select from the possible actions listed on the menu that appears.**

   After you find the person you want, you can select from these options, as shown in Figure 1-13:

   • *Email:* Starts a new e-mail message. See Book III, Chapter 2 for more information about e-mail.

   • *PIN:* Starts a new PIN-to-PIN message, which is a messaging feature unique to BlackBerry. With PIN-to-PIN, you can send someone who has a BlackBerry a quick message. See Book III, Chapter 3 for more details about PIN-to-PIN messaging.

   • *Call:* Uses Phone to dial the number.

   • *SMS:* Starts a new SMS message. SMS stands for short message service, which is used in cellphones for sending and receiving short text messages. See Book III, Chapter 3 for more details about SMS.

- *Send to Messenger Contact:* Adds this contact to your contacts list in BlackBerry Messenger. (Note that this option appears only if you have BlackBerry Messenger installed.)

- *MMS:* Starts a new MMS message. MMS is short for Multimedia Messaging Service, an evolution from SMS that supports voice and video clips. See Book III, Chapter 3 for more details about MMS.

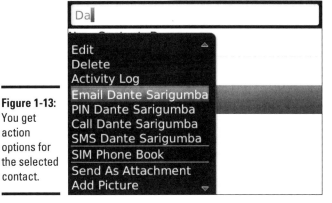

**Figure 1-13:** You get action options for the selected contact.

If you have a finger-fumble and press a letter key in error, press the backspace virtual key on the Storm or Storm 2 and for other models, press the Escape key (the arrow key to the right of the trackball or Menu key) once to return to the original list (the one showing all your contacts), or press the Menu key and select View All.

If you think you're hallucinating when you notice that sometimes the item Email *contact name* or Call *contact name* appears on the menu and sometimes not, just relax. There's nothing wrong with your eyesight or your mind. Contacts is smart enough to know when it's appropriate to show those menu options. If a contact has a phone number, Call *contact name* and SMS *contact name* show up, and the same is true for an e-mail address and personal identification number (PIN). In fact, this list of actions is a convenient way to find out whether you have particular information — a phone number or an e-mail address — for a particular contact.

In a corporate environment, your BlackBerry Enterprise Server administrator might disable PIN-to-PIN messaging because it doesn't go to the corporate e-mail servers and, therefore, can't be monitored. If this is the case, the menu option PIN *contact name* won't appear, even though you entered PIN information for your contacts. Note that you'll still be able to receive a PIN-to-PIN message, but you won't be able to send one.

# *Organizing Your Contacts*

You've been a diligent boy or girl by adding your contacts to Contacts, and your list has been growing at a pretty good clip. It now has all the contact information for your business colleagues, clients, and (of course) family, friends, and relatives. In fact, Contacts has grown so much that it holds hundreds of contacts, and you start to notice that it now takes you more time to find somebody, especially when you can't remember the name right away and are trying to explore the list knowing that when you see the name you'll recognize it.

Imagine that you've just seen an old acquaintance, a sales rep for XYZ, Inc., and you want to greet the person by name. His name is on the tip of your tongue, but you just can't remember it. However, if you saw his name on a list, it would jump right out at you. The trouble is that your list has 300-plus names, which would take you forever to scroll through. It's so long, in fact, that the person would surely come right up to you in the meantime, and you'd have to reveal the fact that you can't remember his name. (How embarrassing.) In this scenario, the tried-and-true Find feature wouldn't be much help. What you need is a smaller pool of names to search through so that you can stumble across the needed name much more quickly.

This isn't rocket science. You're going to want to do one of the following:

+ **Organize your contacts into groups.** Using groups (as every kindergarten teacher could tell you) is a way to arrange something (in your case, contacts) to make them more manageable. How you arrange your groups is up to you because the organizing principle should be based on whatever makes sense (to you, at least) and fits the group you set up. For example, you can place all your customer contacts within a Clients group and all your relatives in a Family group. Then, instead of searching for names of individuals within one humongous list, you can search within a smaller, more manageable group.

+ **Set up your contacts so that you can use some kind of filter.** Another way to organize and streamline how BlackBerry Contacts lists your contacts is to use the Filter feature in combination with BlackBerry's Categories. (*Categories* is just another way that Contacts helps you filter contacts.) Using the Filter feature narrows the Contacts list to such an extent that you have to scroll down and find your contact — no need to type search keywords, in other words.

Whether you use the Group or Filter feature is up to you. You find out how to use both methods in the next sections of this chapter.

## Creating a group

A BlackBerry *group* in Contacts — as opposed to any other kind of group you can imagine — is just a simple filter or category. In other words, a group just arranges your contacts into subsets without affecting the content of your contact entries. In Contacts itself, a group shows up in the contact list just like any other contact. The only wrinkle here is that when you select the group, the contacts associated with that group — and only the contacts associated with that group — appear onscreen.

Need some help visualizing how this works? Go ahead and create a group, following these steps:

**Book II**
**Chapter 1**

*1.* **On the BlackBerry Home screen, select Contacts.**

Contacts opens.

*2.* **Press the Menu key and then select New Group.**

A screen similar to that shown in Figure 1-14 appears. The top portion of the screen is where you type the group name, and the bottom portion is where you add your list of group members.

New Group:
BlackBerry Group Contact
                    * No Contacts *

**Figure 1-14:**
An empty
screen
ready for
creating a
group.

*3.* **In the New Group field, enter the name of the group.**

You can name the group anything, but for the sake of this example, we named the group *Poker Buddies.* After entering the name of the group, you're ready to save it. But hold on a sec — you can't save this group until you associate a member to it. To satisfy such a hard-and-fast rule, proceed to the next step to add a member.

4. **Press the Menu key and then select Add Member from the menu that appears.**

   The main Contacts list shows up in all its glory, ready to be pilfered for names to add to your new group list.

5. **Select the contact you want to add to your new group list, press the Menu key, and then select Continue from the menu that appears.**

   Everybody knows a Rob Kao, so select him. You'll notice that doing so places Rob Kao in your Friends group list, as shown in Figure 1-15. (Rob Kao, a coauthor of the book you're holding, is a very popular fellow.)

   If you simply want to group your Contacts and not necessarily use it as a distribution list, you must follow a rule: The contact needs to have at least an e-mail address or a phone number before you can add it as a member of a group. (Contacts is strict on this point.) If you need to skirt this roadblock, edit that contact's information and put in a fake (and clearly inactive) e-mail address, such as `notareal@emailaddress.no`.

---

New Group: Poker Buddies
BlackBerry Group Contact
Rob Kao (Email)

**Figure 1-15:**
Your new group has one member.

---

6. **Repeat Steps 4 and 5 to add more friends to your list.**

   After you're satisfied, save your group.

7. **Press the Menu key and then select Save Group from the menu that appears.**

   Your Poker Buddies group is duly saved, and you can now see Poker Buddies in your main Contacts list.

Groups is a valuable tool for creating an e-mail distribution list. When adding members to a group, make sure that you select an e-mail address field for your members instead of a phone number. Also, use a naming convention to easily distinguish your group in the list. Appending *-DL* or *-Distribution List* to the group name can quickly indicate a distribution list.

## Using the Filter feature on your contacts

Are you a left-brainer or a right-brainer? Yankees fan or Red Sox fan? An Innie or an Outie? Dividing up the world into categories is something everybody does (no divisions there), so it should come as no surprise that BlackBerry divides your contacts into distinct categories as well. There is a subtle difference between a group and a category. Group is really designed to create a mailing list or a calling list that is unique to the Contacts application. You can say that you assign one or more categories to a contact versus making a contact a member of a group or groups.

By default, two categories are set for you on the BlackBerry:

**Book II**
**Chapter 1**

Managing and
Locating Your
Contacts

✦ Business

✦ Personal

Why stop at two? BlackBerry makes it easy to create more categories. In the following sections, you first find out how to categorize a contact, and then you see how to filter your Contacts list. Finally, you find out how to create categories.

One important aspect of Categories you should be aware of is that they are shared among applications — specifically, among Contacts, MemoPad, and Tasks. This sibling relationship might sound trivial at first, but don't make the common mistake of assuming that what you change in Tasks does not affect other BlackBerry apps. The importance of this comes into play when you delete a category in an application. For example, if you're working in Tasks and you decide to delete a category, you'll soon discover that you've lost that category in Contacts as well — with all its assignments. (The Contacts contact is still intact but will be missing the category assignment.)

### Categorize your contacts

Whether you're creating a contact or editing an existing contact, you can categorize a particular contact as long as you're in Edit mode.

If the trick is getting into Edit mode, it's a pretty simple trick. Here's how you can switch to Edit mode and assign an existing category:

*1.* **On the BlackBerry Home screen, select Contacts.**

Contacts opens.

*2.* **Highlight the contact, press the Menu key, and then select Edit.**

Contacts is now in Edit mode for this particular contact, which is exactly where you want to be.

3. **Press the Menu key and then select Categories.**

   A Categories list appears, as shown in Figure 1-16. By default, you see only the Business and the Personal categories. To find out how to add your own categories, see the next section.

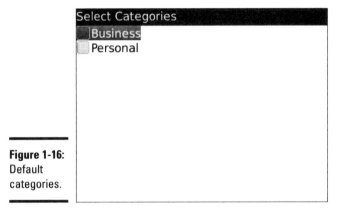

**Figure 1-16:**
Default
categories.

4. **Select the check box next to Personal.**

5. **Press the Menu key and then select Save.**

   You are brought back to the Edit screen for this particular contact.

6. **To complete your changes, press the Menu key and then select Save (again) from the menu that appears.**

You now have one — count 'em, one — contact with Personal as its category, which means you can filter your Contacts list by using a category. Here's how:

1. **On the BlackBerry Home screen, select Contacts.**

2. **Press the Menu key and then select Filter.**

   Your Categories list makes an appearance. If you haven't added any categories in the meantime, all you see here are the default Business and Personal categories.

3. **Select the Personal check box.**

   Your Contacts list shrinks to just the contacts assigned to the Personal category, as shown in Figure 1-17.

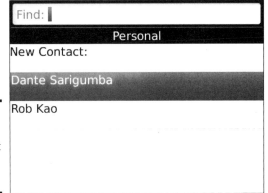

**Figure 1-17:**
The Contacts list after a filter is applied.

As you add more contacts to a category, you can also use the Find feature and enter the first few letters of the name to further narrow the search for a contact. If you need a refresher on how Find works, see the "Looking for Someone?" section, earlier in this chapter.

### Adding a category

Whoever thought the default categories, Business and Personal, were enough for the complexities of the real world probably didn't have many acquaintances. BlackBerry makes it easy to add categories, so you can filter your world as much as you like. Just do the following:

*1.* **On the BlackBerry Home screen, select Contacts.**

*2.* **Press the Menu key and then select Filter.**

You see a view of the default categories. (Refer to Figure 1-16.)

*3.* **Press the Menu key (again) and then select New.**

A pop-up screen appears, asking you to name the new category you want to create.

*4.* **In the Name field, enter a name for your category and then press Enter.**

After you enter the name of the category, it's automatically saved and you see the Filter screen, which lists all the categories, including the one you just created. Just press the Escape key to get back to the main Contacts screen.

If you have a BlackBerry Pearl or BlackBerry Pearl Flip, you might not be able to see special characters, such as a single quotation mark, on the keyboard. To enter a single quote, press the Alt key (it's to the left of the Z key) and the Sym key to display the list of symbols. On the screen that follows, scroll and click the quote.

## Setting preferences

Vanilla, anyone? Some days you'll wish that your Contacts list was sorted differently. For example, there's the day when you need to find the guy who works for ABC Company but has a foreign name that you can hardly pronounce, let alone spell. What's a body to do?

You're in luck. Once again, the good people at RIM have anticipated this scenario and have made available to you Contacts Options, which is a palette of tricks that you can use to navigate some out-of-the-ordinary situations. Contacts Options sports a very simple screen, as Figure 1-18 makes clear, yet it provides you with four important options that change how Contacts behaves:

✦ **Sort By:** Allows you to change the way the list is sorted. You can change the sort field criteria from First Name, Last Name, or Company from the menu choices. You can use the Space key to toggle among the choices. Remember that guy from ABC Company? You can use the Sort By option to sort by company. By doing that, all contacts from ABC Company are listed next to each other, and with any luck, the guy's name will jump out at you.

**Contacts Options**

**Views**

| Sort By: | First Name ▾ |
| --- | --- |
| Separators: | Lines ▾ |

**Actions**

| Allow Duplicate Names: | Yes ▾ |
| --- | --- |
| Confirm Delete: | Yes ▾ |

**Figure 1-18:** Choose your sort type here.

✦ **Separators:** Allows you to change the dividers in the Contacts list. It's purely aesthetics, but check it out — you might like the stripes.

✦ **Allow Duplicate Names:** Self-explanatory. If you turn this on, you can have multiple people who happen to share the same name in your

Contacts list. If it's turned off, you get a warning when you try to add a name that's identical to someone who already exists in your list. Maybe you are just tired and mistakenly try to add the same person twice to your list. Then again, sometimes people just have the same names — perhaps you know three people named John Smith. We recommend keeping the default value of Yes, allowing you to have contacts with the same names.

✦ **Confirm Delete:** Allows you to display a confirmation screen for all contact deletions. You should always keep this feature turned on for normal usage. Because you could accidentally delete somebody from your Contacts list in many ways, this feature is a good way of minimizing those accidents.

**Book II**
**Chapter 1**

How do you change any of these options? The fields behave like any other fields on a BlackBerry application. You simply highlight the field, press the Menu key and select Change Option from the menu that appears. You then see a menu screen that allows you to select the possible option values. For example, Figure 1-19 shows the possible Sort By fields.

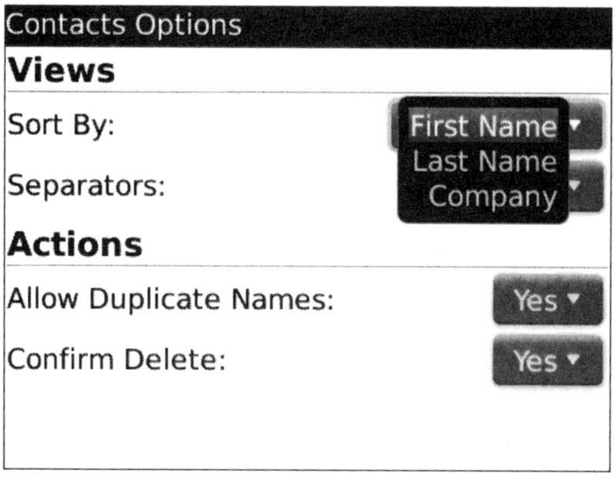

**Managing and
Locating Your
Contacts**

**Figure 1-19:**
The Sort
By field
options.

| Contacts Options | |
| --- | --- |
| **Views** | |
| Sort By: | First Name ▾ |
| | Last Name |
| Separators: | Company ▾ |
| **Actions** | |
| Allow Duplicate Names: | Yes ▾ |
| Confirm Delete: | Yes ▾ |

# Sharing a Contact

Suppose you want to share your contact information with a friend who happens to have a BlackBerry as well. A vCard — virtual (business) card — is your answer and can make your life a lot easier. In BlackBerry Land, a vCard is a contact in Contacts that you send to someone as an attachment to an e-mail. (Keep reading for more on sending vCards.) At the receiving end, the BlackBerry device (being the smart device that it is) recognizes the attachment and informs the BlackBerry owner that he or she has the option of saving it, which makes it available for his or her viewing pleasure in Contacts.

## Sending a vCard

Because a vCard is nothing more than Contacts data attached to an e-mail, sending a vCard is a piece of cake. (Of course, you do need to make sure that your recipient has a BlackBerry device to be able to receive the information.)

Here's how you go about sending a vCard:

*1.* **On the BlackBerry Home screen, select the Messages application.**

The Messages icon may differ from theme to theme, but most themes use an image of a mail envelope, as shown in Figure 1-20.

**Figure 1-20:**
Launch
Messages
here.

*2.* **Press the Menu key and then select Compose Email.**

A screen appears, allowing you to compose a new e-mail. Your next step is to enter the name of the recipient of this e-mail.

*3.* **In the To field, start typing the name of the person in your Contacts that you want to receive this vCard. When you see the name in the drop-down list, highlight it and press the trackball, trackpad, or touch screen.**

You are now presented with an e-mail screen with the name you just selected as the To recipient.

*4.* **Enter the subject and message.**

*5.* **Press the Menu key and then select Attach Contact.**

Contacts opens.

*6.* **Highlight the name of the person whose contact information you want to attach and then press the trackball, trackpad, or touch screen.**

The e-mail composition screen reappears, and an icon that looks like a book indicates that the e-mail now contains your attachment. Now all you have to do is send your e-mail.

7. **Press the Menu key and then select Send from the menu that appears.**

   You just shared the specified contact information. (Don't you feel right neighborly now?)

## Receiving a vCard

If you're the recipient of an e-mail that has a contact attachment, here's how you save it to Contacts:

1. **On the BlackBerry Home screen, select Messages.**

2. **Select the e-mail that contains the vCard.**

   The e-mail with the vCard attachment opens.

3. **Scroll down to the attachment. When the cursor is hovering over the attachment, press the Menu key and select View Attachment from the menu that appears.**

   The vCard makes an appearance onscreen.

4. **Press the Menu key and then select Add to Contacts.**

   The vCard is saved and is now available in your BlackBerry Contacts list.

# Searching for Somebody Outside Your Contacts

Does your employer provide your BlackBerry? Do you use Outlook or Lotus Notes on your desktop machine at work? If your answer to both of these questions is yes, this section is for you. BlackBerry Contacts has a feature that allows you to search for people in your organization, basically through Microsoft Exchange (for Outlook), IBM Domino (for Notes), or Novell GroupWise. Exchange, Domino, and GroupWise serve the same purposes, namely, to facilitate e-mail delivery in a corporate environment and to enable access to a database of names.

For you techies out there, these person databases are called Global Address Lists (GALs) in Exchange, Notes Address Books in Domino, and GroupWise Address Books in GroupWise.

If you want to search for somebody in your organization through a database of names, simply follow these steps:

1. **On the BlackBerry Home screen, select Contacts.**

2. **Press the Menu key and then select Lookup.**

   Some corporations might not have the Lookup feature enabled. Check with your IT department for more information.

Although you access the Lookup feature inside Contacts, it actually is going beyond your device and into your company's Exchange, Domino, or GroupWise database, depending on which e-mail server your company uses.

**3. On the screen that appears, enter the name you're searching for and then press the Enter key.**

You could enter the beginning characters of either a person's surname or first name. You are not searching through your contacts but through your company's database, so this step might take some time to complete depending on the criteria you enter.

For large organizations, we recommend being more precise when doing your search. For example, searching for *Dan* yields more hits than searching for *Daniel.* The more precise your search criteria, the fewer hits you'll get, and the faster the search will be.

While the search is in progress, you'll see the word Lookup and the criteria you entered at the top of the list. For example, if you enter **Daniel**, the top row reads Lookup: Daniel. After the search is finished, BlackBerry displays the number of hits or matches, for example, 20 matches: Daniel.

**4. Select the matches count to display the list of matches.**

A screen that displays the matching names based on your criteria appears. A header at the top of this screen details the matches displayed on the current screen as well as the total hits. For example, if the header reads something like Lookup Daniel (20 of 130 matches), 130 people in your organization have the name *Daniel,* and BlackBerry is displaying the first 20. You have the option of fetching more by pressing the Menu key and selecting Get More Results from the menu that appears.

You could also add the name or names listed in this result to Contacts by using the Add command (for the currently highlighted name) or the Add All command for all the names in the list. (As always, press the Menu key to call up the menu that contains these options.)

When you think you've found the person you're looking for, the next step is to check the information to make sure that you have the correct person.

**5. Select the person whose information you want to review.**

The person's contact information is displayed on a read-only screen. Information might include the person's title; e-mail address; work, mobile, and fax numbers; and the snail-mail address at work. Any of that information gives you confirmation about the person you're looking for. Of course, what shows up depends on the availability of this information in your company's database.

# Synchronizing Facebook Contacts

Do you network like a social butterfly? You must be using one of the popular social networking BlackBerry applications, such as MySpace or Facebook. You must have tons of friends from these networking sites and want to copy their contact information to your BlackBerry. There are ways to achieve this, and individual networking sites will have their own unique way. But if you're in Facebook, you're lucky.

With the latest Facebook application (version 1.7 as of this writing), it's much easier to get Facebook contacts to your BlackBerry. The Facebook app also allows you to synchronize information between your BlackBerry and your friend's information in Facebook.

Book II
Chapter 1

Managing and
Locating Your
Contacts

## Adding a Facebook friend's info to Contacts

Pulling your friend's information from Facebook is quite easy:

1. **Select the Facebook icon from the Home screen.**

   The Facebook application is filed in the Downloads folder.

2. **Press the Menu key and then press the Friends image.**

   Friends is the fourth image in the top.

   Your friends list shows up in the screen similar to the one on the left in Figure 1-21.

**Figure 1-21:**
Select
Facebook
friends
to add to
Contacts
here.

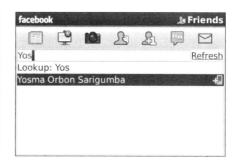

3. **In the Find field, start typing your friend's name.**

   This narrows the list, as shown on the right in Figure 1-21.

**4. Highlight the friend you want to add to Contacts, press the menu key, and select Connect to BlackBerry Contact.**

A dialog box appears, as shown in Figure 1-22, allowing you to choose whether to connect this Facebook friend to an existing contact or add the friend as a new contact to your BlackBerry. In this case, you want to add.

**Figure 1-22:**
Connect a
Facebook
friend to
an existing
contact or
add a new
contact
here.

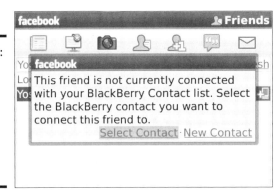

If the same person exists in your BlackBerry and in Facebook, you can simply link the contact here. Also, see the next section to get an automatic update on your BlackBerry whenever your friend changes his or her profile in Facebook.

The term *connect* in the Facebook application means telling the app which contact is associated with a Facebook friend. After the app records this linkage, it knows which contact to update when information in Facebook changes.

**5. Select New Contact.**

A progress screen appears momentarily, telling you that it's getting the contact information from Facebook. When it's finished, a new contact is added to your BlackBerry with the contact info shown on the screen.

**6. Press the Escape key.**

The Facebook app displays a prompt, asking you whether to ask for a phone number. This is a default behavior even if the phone number is already in your BlackBerry.

**7. Select either Yes or No in the prompt to request for phone number.**

You're back to the previous Facebook screen, and an Address Book icon has been added to the right of your friend's name, indicating that this friend is now connected, or linked, to a BlackBerry contact, as shown in Figure 1-23.

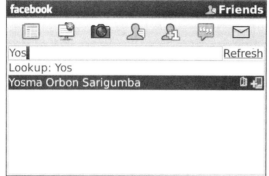

**Figure 1-23:**
The Address Book icon shows that your Facebook friend is connected to a BlackBerry contact.

## *Automatic syncing between Facebook profiles and Contacts*

When running the Facebook application for the first time, you're asked to enable synchronization. Here are Facebook and BlackBerry connections you can choose among:

✦ **BlackBerry Message application:** When enabled, you'll see new Facebook notifications in your Messages application.

✦ **BlackBerry Calendar application:** When enabled, a calendar item is automatically created in your BlackBerry whenever you have a new Facebook event.

✦ **BlackBerry Contacts application:** When enabled, your BlackBerry contacts are periodically updated with the latest Facebook information, including the profile pictures. For this to happen, your BlackBerry contacts also will be sent to Facebook.

If you opted out of these options the first time you ran Facebook, you can still enable them from the Facebook Options screen. The following steps enable Contacts synchronization:

*1.* **Select the Facebook icon from the Home screen.**

The Facebook application is filed under the Downloads folder.

*2.* **Press the Menu key and then select Options.**

The Options screen appears. A lot of information and text are on this screen, and you have to scroll down to see all the options. Feel free to check other options, but for synchronizing contacts, refer to the first two pages of the screen, which look like the ones shown in Figure 1-24.

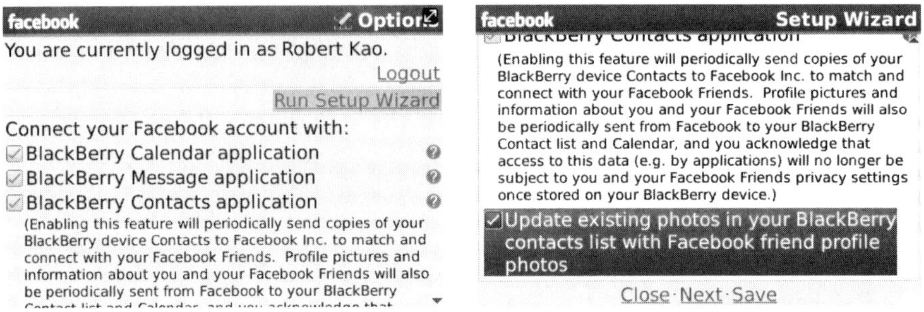

**Figure 1-24:**
Enable
Facebook
friends
synchroni-
zation with
Facebook
Options.

facebook ✎ Options
You are currently logged in as Robert Kao.
Logout
Run Setup Wizard
Connect your Facebook account with:
☑ BlackBerry Calendar application ❷
☑ BlackBerry Message application ❷
☑ BlackBerry Contacts application ❷
(Enabling this feature will periodically send copies of your
BlackBerry device Contacts to Facebook Inc. to match and
connect with your Facebook Friends. Profile pictures and
information about you and your Facebook Friends will also
be periodically sent from Facebook to your BlackBerry

facebook Setup Wizard
☑ BlackBerry Contacts application
(Enabling this feature will periodically send copies of your
BlackBerry device Contacts to Facebook Inc. to match and
connect with your Facebook Friends. Profile pictures and
information about you and your Facebook Friends will also
be periodically sent from Facebook to your BlackBerry
Contact list and Calendar, and you acknowledge that
access to this data (e.g. by applications) will no longer be
subject to you and your Facebook Friends privacy settings
once stored on your BlackBerry device.)
☑ Update existing photos in your BlackBerry
contacts list with Facebook friend profile
photos
Close · Next · Save

3. **Add a check mark to the BlackBerry Contacts application.**

   There's explanatory text right below this check box. If you scroll down, you should see another check box, which allows you to synchronize Facebook profile photos with Contacts photos.

4. **Add a check mark to the option titled Update Existing Photos in Your BlackBerry Contacts List with Facebook Friend Profile Photos (Figure 1-24, right).**

5. **Press the escape key, and select Yes on the Save Changes prompt.**

   Your Contacts will now be periodically updated with Facebook friends.

# Chapter 2: Managing Your Appointments

## In This Chapter

✓ Seeing your schedule from different time frames

✓ Making your Calendar your own

✓ Scheduling a meeting

✓ Viewing an appointment

✓ Deleting an appointment

✓ Sending and receiving meeting requests

To some folks, the key to being organized and productive is mastering time management and using their time wisely (and we're not just talking about reading this book while you're commuting to work). Many have discovered that there is no better way to organize their time than to use a calendar — a daily planner tool. Some prefer digital to paper, so they use a planner software program on their PC — either installed on their hard drive or accessed through an Internet portal (such as Yahoo!). The smartest of the bunch, of course, use their BlackBerry handheld because it has the whole planner thing covered in handy form with its Calendar application.

In this chapter, we show you how to keep your life (personal and work) in order by managing your appointments with your BlackBerry Calendar. What's great about managing your time on a BlackBerry instead of on your PC is that your BlackBerry is always with you to remind you. Just remember that you won't have any more excuses for forgetting that important quarterly meeting or Bertha's birthday bash. To synchronize your desktop calendar with your BlackBerry Calendar, please see Book IV, Chapter 2.

## Accessing BlackBerry Calendar

BlackBerry Calendar is one of the BlackBerry core applications, like Contacts or Phone (read more about the others in Book I, Chapter 2), so it's easy to get to. On the BlackBerry Storm, just press the Menu key and then touch-press the Calendar icon. For all other devices, press the Menu key and then scroll to the Calendar icon and press the trackball or trackpad. Voilà! You have Calendar.

## Choosing Your Calendar View

The first time you open Calendar, you'll likely see the Day view, which is a default setting on the BlackBerry, as shown in Figure 2-1. However, you can change the Calendar view to a different one that works better for your needs.

13 Jan 2010
14:17
M T W T F S S
09:00 Daily Rocks - Day's Priority
10:00
11:00
12:00 Lunch Meeting
13:00
13:00 Blog Interview
14:00
15:00 Jade's Swimming Lessons
16:00

**Figure 2-1:** Day view in Calendar.

- **Day:** This view gives you a summary of your appointments for the day. By default, it lists all your appointments from 9 a.m. to 5 p.m.
- **Week:** This view shows you a seven-day summary view of your appointments. By using this view, you can see how busy you are in a particular week.
- **Month:** The Month view shows you every day of the month. You can't tell how many appointments are in a day, but you can see on which days you have appointments.
- **Agenda:** The Agenda view is a bit different from the other views. It isn't a time-based view like the others; it basically lists your upcoming appointments. And in the list, you can see details of the appointments, such as where and when.

Different views (such as the Week view shown in Figure 2-2) offer you a different focus on your schedule. Select the view you want based on your scheduling needs and preferences. If your life is a little more complicated, you can even use a combination of views for a full grasp of your schedule.

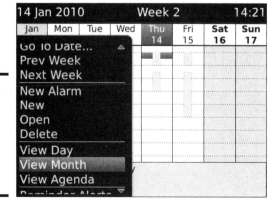

| 14 Jan 2010 | | | Week 2 | | | 14:21 | |
|---|---|---|---|---|---|---|---|
| Jan 2010 | Mon 11 | Tue 12 | Wed 13 | Thu 14 | Fri 15 | Sat 16 | Sun 17 |
| 09:00 | | | | | | | |
| 10:00 | | | | | | | |
| 11:00 | | | | | | | |
| 12:00 | | | | | | | |
| 13:00 | | | | | | | |
| 14:00 | | | | | | | |
| 15:00 | | | | | | | |
| 16:00 | | | | | | | |
| 17:00 | | | | | | | |

Daily Rocks - Day's Priority

09:00 - 10:00

**Figure 2-2:** Change your Calendar view to fit your life.

To switch between different Calendar views, simply follow these steps:

**1. From the Home screen, press the Menu key and then select Calendar.**

Doing so calls up the Calendar application in its default view — more than likely Day view.

**2. Press the Menu key and then select the view of your choice from the menu that appears (shown in Figure 2-3).**

If you start from Day view, your choices are View Week, View Month, and View Agenda.

**Figure 2-3:** The Calendar menu lets you select different views.

| 14 Jan 2010 | | | Week 2 | | | 14:21 | |
|---|---|---|---|---|---|---|---|
| Jan | Mon | Tue | Wed | Thu 14 | Fri 15 | Sat 16 | Sun 17 |

Go To Date...
Prev Week
Next Week
New Alarm
New
Open
Delete
View Day
View Month
View Agenda
Reminder Alerts

## Moving between Time Frames

Depending on what view of Calendar you're in, you can easily move to the previous or next day, week, month, or year. For example, if you're in the

Month view, you can move to the next month (um, relative to the currently displayed month). Likewise, you can also move to the previous month. In fact, if you like to look at things in the long term, you can jump ahead (or back) a year at a time. You can move between time frames by pressing the Menu key, shown in Figure 2-4.

**Figure 2-4:**
Move
among
months or
years in
Month view.

| 4 Mar 2010 | | Week 9 | | 14:23 |
|---|---|---|---|---|
| **M a r c h** | | | | |
| Help | **T** | **F** | **S** | **S** |
| Today | | | | |
| Go To Date... | 25 | 26 | 27 | 28 |
| Prev Month | | | | |
| Next Month | 4 | 5 | 6 | 7 |
| Prev Year | 11 | 12 | 13 | 14 |
| Next Year | 18 | 19 | 20 | 21 |
| New Alarm | 25 | 26 | 27 | 28 |
| New | | | | |
| View Day | 1 | 2 | 3 | 4 |

You have similar flexibility when it comes to the other Calendar views. See Table 2-1 for a summary of what's available.

| Table 2-1 | Moving between Views |
|---|---|
| *Calendar View* | *Move Between* |
| Day | Days and weeks |
| Week | Weeks |
| Month | Months and years |
| Agenda | Days |

You can always go to today's date regardless of what Calendar view you're in. Just press the Menu key and then select Today from the menu that appears.

Furthermore, you can jump to any date you choose by pressing the Menu key and then selecting Go to Date. Doing so calls up a handy little dialog box that lets you choose the date you want. To change the date on a Bold, Curve, or Pearl, scroll the trackball or trackpad to the desired day, month, and year, as shown in Figure 2-5.

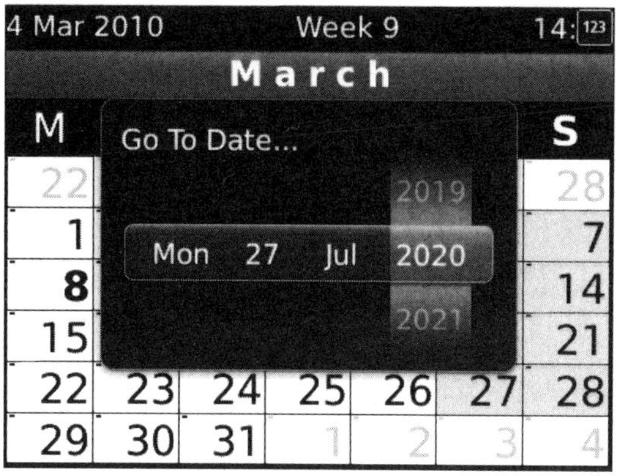

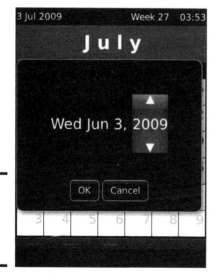

**Figure 2-5:**
Go to
any date
you want
on your
BlackBerry.

To change the date on a Storm, touch-press the date and then press the up
or down arrow above the month, day, and year as shown in Figure 2-6. Once
you are done, touch-press OK.

**Figure 2-6:**
Choose a
date to go
to on the
Storm.

# Customizing Your Calendar

To change the initial (default) view in your Calendar — from Day view
to Month view, for example — Calendar Options is the answer. To get to

Calendar Options, open Calendar, press the Menu key, and select Options from the menu that appears. You see choices similar to the ones shown in Table 2-2.

| Table 2-2 | Calendar Options |
|---|---|
| **Option** | **Description** |
| **Formatting** | |
| First Day of Week | The day that first appears in your Week view. |
| Start of Day | The time of day that defines your start of day in Day view. The default is 9 a.m. If you change this to 8 a.m., for example, your Day view starts at 8 a.m. instead of 9 a.m. |
| End of Day | The time of the day that defines the end of day in Day view. The default is 5 p.m. If you change this to 6 p.m., for example, your Day view ends at 6 p.m. instead of 5 p.m. |
| **Views** | |
| Initial View | Specifies the Calendar view that you first see when opening Calendar. |
| Show Free Time in Agenda View | If yes, this field allows an appointment-free day's date to appear in the Agenda view. If no, the Agenda view does not show the date of days on which you don't have an appointment. |
| Show End Time in Agenda View | If yes, this field shows the end time of each appointment in the Agenda view. If no, the Agenda view shows only the start time of each appointment. |
| **Actions** | |
| Snooze | The snooze time when a reminder appears. The default is 5 minutes. |
| Default Reminder | How far in advance your BlackBerry notifies you before your appointment time. The default is 15 minutes. |
| Enable Quick Entry | Day view only. Allows you to make a new appointment by typing characters. This way, you don't need to press the trackball or trackpad and select New. *Note:* If you enable this, you will not be able to enter appointments by pressing letter keys. You must press the Menu key and select New. |

# Managing multiple calendars

Like your e-mail accounts, you might have multiple calendars. For example, you might have a calendar from your day job and you might have a calendar from your personal life or softball club that you belong to. Whatever the reason, your BlackBerry has a great way for you to manage multiple calendars.

In Calendar Options, you see a screen similar to the following figure. Each colored square represents a different calendar. The folks at RIM use Calendars to represent different calendars in your life, which is the reason you see e-mail addresses next to the colored squares. For example, you can assign the color red to your day job calendar and green to your softball club calendar. When two events conflict at the same time slot, you can better prioritize with the colors.

To change the color of each calendar, follow these steps:

1. **Open Calendar.**

2. **Press the Menu key and select Options Menu.**

   This opens a screen similar to the following figure.

3. **Select a calendar of your choice.**

   The calendar properties screen opens.

4. **Highlight the colored square and then select the desired color.**

5. **Press the Menu key and select Save.**

```
Calendar Options
General Options
Calendars
  betasupport@smrtguard.com
  jave.park@gmail.com
  mltest3@att.blackberry.net
  peekalab@smrtguard.com
  rob.kao@gmail.com
  rob.kao@smrtguard.com
  sales@smrtguard.com
```

# All Things Appointments: Adding, Opening, and Deleting

After you master navigating the different Calendar views (and that should take you all of about two minutes) and you have Calendar customized to your heart's content (another three minutes, tops), it's time (pun intended) to set up, review, and delete appointments. We also show you how to set up a meeting with clients or colleagues.

## Creating an appointment

Setting up a new appointment is easy. You need only one piece of information: when your appointment occurs. Of course, you can easily add related information about the appointment, such as the meeting's purpose, its location, and whatever additional notes are helpful.

In addition to your standard one-time, limited-duration meeting, you can also set all-day appointments. The BlackBerry can assist you in setting recurring meetings as well as reminders. Sweet!

### Creating a one-time appointment

To add a new one-time appointment, follow these steps:

1. **Open Calendar.**

2. **Press the Menu key and then select New.**

   The New Appointment screen appears.

3. **Fill in the key appointment information.**

   Type all the information regarding your appointment in the appropriate spaces, as shown in Figure 2-7. You should at least enter the time and the subject of your appointment.

**New Appointment**

**Subject: Golf Outing**
Location: Alpine Golf Course

☐ All Day Event
Start:                     Tue 8 Jun 2010 11:00
End:                       Tue 8 Jun 2010 17:00
Duration:                     6 Hours 0 Mins
Time Zone:        Casablanca (GMT) ▼
Show Time As:                    Busy ▼
Reminder:                      15 Min. ▼

**Figure 2-7:**
Set an
appointment
here.

4. **Press the Menu key and then select Save.**

   Your newly created appointment is saved.

Your new appointment is now in Calendar and viewable from any Calendar view. Also, keep in mind that you can have more than one appointment in the same time slot. Why? Well, the BlackBerry Calendar allows conflicts in your schedule because it lets you make the hard decision about which appointment you should forgo.

### Creating an all-day appointment

If your appointment is an all-day event — for example, if you're in corporate training or have an all-day doctor's appointment — select the All Day Event check box on the New Appointment screen, as shown in Figure 2-8. With the check box selected, you can't specify the time of your appointment — just the start date and end date (simply because it doesn't make sense to specify a time for an all-day event).

**Figure 2-8:**
Set an all-
day event
here.

Appointment Details
**Subject: Mom's bday**
Location:
☑ All Day Event
Start:                Mon 8 Mar 2010
End:                 Mon 8 Mar 2010
Duration:               1 Day
Time Zone:     Casablanca (GMT) ▾
Show Time As:            Free ▾
Reminder:             15 Min. ▾

### Setting your appointment reminder time

Any appointment you enter in Calendar can be associated with a reminder alert — a pop-up dialog box with either a vibration or a beep, depending on how you set things up in your profile.

Profile is simply another useful BlackBerry feature that allows you to cus-tomize how your BlackBerry alerts you when an event occurs. Examples of events are an e-mail, a phone call, or a reminder for an appointment. (For more on profiles, see Book I, Chapter 3.)

You can set the reminder time for a particular appointment. From the New Appointment screen, simply scroll to the Reminder field and select a reminder time anywhere from None to 1 Week before your appointment time.

By default, whatever reminder alert you set goes off 15 minutes before the event. But you don't have to stick with the default. You can choose your own default reminder time so that you don't have to change the reminder time with each new appointment you enter. Here's how:

1. **Open Calendar.**

2. **Press the Menu key and then select Options.**

   The Calendar Options screen displays.

3. **Select Default Reminder.**

4. **Choose a default reminder time anywhere from None to 1 Week before your appointment.**

So from now on, any new appointment has a default reminder time of what you just set up. Assuming that you have a reminder time other than None, the next time you have an appointment coming up, you see a dialog box like the one shown in Figure 2-9, reminding you of the appointment.

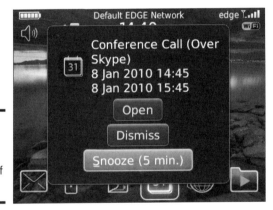

**Figure 2-9:** You get a reminder dialog box if you want.

### Creating a recurring appointment

You can set up recurring appointments based on daily, weekly, monthly, or yearly recurrences. Everyone has some appointment that repeats, such as birthdays or anniversaries (or taking out the trash every Thursday at 7:30 a.m. — ugh).

For all recurrence types, you can define an Every field. For example, say you have an appointment that recurs every nine days. Just set the Recurrence field to Daily and the Every field to 9, as shown in Figure 2-10.

Depending on what you select in the Recurrence field, you have the option to fill in other fields. If you enter Weekly in the Recurrence field, for example, you have the option of filling in the Day of the Week field. (It basically allows you to select the day of the week on which your appointment recurs.)

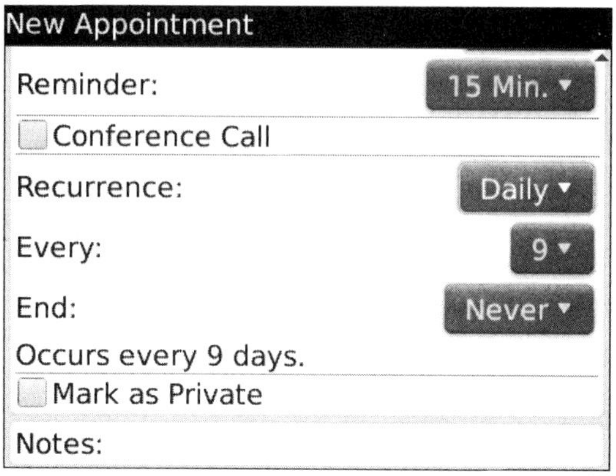

**Figure 2-10:**
An appoint-
ment
recurring
every nine
days.

| New Appointment | |
|---|---|
| Reminder: | 15 Min. ▾ |
| ☐ Conference Call | |
| Recurrence: | Daily ▾ |
| Every: | 9 ▾ |
| End: | Never ▾ |
| Occurs every 9 days. | |
| ☐ Mark as Private | |
| Notes: | |

If you enter Monthly or Yearly in the Recurrence field, the Relative Date check box is available. With this check box selected, you can ensure that your appointment recurs relative to today's date. For example, if you choose the following, your appointment occurs every two months on the third Sunday until July 31, 2012:

**Start:** Sunday, June 21, 2009 at 12 p.m.

**End:** Sunday, June 21, 2009 at 1 p.m.

**Recurrence:** Monthly

**Every:** 2

**Relative Date:** Selected

**End:** Saturday, July 31, 2012

On the other hand, if all options in this example remain the same except that Relative Date is not selected, your appointment occurs every two months, on the 21st of the month, until July 31, 2012.

If all this "relative" talk has you dizzy, don't worry: The majority of your appointments won't be as complicated as this.

## Opening an appointment

After you set an appointment, you can view it in a couple of ways:

✦ If you've set up reminders for your appointment and the little Reminder dialog box appears onscreen at the designated time before your appoint-ment, you can view your appointment by clicking the box's Open button. In the same dialog box, you can Snooze the reminder. (Refer to Figure 2-9.)

✦ You can open the appointment from Calendar by going to the exact time of your appointment and viewing it there.

While looking at an appointment, you have the option of making changes (a new appointment time and new appointment location) and then saving them.

## Deleting an appointment

Deleting an appointment is straightforward. When in Day or Week view, simply scroll to the appointment that you want to delete, press the Menu key, and select Delete from the menu that appears.

If the appointment that you're deleting is part of a recurring appointment, a dialog box pops up asking whether you want to delete all occurrences of this appointment or just this particular occurrence, as shown in Figure 2-11. After you make your choice, your appointment is history.

**Figure 2-11:**
You can delete all occurrences or just the single instance of a recurring appointment.

# Appointments versus Meetings

Technically, any event in your Calendar counts as an appointment, whether it's a reminder for your best friend's birthday or a reminder of a doctor's appointment for a checkup. However, when you invite people to an appointment or you get invited to one, regardless of whether it's a face-to-face meeting or a phone conference, that appointment becomes a *meeting*.

## Sending a meeting request

Sending a meeting request to others is similar to creating a Calendar appointment. Follow these steps:

*1.* **Open Calendar.**

*2.* **Press the Menu key and then select New.**

*3.* **Fill in the key appointment information (subject, location, and time).**

*4.* **Press the Menu key and then select Invite Attendee.**

You're taken to Contacts to select your meeting attendee(s).

5. **From Contacts:**

   - *If you have contacts in Contacts:* Highlight the contact you want and press it.

     Current BlackBerry models use one of the three navigation methods — trackball, trackpad, or touch screen. One thing they all have in common is that you press them to make a selection. Throughout this book, whenever we provide steps that simply say "press" we mean to press the trackball, trackpad, or touch screen depending on your BlackBerry model.

   - *If you don't yet have contacts or if the one you want isn't in Contacts:* Select the Use Once option to enter the appropriate e-mail address and press the Enter key to finish and return to Calendar.

6. **After returning from Contacts, you see the attendees in your Calendar meeting notice.**

7. **Press the Menu key and then select Save.**

   This action actually sends an e-mail to your meeting attendees, inviting them to your meeting.

## Responding to a meeting request

Whether for work or a casual social event, you've likely received a meeting request by e-mail, asking you to respond to the meeting by choosing one of three options: Accept, Tentative, or Decline. (If it's from your boss for an all-staff meeting and you just can't afford to decline again because it's so close to Christmas bonus time, that's an Accept.)

You can accept any meeting request from your managers or colleagues on your BlackBerry just as you would on your desktop PC. In the PC world, you respond to an e-mail request for a meeting by clicking the appropriate button in your e-mail client (Microsoft Outlook, for the vast majority of you). In the BlackBerry world, a meeting request also comes in the form of an e-mail; upon reading the e-mail, just choose Accept, Tentative, or Decline in the Messages application. Your response is sent back in an e-mail.

After you respond to the meeting request, the meeting is added to your Calendar automatically, if you accepted it. If you have a change of heart later, you can change your response (yes, you can later decline that useless meeting after all) in Calendar, and the declined event disappears from your Calendar.

## Setting your meeting dial-in number

You may have colleagues and friends all over the country or even on another continent. Group phone meetings may require a

- ✦ Dial-in number
- ✦ Moderator code (if you are the moderator)
- ✦ Participation code

**Book II
Chapter 2**

**Managing Your
Appointments**

To set your phone conference dial-in details, follow these steps:

1. **Open Calendar.**

2. **Press the Menu key and then select the Options menu item.**

3. **Select Conference Call Options.**

   A screen similar to Figure 2-12 appears.

4. **Enter the appropriate numbers.**

5. **Press the Menu key and select Save.**

   Your conference call number is saved.

**Figure 2-12:**
Setting up
conference
call dial-in
details.

The next time you create a new appointment, if you tap the Conference Call check box in the Appointment screen, you see the conference number, as shown in Figure 2-13.

**Figure 2-13:**
Conference
call
information
displayed
in the
Appointment
Details
screen.

# Chapter 3: Creating To-Do's and Safely Storing Your Passwords

## In This Chapter

✔ **Getting to know the Tasks application**

✔ **Adding, changing, and deleting tasks**

✔ **Managing your Tasks list**

✔ **Setting up recurring tasks**

✔ **Categorizing tasks**

✔ **Using Password Keeper**

*U*sing your BlackBerry as an organizational tool is one of the key themes in this book. And speaking of organization, what better proof of your impressive organizational skills could exist than the fact that you use task, or to-do, lists? Knowing what you need to do today, tomorrow, the entire week, or perhaps the whole month makes you more efficient on your job and in your personal life. The fact is that you not only need to know what your tasks are, but you also need to prioritize them — and reprioritize them, if necessary. And with your BlackBerry as your able assistant, you can.

In this chapter, we introduce you to your Blackberry smartphone's Tasks application. Stick with us as we explore this valuable tool, jumping from creating and maintaining your to-do list to setting alerts for a recurring task in a single bound. We also throw in a few tips and tricks to make maintaining and searching your Tasks list easier and faster. Of course, we can't advise you on how to actually *do* the task after your BlackBerry calls your attention to it, but with your BlackBerry in your palm, you're at least a step closer to clearing your desk. Finally, we also give you the scoop on keeping your passwords safe by using the Password Keeper application.

Even if you keep a to-do list on your desktop (Outlook or Lotus Notes, anyone?), consider switching to all BlackBerry, all the time. You'll love the greater flexibility that comes with greater mobility.

# Accessing Tasks

You can find tasks in the Applications folder on the Home screen, as shown in Figure 3-1. Look for the icon of a clipboard with a check mark.

**Figure 3-1:**
Going from
Applications
to Tasks.

The icons used by other providers might be different from what you see here. If this is the case for your BlackBerry, just remember that Tasks is always in Applications. If you can locate the Applications icon, it's just a matter of selecting it to locate the Tasks icon.

# Recording a New Task

The first step when building a to-do list is to start recording one. Don't groan and roll your eyes, dreading how long this will take. This is easy — so just relax and you'll be finished in a snap.

Follow these simple steps:

*1.* **Select Tasks.**

The Tasks application opens. Similar to Contacts and MemoPad, the screen that appears is divided into two parts: The top shows the Find field, followed by the list of tasks or *\*No Tasks\**.

*2.* **Select Add Task.**

Alternatively, you can press the Menu key and then select New (as shown on the left side of Figure 3-2).

The New Task screen appears, as shown on the right side of Figure 3-2, ready and willing to document your new task. This simple screen features easy-to-understand fields that describe the task you're about to enter.

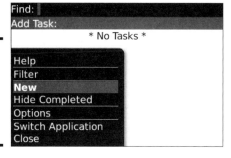
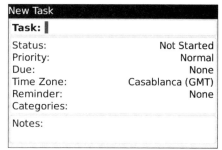

**Figure 3-2:**
Select
New and
an empty
task screen
appears.

3. **Move to each field and enter information for your task.**

    Some fields are for text that you enter yourself, and some fields hold items you select from a menu specific for that field. In other words, for text fields, you have to type the stuff you want; but for nontext fields, you select the field, press again, and then make your choice from the selection that appears. (Pretty convenient, huh?) We explain all these fields in the next section of this chapter.

4. **After filling in the relevant fields, press the Menu key and then select Save.**

    Doing so saves your task, and you see the task added to the Tasks list. Remember that unless you change the sort field (see "Organizing Your Tasks List" later in this chapter to find out how to change the sort field), the Tasks list is sorted by task subject.

Book II
Chapter 3

Creating To-Do's
and Safely Storing
Your Passwords

# Navigating the Tasks Fields

The New Task screen (refer to Figure 3-2) is straightforward and contains few fields. Although the few fields there are self-explanatory, we (being the thorough guys that we are) describe each one here. Task (or Subject), Status, Priority, and Due Date can be used to sort your Tasks list. See the "Organizing Your Tasks List" section, which tells you how to change the sort order.

## Task field

Use this field to log the subject of your task or a short description of your task. Make this field as descriptive as you can: The subject you type here should be specific enough that you can differentiate this task from the rest in your list. For example, if you make several presentations to clients, you don't want to call your task Prepare Presentations. You want to be specific so that you can distinguish it from other tasks. Perhaps name it Product X Benefit Forecast to XYZ CFO. You can search this field from the main Tasks screen.

## Status field

Fill in the Status field to indicate the current state of your task. Status is a *selection field* — that is, after highlighting the field, you press the trackball, trackpad, or touch screen and select a value from the list.

The choices give you a good idea of the field's purpose. The following are the possible values you can choose from:

✦ **Not Started:** You haven't started this task yet. Because this choice is the most common, Not Started is the default choice when creating a task.

✦ **In Progress:** You are in the midst of the task.

✦ **Completed:** You are finished with the task.

✦ **Waiting:** Your task is ongoing and depends on another task or another event. For example, you're waiting for Joe in Accounting to get you a spreadsheet so that you can include it in your report and complete the task.

✦ **Deferred:** Your task is on hold. Maybe you just don't need to work on this task at the moment, or you need more information before you decide whether this task is worth doing. Either way, you want to keep the task listed so that you can track it or resurrect it. Perhaps this task isn't a big deal today, but it could become important in a month or two. By tagging a task as Deferred, you keep yourself aware of a task that might or might not ramp up.

## Priority field

In the Priority field, you can specify the timeliness or urgency of the task. Like the Status field, the choices here are selections you make from a menu. You can choose from one of the following values:

✦ **High:** This is the highest possible setting. You should consider the most urgent task to be of high priority.

✦ **Normal:** This is the default value, which applies to most tasks. In reality, a Normal task can jump to become a high priority when it's not finished in time, but you have to decide and assign that yourself.

✦ **Low:** Just as you'd surmise, a Low rating tags a task as being less critical — you can put this one off until you're finished with the High and Normal tasks. You can rate all your nice-to-have tasks with this priority. *Hint:* When you're finished with your High and Normal tasks, reprioritize your Low tasks.

## Due field

Consider the Due field your task completion deadline. Here, you can enter a due date for your task. The default here is None; to change the value to a specified due date, follow these steps:

*1.* **Select the field.**

A pop-up menu appears onscreen offering two options: None and By Date.

*2.* **Select the By Date option.**

A date field appears on the next line, as shown in Figure 3-3. The value of the date defaults to the current date. If the current date is not your intended due date, proceed to Step 3 to change the value of this date.

**Book II**
**Chapter 3**

Creating To-Do's
and Safely Storing
Your Passwords

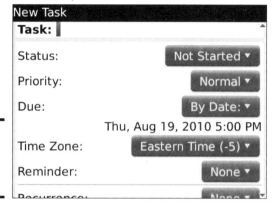

**Figure 3-3:**
Set a task's
due date
here.

*3.* **Select the entire date (Storm only) or a specific portion of the date (trackball/trackpad devices) that you want to change.**

The date or portion of the date that you highlighted is now editable. The date portions that you can change are year, month, day, and time.

Although this is a date field — like any date field in Tasks, for that matter — setting its value does not create an entry in Calendar. You can, however, set it to remind you on the due date if you set the reminder field. See the "Reminder field" section for more on the Reminder field.

*4.* **Change the date value.**

For trackball/trackpad devices, scroll to the specific date value you want and then press the trackball or trackpad to accept the change. For the Storm, use a finger swipe to change the date value and then touch-press OK.

At this point, you should have the correct value of the date component you want.

Say you modified the day but you want a specific time on that day. Repeat Steps 3 and 4, but this time highlight (and edit) the time component of the date.

While you are specifying the due date, you might see the Recurrence and No Recurrence fields. These fields show up only when you select By Date in the

Due field. Intrigued? Check out the "Creating Recurring Tasks" section, later in this chapter, for details about these fields.

## Time Zone field

The Time Zone field holds the time zone related to the date fields used for this task: Due (refer to the preceding section) and Reminder (see the following section). If these fields have values of None, this field is irrelevant.

You can specify a time zone different from your locale. For example, if you live in New York and you anticipate completing this task in Mexico City, you can specify the Mexico City time zone. Then, all the times in this task become relative to Mexico City.

## Reminder field

From the Reminder field (a date field), you can set an alarm or a notice on the date and time you specify. You can set it just like you would set the Due field. (Refer to the earlier section "Due field.") Setting a reminder is useful, especially for those important tasks that you can't afford to forget (such as buying a birthday gift for your significant other).

Again, just like any date field in Tasks, setting its value does not make it show up in your Calendar. The type of reminder you get is based on your active profile. (See Book I, Chapter 3 for details on how to customize notifications in your profile.)

When the reminder date is met, your BlackBerry notifies you and displays a Reminder screen. On this screen, you will see the name of the task and possible actions: Open (opens the task), Mark Completed (shows up for a task with a status other than Completed), and Dismiss (closes the screen).

## Categories field

Use the Categories field to assign a specific category that you can use to filter your Tasks list later. By default, this field is blank. However, you can easily assign a value to it from the Categories screen available through the context menu. Using the Categories field is important for organizing your list, so we describe it fully in the section "Organizing Your Tasks List," later in this chapter.

## Notes field

The Notes field is a free-for-all text field. You can put anything here you want, such as a detailed description of the task or any other info that relates to this task.

## Updating Your Tasks

When it's time to update your Tasks list — say, after finishing a high-priority task or when you want to change the due date for a specific task — the Tasks application won't stand in your way. You can quickly go back to your Tasks list and update those records.

To update a specific task, follow these steps:

*1.* **Select Tasks.**

The Tasks application opens to the Find screen, which displays your current Tasks list.

*2.* **In the Tasks list, select the task you need to edit.**

The screen that opens is the same one you used to create the high-lighted task, although obviously this display has fields filled with the information you already entered.

*3.* **Update the fields.**

Go through each of the fields that you want to edit.

Tasks and Notes fields are text fields that you can edit from here. To update the other fields, select the fields to make them editable.

*4.* **Press the Menu key and then select Save.**

Your task is saved, and you can see the updated task in the Tasks list.

## Deleting a Task

Just like folks make a ritual of spring-cleaning when winter fades, the same is true for your tasks. When a task is completed and keeping it just takes up space, simply delete it through your Tasks application.

In simple steps, here's how:

*1.* **Select Tasks.**

The Tasks application opens, displaying your Tasks list.

*2.* **Highlight the task you want to delete.**

*3.* **Press the Menu key and then select Delete.**

You see the standard Confirmation screen.

*4.* **If you're sure that this task is doomed for the dustbin, select Delete on the Confirmation screen.**

The task is deleted, and your Tasks list is updated.

## Organizing Your Tasks List

As time goes by, your Tasks list is sure to grow — which means that the time it takes to find a task in your list is sure to grow as well. One way to stay organized is to make it a habit to delete finished tasks from your list, as we detail in the preceding section. (The shorter the list, the better, we always say.) We recommend weeding out your Tasks list every time a project or a goal is completed. After all, when a particular project is completed, you probably don't need to go back to the tasks you did for it.

If you're someone who just loves to document everything you accomplish (or you work in an environment where you're expected to keep a listing of tasks completed — can you say quarterly employee review?), you might not relish the idea of deleting any entry from your Tasks list. In that case, regularly archive a copy of your entire Tasks list before you do any weeding, and you'll have a complete record of every stitch of hard work you contributed to a project. To archive, synchronize your BlackBerry with your desktop and store the data in whatever time-management software you use on your desktop. BlackBerry can synchronize to personal as well as enterprise time-management software. (For details on how to synchronize, see Book IV, Chapter 2.)

After synchronization, you can print the Tasks list related to this completed project (via your desktop application), which you can file. Having a hard copy of those completed tasks can give you peace of mind as you delete tasks from your BlackBerry. The best of both worlds, right? You clean up your Tasks list (making it easier for you to do a search), and you have an archive (in case you need a historical reference).

Another way to stay organized is to sort your Tasks list. Even after you weed out tasks you've completed, you may still have difficulty finding a task from your list. Fear not; help is available. Chances are that when you look for a task, you know something identifiable about it, such as its priority or due date. Because you assign information about your task when you create it (see the earlier sections "Navigating the Tasks Fields" and "Updating Your Tasks"), you can use that information as part of a task sort in your BlackBerry. By default, your Tasks list is sorted by the name or subject of your task, but you can also sort by priority, due date, or status. For example, if you know the due date of your task, you can sort your list by due date and thus quickly find the task at hand.

"Sounds great," you say, "but how do I change the sort criteria?" We're glad you asked. Because changing how you sort basically involves customizing your BlackBerry device, we tell you how to do that first and then tell you how to sort.

## Customizing tasks

You can make further Tasks customizations through the Options screen: sorting by criteria and toggling the deletion confirmation screen. Locating the Options screen from the Tasks application is easy. Press the Menu key and select Options. The Tasks Options screen appears, displaying two sections, Views and Actions, as shown in Figure 3-4. You can set the following here:

✦ **Sort By:** Here you can change how the list is sorted. The default task listing is the alphabetical order of the subject from *A* to *Z* (no reverse). To change to a different sort field, follow these steps:

  a. *On the Tasks Options screen, press the Sort By field.*

  A list displays your sorting options. You can choose from Subject, Priority, Due Date, or Status.

  b. *Press your choice.*

  Make sure you select save when you exit this screen; otherwise your changes here will be discarded.

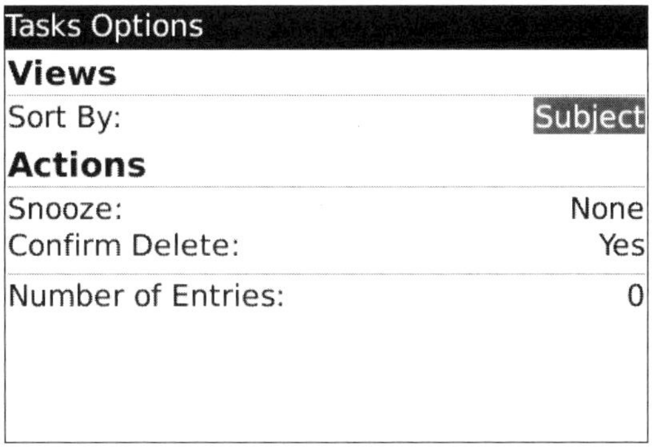

**Book II**
**Chapter 3**

**Creating To-Do's and Safely Storing Your Passwords**

| Tasks Options | |
|---|---|
| **Views** | |
| Sort By: | Subject |
| **Actions** | |
| Snooze: | None |
| Confirm Delete: | Yes |
| Number of Entries: | 0 |

**Figure 3-4:** Change your Tasks sort options here.

✦ **Snooze:** For tasks with reminders, this option allows you to snooze the alarm. The default value is None, but you can set it to 1, 5, 10, 15, or 30 minutes.

✦ **Confirm Delete:** This option allows you to control whether you want the application to display a confirmation screen upon deleting an item from your list. In other BlackBerry applications, this option is common. These confirmations appear as a safety feature so that you don't accidentally press Delete and lose something important. You can always turn this feature off in any of your BlackBerry applications, but we generally recommend that you keep the value of this field set to Yes, meaning that it prompts you for every delete.

There are exceptions to every rule, however. When you begin to weed out a bunch of outdated tasks from your Tasks list, toggle this feature off so that you don't get 27 prompts in a row when you're deleting 27 items. (But toggle the feature back on when you're finished weeding.)

To turn off this feature, just do the following:

a. *On the Tasks Options screen, select the Confirm Delete field.*

The screen shows your two choices: Yes and No. (See Figure 3-5.) *No* means you want to toggle off the Confirmation screen.

b. *Select No.*

The Tasks Options screen updates to show No in the Confirm Delete field.

c. *Press the Menu key and select Save.*

The Tasks application applies the change you made.

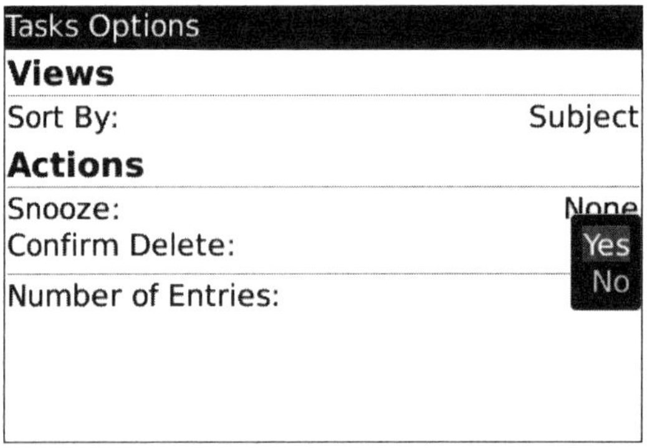

**Figure 3-5:**
Toggle
delete
confirmation
here.

| Tasks Options | |
|---|---|
| **Views** | |
| Sort By: | Subject |
| **Actions** | |
| Snooze: | None |
| Confirm Delete: | Yes |
| | No |
| Number of Entries: | |

Another field you can see from the Tasks Options screen is Number of Entries. This field is just informational, showing you how many tasks you have in your Tasks application.

## Creating a category

Sometimes sorting on a specific criterion such as Subject or Due Date still might not give you a quick answer to what you're looking for. For example, if you want to know how many more personal tasks you still have to do as opposed to the more business-oriented stuff you have going, sorting is of no help, right? What you really need is a way to filter your list based on certain groups that you define. (Personal versus Business would be a good start.) The good people at RIM anticipated such a need and introduced Categories.

So what exactly is a category, and why is it important? A *category* is a way for you to group your tasks in a manner that you can come back to. The grouping is the category, and the listing with a certain category is a *filter.* To make a task part of a group or category, you simply assign it a category when you record a task, or you can update the task and assign it a category then. (Refer to earlier sections in this chapter for how to do both.)

One important aspect of Categories you should be aware of is that they are shared among applications — specifically, among Contacts, MemoPad, and Tasks. This sibling relationship might sound trivial at first, but don't make the common mistake of assuming that what you change in Tasks does not affect other BlackBerry apps. The importance of this comes into play when you delete a category in an application. For example, if you're working in Tasks and you decide to delete a category, you'll soon discover that you've lost that category in Contacts as well — with all its assignments. (The Contacts contact is still intact but will be missing the category assignment.)

Book II
Chapter 3

Creating To-Do's
and Safely Storing
Your Passwords

To make use of this feature, start by creating a category by following these steps:

1. **Select Tasks.**

   The Tasks application opens, displaying your Tasks list on the Find screen.

2. **Press the Menu key and then select Filter.**

   The Select Categories screen appears (see Figure 3-6), listing all the available categories. The two default entries on the list are Business and Personal. By all means, feel free to use these default categories. Consider, though, that these categories are broad and might not be helpful if you have a lot of tasks. (Imagine going to a grocery store with only two sections: perishable and nonperishable.) Our advice to you: Go the extra mile and create some categories to work with.

   Strive to define groups or categories that are meaningful in your line of work and not so broad.

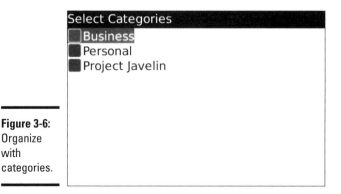

**Figure 3-6:**
Organize
with
categories.

3. **Press the Menu key and then select New.**

   The New Category screen appears, allowing you to define a new category. Imagine that.

4. **In the New Category screen, enter your category name in the text field and then press Enter.**

   Doing so establishes your category and lists it as an option on the Select Categories screen. Refer to Figure 3-6.

You can define as many categories as necessary up front. This way, you won't have to go back and create them. After you have the categories you want, assign your tasks to those categories.

## Assigning categories to your tasks

Here's how to assign a category to an existing task while you're in the Tasks application:

1. **Select the task in the Tasks list.**

   The Edit Task screen appears, ready for any changes you'd like to make.

2. **Select the Category field.**

3. **In the list that appears, select the category.**

   The category you selected is entered into the field.

4. **Press the Menu key and then select Save.**

   The task is now associated with the category you selected.

## Filtering the list

As you can read in the two preceding sections, you create categories and assign them to your tasks. To see the tasks associated with a certain category, just filter the list, which is way easy. Like all other options, Filter is available through the menu while you are in the Tasks application. Selecting that menu option gives you a screen that allows you to select any category listed. After you choose a category, the list is filtered so that only those tasks assigned to the chosen category are visible.

So much for the overview. Here are the nitty-gritty details:

1. **Select Tasks.**

   The Tasks application opens, displaying your Tasks list.

2. **Press the Menu key and then select Filter.**

   The Select Categories screen appears, listing all the available categories. (Refer to Figure 3-6.)

3. **Select the category you want.**

   A filtered list appears, containing just the tasks associated with the chosen category. Note that the list has the category name as the heading, as shown in Figure 3-7.

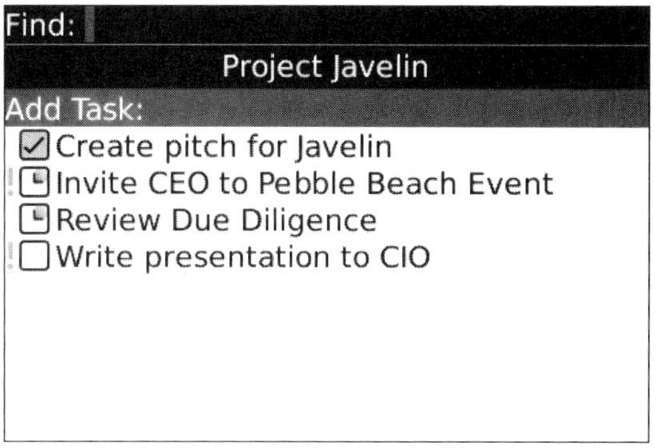

---

Find:
| Project Javelin |
| --- |

Add Task:
- ☑ Create pitch for Javelin
- ⊡ Invite CEO to Pebble Beach Event
- ⊡ Review Due Diligence
- ☐ Write presentation to CIO

**Figure 3-7:**
A filtered task list.

After you have the necessary categories, divide your tasks into appropriate categories to make running a task sort easier.

# Creating Recurring Tasks

A more advanced feature in Tasks is to create a *recurring task* — you know, one that repeats periodically. Maybe the recurring task is deadline driven, such as filing your taxes or paying bills, or buying presents for your significant other for his or her birthday, for Valentine's Day, or for an anniversary. (Other times, a recurring task might be something a bit more mundane, such as turning in a weekly status report to your boss.)

Making a task recur is simple, although it might not be obvious at first. (Rest assured; we show you the trick.) Basically, you either create a task from scratch or you use an existing task.

If you don't see the Recurrence field while looking at your existing task, the task doesn't have a due date — and it must have a due date to make it a recurring task.

Say you want a task to repeat every year. With the Tasks application running, do the following to make the task recur on a yearly basis:

1. **In the Tasks list, select the task you want to recur.**

   The Edit Task screen for the specified task appears.

2. **If the Due field on the Edit Task screen reads None, select the field, and then select By Date from the menu.**

   The screen displays a Date field — short for Due Date. The default date is the current date, so if you haven't yet changed the default date to your due date, do that now. (The "Due field" section, earlier in this chapter, has all the details on how to do this.)

   If you scroll down the screen, notice that the Recurrence field appears with a default value of None. Recurrence always appears after a due date is specified, so you will see it the next time you open this task.

3. **Select the Recurrence field.**

   Daily, Weekly, Monthly, and Yearly selections appear. These designate when your task recurs.

4. **Select Yearly from the list.**

   At this point, your task recurs yearly, beginning on the date you specified in the Due field (in Step 2).

*Note:* The End field below the Recurrence field is the end of the recurrence and comes in handy when you specify a relative reminder. For example, consider the following:

**Due:** Wed, Dec 15, 2010 5:00 p.m.

**Reminder:** Relative

   15 Min.

**Recurrence:** Yearly

**Every:** 1

**End:** Sun, Dec 15, 2013

**Relative Date:** (Checked)

Occurs every year until Dec 15, 2013

Your task will recur every December 15 of each year until 2013. You won't see any difference in the behavior of this task versus a nonrecurring task. However, at 4:45 p.m. every December 15, you'll get a reminder. By default, your BlackBerry will vibrate to alarm you. (See Book I, Chapter 3 for details on customizing your Tasks notification in Profiles.).

Also, after the End field is a Relative Date check box field. To select/deselect this field, highlight the field, and then press the Space key. Mark the Relative Date field if the date is not exact but *relative* on the day: For example, you want the task to recur every third Monday of December rather than every December 15. This field is associated to Recurrence and will appear only when it's appropriate. In fact, you will not see this field for daily or weekly — only for monthly and yearly recurrence.

# Using Password Keeper

Suppose you're in front of an Internet browser, trying to access an online account. For the life of you, you just can't remember the username that matches your password. It's your third login attempt, and if you fail this time, your online account is locked. Then you have to call the customer hotline and wait hours before you can speak to a representative. Argghh! We've all done it. Luckily, the folks at RIM created an application just for people like you (and us) so that you're never locked out of your own online accounts again — and yet can keep intruders at bay.

Password Keeper is the simple yet practical BlackBerry application that makes your life that much easier.

## Accessing Password Keeper

No matter which BlackBerry smartphone you have, Password Keeper is always filed in Applications (as shown in Figure 3-8).

**Figure 3-8:**
Password
Keeper
in the
Applications
folder.

Not all BlackBerry devices come with Password Keeper out of the box. You may need to specifically request that your service provider install it on your BlackBerry. It is free, though.

## Setting a password for Password Keeper

The first time you access Password Keeper, you're prompted to enter a password. *Be sure to remember the password you choose* because this is the password to all your passwords. Forgetting this password is like forgetting the combination to your safe. There is no way to retrieve a forgotten Password Keeper password, unlike with many Web sites. In addition, you're prompted to enter this master password every time you access Password Keeper.

We know it sounds perverse to set up yet another password to help you manage your passwords, but trust us. Having to remember just one password is a lot easier than having to remember many.

## Creating credentials

Okay, so you're ready to fire up your handy-dandy Password Keeper application. Now, what kinds of things does it expect you to do for it to work its magic? Obviously, you're going to need to collect the pertinent info for all your various password-protected accounts so that you can store them in the protected environs of Password Keeper. So, when creating a new password entry, be sure you have the following information. (See Figure 3-9.)

+ **Title:** This one's straightforward. Just come up with a name to describe the password-protected account — My Bank Account, for example.

+ **Username:** This is where you enter the username for the account.

+ **Password:** Enter the password for the account here.

+ **Website:** Put the Web site address (its URL) here.

+ **Notes:** Not exactly crucial, but the Notes field does give you a bit of room to add a comment or two.

Make sure you press Save on the prompt upon exiting this screen; otherwise your changes here will be discarded.

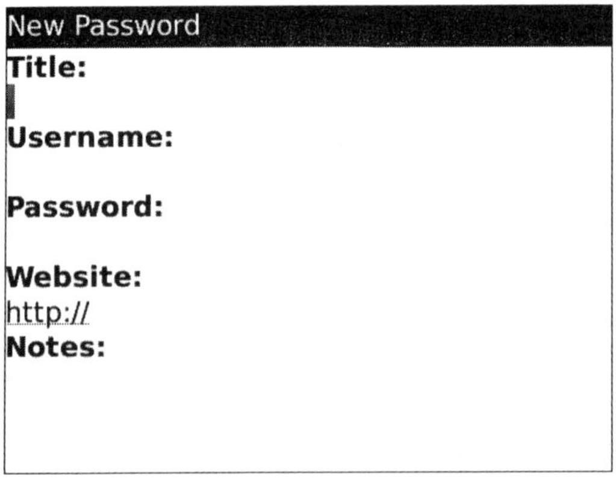

New Password
Title:

Username:

Password:

Website:
http://
Notes:

**Figure 3-9:**
Set your
password
here.

The only required field is Title, but a title alone usually isn't of much use to you. We suggest that you fill in as much other information here as possible, *but at the same time be discreet about those locations where you use your username and password* — so don't put anything in the Web site field or use

My eBay Account as a title. That way, if someone *does* somehow gain access to your password to Password Keeper, the intruder will have a hard time figuring out where exactly to use your credentials.

## Random password generation

If you're the kind of person who uses one password for everything but knows deep in your heart that this is just plain wrong, wrong, wrong, random password generation is for you. When creating a new password for yet another online account (or when changing your password for an online account you already have), fire up Password Keeper, press the Menu key, and then select Random Password from the menu that appears, as shown in Figure 3-10. Voilà! A new password is automatically generated for you.

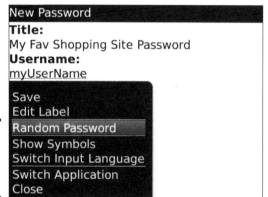

**Figure 3-10:**
Generate
a random
password.

Using random password generation makes sense in conjunction with Password Keeper because you don't have to remember the randomly generated password that Password Keeper came up with for any of your online accounts — that's Password Keeper's job.

## Using your password

The whole point of Password Keeper is to let your BlackBerry's electronic brain do your password remembering for you. So, imagine this scenario: You can no longer live without owning your personal copy of the *A Chipmunk Christmas* CD, so you surf on over to your favorite online music store and attempt to log in. You draw a blank on your password, but instead of seething, you take out your BlackBerry, open Password Keeper, and do a search. Like Contacts (see Book II, Chapter 1), you just type the first letters of your account title in the Find field to search for the title of your password. After you find the title, simply press the trackball, trackpad, or touch screen and the screen for your account appears, conveniently listing the password. All you have to do now is enter the password into the login screen for the online music store, and Alvin, Simon, and Theodore will soon be wending their way to your address, ready to sing "Chipmunk Jingle Bells."

Yes, you *do* have the option of copying and pasting your password from Password Keeper to another application — BlackBerry Browser, for instance. Just highlight the password name, press the Menu key, and select Copy to Clipboard from the menu that appears. Then navigate to where you want to enter the password, press the Menu key, and select Paste from the menu. Keep in mind, for the copy-and-paste function to work for passwords from Password Keeper, you need to enable the Allow Clipboard Copy option in the Password Keeper options. (See the upcoming Table 3-1.) You can copy and paste only one password at a time.

After you paste your password into another application, clear the Clipboard by pressing the Menu key and choosing Clear Clipboard. The Clipboard keeps your last copied password until you clear it.

## Password Keeper options

The Password Keeper Options menu, which is accessible by pressing the Menu key while in Password Keeper, allows you to control how Password Keeper behaves. For example, you can set what characters can make up a randomly generated password. Table 3-1 describes all these options.

| Table 3-1 | Password Keeper Options |
|---|---|
| *Name of Option* | *Description* |
| Random Password Length | Select between 4 and 16 for the length of your randomly generated password. |
| Random Includes Alpha | If True, a randomly generated password includes alphabetic characters. |
| Random Includes Numbers | If True, a randomly generated password includes numbers. |
| Random Includes Symbols | If True, a randomly generated password includes symbols. |
| Confirm Delete | If True, all deletions are prompted with a confirmation screen. |
| Password Attempts | Select between 1 and 20 attempts to successfully enter the password to Password Keeper. |
| Allow Clipboard Copy | If True, you can copy and paste passwords from Password Keeper. |
| Show Password | If True, the password appears on the View screen. If False, asterisks take the place of the password characters. |

## Changing your Password Keeper password

If you want to change your master password to Password Keeper — the password for opening Password Keeper itself — simply follow these steps:

1. **Select Password Keeper.**

   The initial login screen for the Password Keeper application appears.

2. **Enter your current password to access Password Keeper.**

3. **With the Password Keeper application open, press the Menu key and then select Change Password.**

   The Password Keeper screen opens, allowing you to enter a new password, as shown in Figure 3-11.

**Figure 3-11:**
Change your
Password
Keeper
password
here.

4. **Enter a new password, confirm it by entering it again, and then click OK.**

## Changing info within Password Keeper

Does one of your systems require you to change passwords at a regular interval? No worries. You can quickly generate or change the password — or any information, for that matter — inside Password Keeper. Follow these steps:

1. **Select Password Keeper.**

   The initial login screen for the Password Keeper application appears.

2. **Enter your current password to access Password Keeper.**

3. **Press the password entry you want to change.**

   The same screen that allows you enter a new password entry appears, only this time the screen is populated with the previous information you entered. All the fields are editable.

4. **Change the information you want to change.**

5. **Press the Escape key and select Save on the screen that follows.**

# Chapter 4: Taking Control of the Phone

## In This Chapter

✓ Accessing the BlackBerry Phone application

✓ Making and receiving calls

✓ Managing your calls with call forwarding and more

✓ Customizing your BlackBerry Phone setup

✓ Conferencing: Connecting with more than one person at once

✓ Talking hands-free on your BlackBerry phone

✓ Multitasking with your BlackBerry phone

*T*he BlackBerry phone operates no differently than any other phone you've used. So why bother with this chapter? Although your BlackBerry phone operates like any other phone, it has capabilities that far outreach those of your run-of-the-mill cellphone. For example, when was the last time your phone was connected to your to-do list? Have you ever received an e-mail and placed a call directly from that e-mail? We didn't think so. But with your BlackBerry, you can do all these things and more.

In this chapter, we first cover phone basics and then show you some of the neat ways BlackBerry Phone intertwines with other BlackBerry applications and functions. We'll give you everything you need to take control of your BlackBerry Phone capabilities.

## Using the BlackBerry Phone Application

Accessing the Phone application from the BlackBerry is a snap. You can press the green Send button located right below the display screen to get into the Phone application.

You can also get to the Phone application by pressing any of the numeric keys. To do this, however, you must make sure that the Dial from Home Screen option is enabled in Phone Options. If you're a frequent phone user, we recommend that you enable this option.

Throughout this section, we assume that you have the Dial from Home Screen option enabled if you're using a non-SureType or non–touch screen model.

On a BlackBerry Pearl or Storm, you don't need to go through the following steps. Because of the way these devices are designed, Dial from Home Screen isn't an option.

To enable dialing from the Home screen for other models, follow these steps:

1. **On your BlackBerry, press the green Send key.**

   Phone opens, showing the dial screen as well as your call history list.

2. **Press the Menu key and then select Options.**

3. **Select General Options.**

4. **Highlight the Dial from Home Screen option, press the trackball or trackpad, and then select Yes.**

   You can now make a phone call by pressing the numeric keys when you're at the Home screen.

5. **Press the Menu key and then select Save.**

# Making and Receiving Calls

The folks at RIM have created an intuitive user interface to all the essential Phone features, including making and receiving calls.

## Making a call

To make a call, start from the Home screen and type the phone number you want to dial. As soon as you start typing numbers, the Phone application opens. When you finish typing the destination number, press the green Send key.

### Calling from Contacts

Because you can't possibly remember all your friends' and colleagues' phone numbers, calling from Contacts is convenient and useful. To call from Contacts, follow these steps:

1. **Open the Phone application.**

2. **Press the Menu key.**

   The Phone menu appears, as shown in Figure 4-1.

3. **Select Call from Contacts.**

   Contacts opens. From here, you can search as usual for the contact you'd like to call.

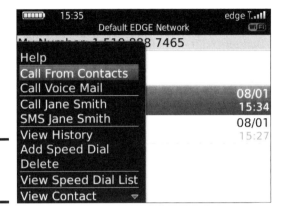

**Figure 4-1:**
The Phone menu.

4. **From Contacts, highlight your call recipient, press the Menu key, and then select Call.**

   This makes the call.

### Dialing letters

One of the nice features of BlackBerry Phone is that you can dial letters, and BlackBerry will figure out the corresponding number. For example, to dial 1-800-11-LEARN, do the following on your BlackBerry:

1. **From the Home screen or the Phone application, dial 1-8-0-0-1-1.**

   As you type the first number, the Phone application opens (if it isn't open already) and displays the numbers you dialed.

2. **Press and hold the Alt key and then dial (press) L-E-A-R-N.**

   The letters appear onscreen as you type.

3. **Press the green Send key.**

   The call is initiated.

This does not apply to BlackBerry Storm, as you can't type letters on the Storm's virtual dial pad.

### Receiving a call

Receiving a call on your BlackBerry is even easier than making a call. When you get an incoming call, you can simply press the green Send key to pick up the call. In addition, you can receive calls using your BlackBerry's automated answering feature.

Automated answering is triggered whenever you take your BlackBerry out of your holster; in other words, just taking out the BlackBerry forces it to automatically pick up any call, so you can start talking right away. The disadvantage of this is that you don't have time to see who is calling you (on your caller ID). *Note:* To disable autoanswering, you can adjust the Phone General Options and set "Auto Answer Calls" to Never.

What's the disadvantage of disabling autoanswer? Manual answering prompts you to answer or ignore an incoming call (see Figure 4-2). This way, you can see on your caller ID who is calling before you pick up or decide to ignore the call.

**Figure 4-2:**
An incoming call on a BlackBerry.

## Missed calls notification

So, you missed that call from that important client. What made it worse is that you didn't notice the missed call because you didn't see the little Missed Call icon. This happened because you pay attention only to what is in your e-mail message box. What can you do to make sure that you return that call?

Set your BlackBerry device to send you an e-mail notification when you miss a call, so that you are sure to return your missed calls (if you choose to, that is).

To have your missed calls appear in your inbox, follow these steps:

*1.* **Open the Phone application.**

*2.* **Press the Menu key.**

The Phone menu appears. (Refer to Figure 4-1.)

3. **Select Options.**

   The phone options screen appears, listing the different categories of options.

4. **Select Call Logging.**

   The Call Logging screen opens.

5. **Scroll to Missed Calls and press the trackball, trackpad, or touch screen.**

   You can also select All Calls, which means that incoming and outgoing calls will be displayed in your e-mail inbox.

6. **Press the Menu key and select Save.**

From now on, any missed call will show up in your inbox as a row that displays the time and number of missed calls.

# Muting, Holding, and Turning It Up

When you're on the phone, situations might arise where you'd want to mute your conversation, place a call on hold, or change the call volume. No problem. BlackBerry makes such adjustments easy.

## Muting your call

You might want to use the Mute feature while on a conference call (see the upcoming section "Arranging Conference Calls") when you don't need to speak but do need to hear what is being discussed. Maybe you're on the bus or have kids in the background, making your surroundings noisy. By using Mute, these background noises are filtered out from the conference call.

To mute your call, follow these steps:

1. **While in a conversation, press the Menu key.**

   The Phone menu appears in all its glory.

2. **Select Mute.**

   You hear a tone, indicating that your call is being muted.

Follow these steps to unmute your call:

1. **While a call is on mute, press the Menu key.**

   The Phone menu makes another appearance.

2. **Select Turn Mute Off.**

   You hear a tone, indicating that your call is now unmuted.

### Placing your call on hold

Unlike muting a call, placing a call on hold prohibits both you and your caller from hearing one another. To put a conversation on hold, follow these steps:

1. **While in a conversation, press the Menu key.**

   The Phone menu appears yet again.

2. **Scroll to Hold and press the trackball, trackpad, or touch screen.**

   Your call is now on hold.

Follow these steps to unhold your call:

1. **While a call is on hold, press the Menu key.**

   A new menu appears.

2. **Scroll to Resume and press the trackball, trackpad, or touch screen.**

   You can continue your conversation.

### Adjusting the call volume

Adjusting the call volume, a simple yet important action on your BlackBerry phone, can be performed by simply pressing the volume up or volume down key on the side of your BlackBerry.

## Customizing the BlackBerry Phone

For your BlackBerry phone to work the way you like, you have to first set it up the way you want it. In the following sections, we go through some settings that can make you the master of your BlackBerry phone.

### Setting up your voice mail access number

This section shows you how to set up your voice mail access number to check your voice mail. Unfortunately, the instructions for setting up your voice mailbox vary, depending on your service provider. However, most service providers are more than happy to walk you through the steps to get your mailbox set up in a jiffy.

To set up your voice mail access number, follow these steps:

1. **Open the Phone application.**

2. **Press the Menu key and then select Options.**

   A list of phone options appears.

3. **Select Voice Mail.**

The voice mail configuration screen opens.

4. **Scroll to access the number field and enter your voice mail access number.**

If this field is empty and you don't know this number, contact your service provider and ask for your voice mail access number.

5. **Press the Menu key and then select Save.**

## Using call forwarding

On the BlackBerry, you have two types of call forwarding:

✦ **Forward all calls.** Any calls to your BlackBerry are forwarded to the number you designate. Another name for this feature is *unconditional forwarding.*

✦ **Forward unanswered calls.** Calls that meet different types of conditions are forwarded to different numbers as follows:

- *If busy:* You don't have call waiting turned on, and you're on the phone.

- *If no answer:* You don't hear your phone ring or somehow are unable to pick up your phone. (Perhaps you're in a meeting.)

- *If unreachable:* You're out of network coverage and cannot receive any signals.

Out of the box, your BlackBerry forwards any unanswered calls, regardless of conditions, to your voice mail number by default. However, you can add new numbers to forward a call to.

You need to be within network coverage before you can change your call forwarding options. After you're within network coverage, you can change your call forwarding settings by doing the following:

1. **Open the Phone application, press the Menu key, and select Options.**

A list of phone options appears.

2. **Select Call Forwarding.**

Your BlackBerry now attempts to connect to the server. If it's successful, you'll see the Call Forwarding screen.

If you don't see the Call Forwarding screen, wait until you have network coverage and try again.

3. **From the Call Forwarding screen, press the Menu key and then select Edit Numbers.**

   A list of numbers appears. If this is the first time you're setting call forwarding, mostly likely only your voice mail number is in this list.

4. **To add a new forwarding number, press the Menu key and then select New Number.**

   A pop-up menu appears, prompting you to enter the new forwarding number.

5. **In the pop-up window, enter the number you want to forward to and then press the trackball, trackpad, or touch screen.**

   The new number you entered now appears on the call forward number list. You can add this new number to any call forwarding types or conditions.

6. **Press the Escape key.**

   (The Escape key is the arrow key to the left of the end key.) You return to the Call Forwarding screen.

7. **Scroll to the If Unreachable field and press the trackball, trackpad, or touch screen.**

   A drop-down list appears, listing numbers from the call forwarding number list, including the one you just added.

8. **Select the number you want to forward to and then press the trackball, trackpad, or touch screen.**

   Doing so places the selected number into the If Unreachable field. You can see this on the Call Forwarding screen.

9. **Confirm your changes by pressing the Menu key and then selecting Save.**

## Configuring speed dial

Speed dial is a convenient feature on any phone because it lets you call a number by pressing just one key. And after you get used to having it on a phone system, it's hard not to use it on other phones, including your BlackBerry phone.

### Viewing your speed dial list

To view your speed dial list, follow these steps:

1. **Open the Phone application.**

2. **Press the Menu key and then select View Speed Dial List.**

   A list of speed dial entries displays, as shown in Figure 4-3. If you haven't set up any speed dials, this list will be empty.

**Book II**
**Chapter 4**

Taking Control
of the Phone

**Figure 4-3:**
The speed
dial list.

Notice in Figure 4-3 that the * key and # key have hard-set functions. This is the same on all trackball and trackpad BlackBerry models (QWERTY and SureType). Basically this is the shortcut for locking your BlackBerry screen (* key) and changing your profile to Quiet (# key). This does not apply to the BlackBerry Storm.

### Adding a number to speed dial

Setting up speed dial numbers is as easy as using them. It takes a few seconds to set them up, but you benefit every time you use this feature.

To assign a number to a speed dial slot, follow these steps:

*1.* **Open the Phone application.**

*2.* **Press the Menu key, and choose Options⇨View Speed Dial List.**

   Your list of speed dial numbers appears.

*3.* **Highlight an empty speed dial slot, press the Menu key, and then select New Speed Dial.**

   The BlackBerry Contacts application appears so that you can select a contact's phone number.

*4.* **Select a contact, and then press the trackball, trackpad, or touch screen.**

   If more than one number is associated with the selected contact, you're prompted to select which number to add to the speed dial list.

   The number appears in the speed dial list.

### Using speed dial

After you have a few speed dial entries set up, you can start using them. To do so, while on the Home screen or in the Phone application, press and hold a speed dial key until the call initiates. The call is initiated to the number associated with that particular speed dial key.

## Arranging Conference Calls

To have two or more people on the phone with you — the infamous conference call — do the following:

*1.* **Use the Phone application to place a call to the first participant.**

*2.* **While the first participant is on the phone with you, press the Menu key and then select New Call.**

The first call is automatically put on hold and a New Call screen appears, as shown in Figure 4-4, prompting you to place another call.

**Figure 4-4:** A meeting participant is on hold.

*3.* **Place a call to the second participant by dialing a number, pressing the Menu key, and then selecting Call.**

You can dial the number by using the number pad, or you can select a frequently dialed number from your call log. To place a call from your Contacts, press the Menu key from the New Call screen and select Call from Contacts. Your BlackBerry then prompts you to select a contact to dial.

The call to the second meeting participant is just like any other phone call (except that the first participant is still on the other line).

4. **While the second participant is on the phone with you, press the Menu key and then select Join Conference, as shown in Figure 4-5.**

   The first participant is reconnected with you, along with the second participant. Now you can discuss away with both participants at the same time.

**Figure 4-5:**
Join two people in a conference call.

Another name for having two people on the phone with you is *three-way calling,* which is not a new concept. If you want to chat with four people or even ten people on the phone at the same time, you certainly can. Simply repeat Steps 2 through 4 until all the participants are on the phone.

## Talking privately to a conference participant

During a conference call, you might want to talk to one participant privately. This is called *splitting* your conference call. Here's how you do it:

1. **While on a conference call, press the Menu key and then select Split Call.**

   A pop-up screen appears, listing all the participants of the conference call, as shown in Figure 4-6.

2. **From the pop-up screen, select the participant with whom you want to speak privately.**

   All other participants are placed on hold and you're connected to the participant you selected. On the display screen, you can see to whom you are connected. This confirms that you selected the right person to chat with privately.

3. **To talk to all participants again, press the Menu key and then select Join Conference.**

   Doing so brings you back to the conference call with everyone.

**Figure 4-6:**
Split your
conference
call.

## Alternating between phone conversations

Whether you're in a private conversation during a conference call or you're talking to someone while you have someone else on hold, you can switch between the two conversations by swapping them. Follow these steps:

*1.* **While talking to someone with another person on hold, press the Menu key and then select Swap.**

   Doing so switches you from the person with whom you're currently talking to the person who was on hold.

*2.* **Repeat Step 1 to go back to the original conversation.**

## Dropping that meeting hugger

If you've been on conference calls, you can identify those chatty "meeting huggers" who have to say something about everything. Don't you wish that you could drop them off the call? Well, with your BlackBerry, you can (as long as you are the meeting moderator or the person who initiates the call). Follow these steps to perform the drop-kick:

*1.* **While on a conference call, press the Menu key and then select Drop Call.**

   A pop-up screen appears, listing all conference call participants.

*2.* **Select the meeting hugger you want to drop.**

   That person is disconnected.

*3.* **Everyone else now can continue the conversation as usual.**

# Communicating Hands-Free

Because more and more places prohibit the use of mobile phones without a hands-free headset, we thought we'd go through the hands-free options you have on your BlackBerry.

## Using the speaker phone

The Speaker Phone function is useful under certain situations, such as when you're in a room full of people who want to join your phone conversation. Or you might be all by your lonesome in your office but are stuck rooting through your files — hard to do with a BlackBerry scrunched up against your ear. (We call such moments *multitasking* — a concept so important we devote an entire upcoming section to it.)

To switch to the speaker phone while you're on a phone call, press the Menu key and then select Activate Speaker Phone.

**Book II**
**Chapter 4**

Taking Control
of the Phone

## Pairing your BlackBerry with a Bluetooth headset

Because BlackBerry smartphones come with a wired hands-free headset, you can start using yours by simply plugging it into the headset jack on the right side of the BlackBerry. You adjust the volume of the headset by pressing up or down on the volume keys, the same as you would to adjust the call volume without the headset.

Using the wired hands-free headset can help you avoid being a police target, but if you're multitasking on your BlackBerry, the wired headset can get in the way and become inconvenient.

This is where the whole Bluetooth wireless thing comes in. You can purchase a BlackBerry Bluetooth headset to go with your Bluetooth-enabled BlackBerry. At the time this book was written, the following headsets were supported by the Bluetooth feature on BlackBerry smartphones:

| | |
|---|---|
| Flamingo Bluetooth | Jabra BT250v |
| Jabra JX10 | Jabra BT125 |
| Logitech Mobile Pro | Motorola Bluetooth H500 |
| Motorola Bluetooth HS820 | Motorola Bluetooth HS850 |
| nXZEN 5000 | Plantronics M3000 Bluetooth |
| Plantronics Discovery 640 Bluetooth | Plantronics Explorer 320 Bluetooth |
| Plantronics Voyager 510 Bluetooth | scala-500 |
| Sony Ericsson Akono Bluetooth HBH-602 | |

After you purchase a BlackBerry-compatible Bluetooth headset, you can pair it with your BlackBerry. Think of *pairing* a Bluetooth headset with your BlackBerry as registering the headset with your BlackBerry so that it recognizes the headset.

First things first: You need to prep your headset for pairing. Now, each headset manufacturer has a different take on this, so you'll need to consult your headset documentation for details. With that out of the way, continue with the pairing as follows:

*1.* **From the Home screen, go to Options (the wrench icon).**

*2.* **Choose Bluetooth.**

*3.* **Press the Menu key to display the Bluetooth menu.**

You see the Enable Bluetooth option. If you see the Disable Bluetooth option, skip to Step 5.

*4.* **Highlight Enable Bluetooth and then press the trackball, trackpad, or touch screen.**

Bluetooth is now enabled on your BlackBerry.

*5.* **Press the Menu key to display the Bluetooth menu and then select Add Device.**

You see the Searching for Devices progress bar, um, progressing, as shown in Figure 4-7. When your BlackBerry discovers the headset, a Select Device dialog box appears with the name of the headset.

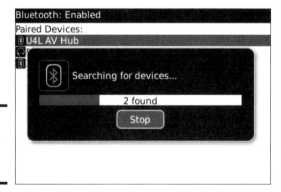

**Figure 4-7:** Searching for a headset.

*6.* **From the Select Device dialog box, select the Bluetooth headset.**

A dialog box appears to prompt you for a passkey code to the headset.

7. **Enter the passkey and press the trackball, trackpad, or touch screen.**

   Normally, the passkey is 0000, but refer to your headset documentation. After you successfully enter the passkey, you see your headset listed in the Bluetooth setting.

8. **Press the Menu key to display the Bluetooth menu and then select Connect.**

   Your BlackBerry now attempts to connect to the Bluetooth headset.

9. **When you see a screen similar to Figure 4-8, you can start using your Bluetooth headset.**

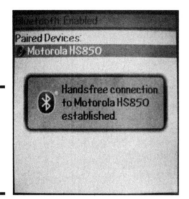

**Figure 4-8:**
Now
you can
use your
Bluetooth
headset.

## Using voice dialing

With your headset and the Voice Dialing application, you can truly be hands-free from your BlackBerry. You may be thinking, how do I activate the Voice Dialing application without touching my BlackBerry? Good question. The majority of hands-free headsets (Bluetooth or not) come with a multipurpose button.

Usually, a multipurpose button on a hands-free headset can mute, end, and initiate a call. Refer to the operating manual of your hands-free headset for more info.

After your headset is active, press its multipurpose button to activate the Voice Dialing application. You will be greeted with a voice stating, "Say a command." At this point, simply say "Call" and state the name of a person or say the number. (For example, say "Call President Obama" or "Call 555-2468.") The Voice Dialing application is good at recognizing the name of the person and the numbers you dictate. However, we strongly suggest that you try out the voice dialing feature before you need it. See Book II, Chapter 5 for more on voice dialing and other verbal commands that your BlackBerry recognizes.

# Multitasking while on the Phone

One of the great things about the BlackBerry is that you can use it for other tasks while you're on the phone. For example, you can take notes or make a to-do list. Or you can look up a phone number in BlackBerry Contacts that your caller is asking you for. You can even compose an e-mail while on a call.

It makes sense to multitask while you're using a hands-free headset or a speaker phone. Otherwise, your face would be stuck to your BlackBerry, and you couldn't engage in your conversation and multitask at the same time.

## Accessing applications while on the phone

After you've donned your hands-free headset or turned on a speaker phone, you can start multitasking by doing the following:

1. **While in a conversation, from the Phone application, press the Menu key and then select Home Screen.**

   Alternatively, you can simply press the Escape key (the arrow key to the right of the trackball, trackpad, or Menu key) while in the Phone application to return to your Home screen.

   You return to the Home screen without terminating your phone conversation.

2. **From the Home screen, you can start multitasking.**

With the exception of the GSM 3G (HSDPA) network, you can compose e-mails during a phone conversation, but you can't send an e-mail until you finish a phone conversation. In addition, you can't surf the Web while on the phone. With the GSM 3G network, you can receive and send e-mails, as well as browse the Web. (Contact your service provider if you aren't sure which network you have.)

While on the phone and multitasking, however, you can still access the Phone menu from other applications. For example, from your to-do list, you can end a call or put a call on hold.

## Taking notes while on the phone

To take notes of your call, follow these steps:

1. **During a phone conversation, press the Menu key and then select Notes.**

   The Notes screen opens.

2. **Type notes for the conversation, as shown in Figure 4-9.**

   When the call ends, the notes are automatically saved for you.

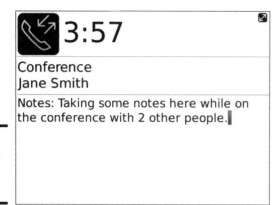

**Figure 4-9:**
Take notes while on a phone call.

Book II
Chapter 4

Taking Control
of the Phone

## Accessing phone notes

From the call history list (see Figure 4-10), you can access notes that you've made during a call or a conference call. In addition, you can also edit and add new notes.

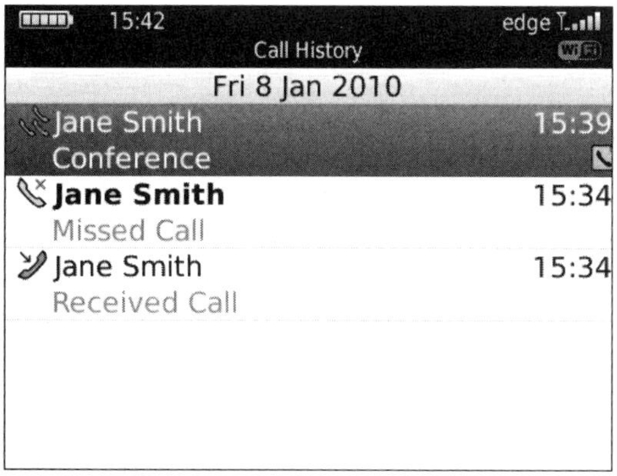

**Figure 4-10:**
Call history, where you can see conversation notes.

**Forwarding phone notes**

You can forward your phone notes just like any e-mail. While on the Call History screen (refer to Figure 4-10), press the Menu key and then select Forward.

You can add notes not only while you're on the phone but also afterward. While you are viewing a call history, press the Menu key. Then select Add Notes if you have no notes for the call, or select Edit Notes if you already have notes for the call.

# Chapter 5: Using Your Voice to Do More

## In This Chapter

✔ **Calling a contact with Voice Dialing**

✔ **Calling a number with Voice Dialing**

✔ **Checking your battery power and coverage level**

✔ **Recording, playing, and sending voice notes**

*H*ave you ever wanted to call somebody while you were driving, but didn't want to take your eyes off the road to look them up in your contact list and dial them? Or maybe you weren't driving but wished there were a faster way to dial than opening Contacts and searching? Tired of clicking through several screens or holding down several keys at once to check your battery level, signal strength, or other device status items? Ever wanted to add the personal touch of your voice to an e-mail message, or record a voice note to yourself so you don't forget something? If so, you're in luck! BlackBerry Voice Dialing lets you place calls with the push of a button and the sound of your voice. BlackBerry Voice Notes let you record and play back messages or add your voice to a message any time you want.

Nowadays, most cellphones — smartphone or not — include basic voice dialing functionality. However, the BlackBerry Voice Dialing application has a few options that those other devices may lack. In this chapter, we teach you the ins and outs of voice commands and voice notes so you can master these useful hands-free BlackBerry applications.

## Using Voice Dialing

Accessing Voice Dialing from the BlackBerry is a breeze. Go to the Applications folder, highlight the Voice Dialing icon (shown in Figure 5-1), and press the trackball, trackpad, or touch screen. Your BlackBerry greets you with the verbal prompt, "Say a command."

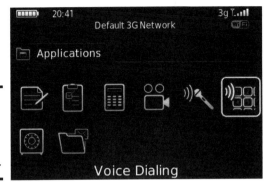

**Figure 5-1:**
The
BlackBerry
Voice
Dialing icon.

The default action for the left convenience key — the button on the left side of the BlackBerry smartphone — opens the Voice Dialing application. If your left convenience key doesn't open the Voice Dialing application, you can set it up using these steps:

1. **From the BlackBerry Home Screen, scroll to the Options icon and press the trackball or touch screen.**

2. **Select the Screen/Keyboard setting.**

   The Screen/Keyboard screen appears with its various customizable fields.

3. **Scroll to the Left (or Right) Side Convenience Key Opens option and press the trackball, trackpad, or touch screen to choose it.**

4. **Choose Default (Voice Dialing) from the drop-down list.**

5. **Press the Menu key and choose Save.**

   Now you just have to press the convenience key to start Voice Dialing.

## Calling a contact with Voice Dialing

Calling a contact with voice dialing is easy as ABC and 123. Just follow these steps, and you'll soon be voice dialing your contacts like a pro:

1. **From the Home screen, select the Voice Dialing icon.**

   Or if you set a convenience key to activate voice dialing, press the convenience key.

   Voice Dialing opens and prompts you to "Say a command."

**2. Say "Call" and the contact name.**

For example, if you wanted to call Tim Calabro, you would say, "Call Tim Calabro."

If the BlackBerry smartphone finds an exact match in Contacts with only one number, it calls the contact.

Sometimes, the Voice Dialing application needs more input from you. When that happens, respond as follows:

✦ **BlackBerry needs you to indicate which number to dial.** If your BlackBerry smartphone finds an exact match with more than one number — maybe you have home, mobile, and work numbers saved for the contact you'd like to call — it asks, "Call *contact name*, which number?" (Figure 5-2 shows what your BlackBerry displays.) Say the name of the number you would like to call. For example, if you want to call the mobile number say, "Mobile."

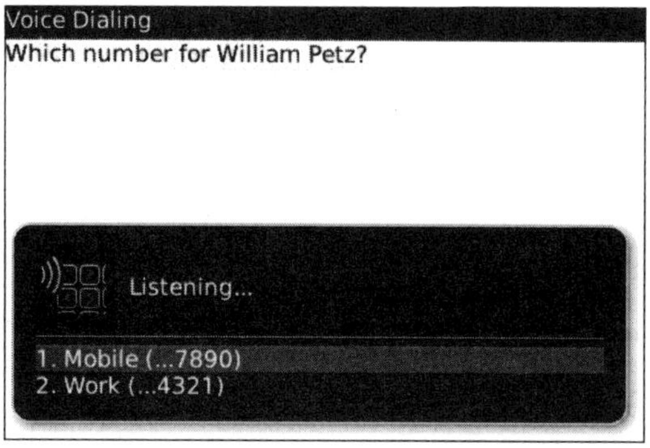

> Voice Dialing
> Which number for William Petz?
>
> )))▢▢( Listening...
>
> 1. Mobile (...7890)
> 2. Work (...4321)

**Figure 5-2:**
Your BlackBerry asks which contact number you want to call.

✦ **BlackBerry needs you to confirm the contact's name.** If your BlackBerry smartphone finds more than one possible match, it asks, "Did you say call *contact name*?" (with the contact name that it thinks you said). Figure 5-3 shows the BlackBerry screen that would appear.

• *If the contact name matches, say "Yes."*

• *If the contact name does not match, say "No." The BlackBerry goes to the next possible match and asks, "Did you say call* contact name*?" for the next name in the list.*

• *If you didn't hear the name, say "Repeat" to hear it again.*

• *Say "Cancel" if you want to cancel the request.*

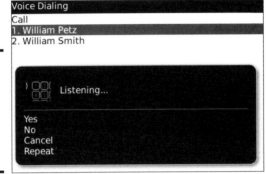

**Figure 5-3:**
You have several options when voice dialing a contact.

## Calling a number with Voice Dialing

Want to use Voice Dialing to call a number that isn't in your Contacts? No problem! Follow these steps:

*1.* **From the Home screen, select the Voice Dialing icon.**

Voice Dialing opens and prompts you to "Say a command."

*2.* **Say "Call" and the number.**

For example, if you want to call 123-456-7890, say "Call 1234567890."

The BlackBerry smartphone displays your options, as shown in Figure 5-4, and prompts you for an answer by asking "Did you say call 1234567890?"

*3.* **Confirm your request:**

• If the number matches say "Yes".

• If the number does not match say "No". The BlackBerry will go to the next possible match and ask "Did you say call this number?" for the next number in the list.

• If you didn't hear the number, say "Repeat" to hear it again.

• Say "Cancel" if you want to cancel the request.

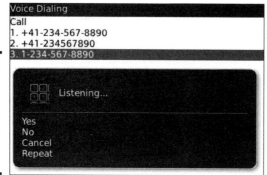

**Figure 5-4:**
You have
several
options
when voice
dialing a
number.

# Delivering Other Verbal Commands

Although the icon says Voice Dialing, it can be used for other voice commands as well. You can use it to turn voice prompts on or off, check the status of your BlackBerry smartphone, and even find out your BlackBerry's phone number.

## Checking your device information

You can have your BlackBerry tell you device information such as battery level, network coverage, and phone number without you visually checking for it. See Figure 5-5.

1. **From the Home screen, select the Voice Dialing icon.**

   Voice Dialing opens and prompts you to "Say a command."

2. **Say "Check" and the item you would like to check. The check status options are as follows:**

   - *Check battery:* Displays and states your battery level.

   - *Check coverage:* Displays and states your coverage level.

   - *Check my phone number:* Displays and states your BlackBerry Smartphone number.

   - *Check signal strength:* Displays and states your signal strength.

   - *Check status:* Displays your battery level, coverage, and signal strength.

**Figure 5-5:**
You can
use your
voice to
check your
BlackBerry
smart-
phone's
status.

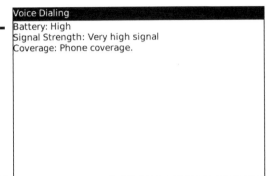

```
Voice Dialing
Battery: High
Signal Strength: Very high signal
Coverage: Phone coverage.
```

# Using Voice Notes Application

Recording, playing, and sending voice notes from the Voice Notes
Application on your BlackBerry are simple.

With the Voice Notes application, you can remind yourself to pick up some
groceries or run an errand. Perhaps you want to send your parents a voice
message from their grandchildren. No matter what the reason, you most
likely can find a reason to use voice notes. Voice notes are saved as a .amr
file on your BlackBerry memory or microSD card. Unless you are recording
a 2-hour concert, the voice notes will not take up large memory space or
waste battery on your BlackBerry.

In the following sections, we show you how to record, play, and send notes.

## Recording a voice note

To record a voice note, follow these steps:

*1.* **From the Home screen, select the Voice Dialing icon.**

*2.* **Press the trackball or trackpad or touch-press the Record button to
begin recording.**

Speak your note into the phone or headset.

*3.* **When you're done recording, press the trackball or trackpad or touch-
press the Pause button to pause recording.**

By pausing the recording, you are given the choice to stop and finish the
recording, continue the recording (same file) later, and more. (See the
next step.)

The voice note is automatically saved (on your microSD card or on your BlackBerry internal memory) with the filename VN0000# – *today's date*. The first voice note of the day is VN00001 – *today's date*, the second is VN00002 – *today's date*, and so on.

4. **Choose what to do with your recording:**

   Highlight the option at the bottom of the screen you would like and then press the trackball, trackpad, or touch screen to continue. All of the options are shown in Figure 5-6.

   • *Continue Recording:* Continue recording the voice note.

   • *Stop:* Close the recording screen. You're taken to the list of existing voice notes.

   • *Play:* Play back the voice note.

   • *Send:* You have the option to send the voice note in an e-mail, MMS, or BlackBerry Messenger message. See the "Sending a voice note" section later in this chapter for more information on sending voice notes.

   • *Rename:* Rename the voice note. Just type the new filename, and then highlight Save and press the trackball, trackpad, or touch screen to rename the file.

   • *Delete:* Delete the voice note.

**Figure 5-6:**
You have
several
options in
the Voice
Note
Recorder
screen.

Continue  Stop  Play  Send  Rename  Delete

5. **Press the Escape key (to the left of the End key) twice to return to the BlackBerry Home screen. Choose Save if you're prompted.**

## Playing a voice note

So you've recorded and saved a few voice notes, but now you're wondering how you replay them. Don't worry; here are the steps for playing a voice note:

1. **From the Home screen, select the Media icon.**

2. **Highlight the Voice Notes icon and press the trackball, trackpad, or touch screen.**

3. **Highlight the voice note you would like to play and press the trackball, trackpad, or touch screen.**

   The voice note opens and begins playing.

4. **Choose what to do with your voice note.**

   Highlight the option you would like, and then press the trackball, trackpad, or touch screen to continue. Figure 5-7 shows all of the Voice Note player options.

   - *Pause/Play:* Pause the voice note. Highlight and press Play to resume playing where you left off.

   - *Stop:* Stop playing the voice note. Highlight and press Play to start playing the voice note from the beginning again.

   - *Skip Forward:* Skip to the next voice note.

   - *Skip Back:* Skip back to the previous voice note.

**Figure 5-7:** You have several options when playing a voice note.

5. **Press the Escape key (to the left of the End key) three times to return to the BlackBerry Home screen.**

## Sending a voice note

What fun is recording and playing a voice note if you can't share it with others? Follow these steps to share your voice notes with others:

1. **From the Home screen, select the Media icon.**

2. **Highlight the Voice Notes button and press the trackball, trackpad, or touch screen.**

3. **Highlight the voice note that you want to send and then press the Menu key.**

4. **Choose Send As Email, Send As MMS, or Send to Messenger Contact.**

5. **Choose the person you'd like to send the voice note to.**

   How you choose the receiver depends on the method you use to send the voice note.

   • *E-Mail:* Fill out the To, CC, and Message fields, and then press the Menu key and choose Send.

   • *MMS:* Fill out the To, CC, and Message fields, and then press the Menu key and choose Send.

   • *Messenger:* Highlight the Messenger contact, press the trackball, trackpad, or touch screen, and then choose OK.

6. **Press the Escape key (to the left of the End key) twice to return to the BlackBerry Home screen.**

# Book III

# Collaborating, Communicating, and Getting Online

The 5th Wave    By Rich Tennant

Cell Phones

"This model comes with a particularly useful function — a simulated static button for breaking out of long winded conversations."

# Contents at a Glance

# Chapter 1: Surfing the Web with a BlackBerry

## In This Chapter

✔ Using BlackBerry Browser to surf the Web

✔ Mastering browser shortcuts and navigation tips

✔ Using bookmarks

✔ Customizing and optimizing Browser

✔ Downloading and installing applications from the Web

✔ Using Browser in an enterprise environment

*I*t's hard to believe that just over 13 years ago, more folks did not have access to the Internet than did. Today, you can surf the Web anytime, anywhere, and you can do it by using a traditional desktop or laptop computer, or even a tiny mobile device such as a PDA or a smartphone. Having said that, it should be no surprise that your BlackBerry has a Web browser of its own.

In this chapter, we explore ways to use BlackBerry Browser effectively. We offer shortcuts that improve your Web-browsing experience. We also throw in timesaving tips, including the coolest ways to customize BlackBerry Browser to make pages load faster, and a complete neat-freak's guide to managing your bookmarks.

Your network service provider might also have its own custom browser for you to use. We compare these proprietary browsers with the default BlackBerry Browser so that you can decide which best suits your needs.

## Getting Started with BlackBerry Browser

BlackBerry Browser comes loaded on your smartphone and accesses the Web using a cellular data connection or, in the case of the Wi-Fi browser, a Wi-Fi data connection. When you run your BlackBerry on a BlackBerry Enterprise Server (BES), the application is called *BlackBerry Browser;* otherwise, it's called *Internet Browser.* We just use *Browser* to make things easier.

There are four variations of Browser. The first three variations listed here are available on every BlackBerry model, but the Wi-Fi browser is available only on models that support Wi-Fi. The variations are

+ One that's connected to your BES.

+ One that goes directly to your service provider's network. This might be called by the network service provider's brand name.

+ A WAP browser. *Wireless Application Protocol,* or *WAP,* was popular in the '90s when mobile device displays were very limited and could display only five to six rows of text. This technology lost its appeal with the advent of high-resolution screens.

+ A browser that uses the Wi-Fi connection.

If you are a corporate BlackBerry user, your company administrator might disable all browsers except for the one that connects through the company's BES.

The following sections get you started using the Browser. After you get your feet wet, we promise that you'll be chomping at the bit to find out more!

## Accessing Browser

Browser is one of the main applications of your device, with its globe icon visible right on the Home screen (shown in Figure 1-1). In most cases, you open Browser by scrolling to this icon and then pressing the trackball, trackpad, or touch screen. Opening Browser by selecting its icon on the Home screen gives you a start page with a list of recently visited bookmarks as well as recent browsing history. If you haven't yet added bookmarks, the default Browser screen looks like Figure 1-2. You can find out more about adding bookmarks later in this chapter.

**Figure 1-1:**
You can open Browser from the Home screen.

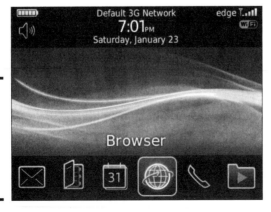

Default 3G Network    edge T.ıll

7:01PM

Saturday, January 23

Browser

**Figure 1-2:**
Browser
with the
default
empty
Bookmarks
screen.

You can also access Browser from any application that displays a Web address. For example, from Contacts, you can open Browser by highlighting and clicking the link on the Web Page field. If you get an e-mail that contains a Web address, just highlight the link and press the trackball, trackpad, or touch screen. Browser automatically opens and loads the Web page. When you click a Web address within another application, the Web page associated with that address opens. (In Figure 1-2, we're opening Browser from the Messages application.)

If you want to access Browser from another application, you don't have to close that application to jump to Browser. Just press the Menu key and choose Switch Application to open a pop-up screen with application icons. Use your trackball, trackpad, or touch screen to highlight the Browser icon, and then press the trackball, trackpad, or touch screen to launch Browser. (See Figure 1-3.)

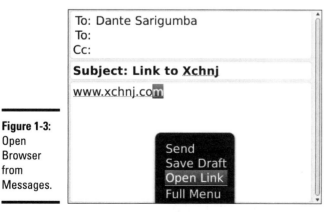

**Figure 1-3:**
Open
Browser
from
Messages.

## Hitting the (air) waves

After you locate Browser, you're ready to surf the Web. Here's how:

1. **On your BlackBerry smartphone, click the Browser icon.**

   Unless the configuration is changed, your smartphone displays the start page when you open Browser.

2. **Select the bookmark or type the Web address.**

   • *Bookmark:* Highlight the bookmark and then press the trackball, trackpad, or touch screen to launch the Web page. If the bookmark isn't under your recent bookmarks, press the Menu key and choose Bookmarks from the menu to see your full list of bookmarks.

   • *URL field:* Enter a Web address at the top of the screen that appears (as shown in Figure 1-4) and then press Enter or Go.

   The Web page appears. While the page is loading, progress is indicated at the bottom of the screen.

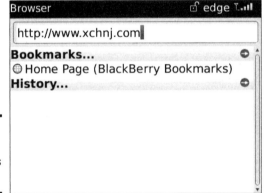

**Figure 1-4:**
Opening a
Web page is
simple.

## Navigating Web pages

Using Browser to navigate to a Web page is easy. Note that hyperlinks are highlighted onscreen. To jump to a particular hyperlink, scroll to the highlighted link and press the trackball, trackpad, or touch screen.

Here are a few shortcuts you can use while navigating a Web page:

✦ Quickly move up or down one full display page at a time by pressing Space ( move down one page) or Shift+Space (move up one page). On the Blackberry Storm, quickly flick your finger up (move down one page) or down (move up one page) without pressing the touch screen.

✦ Press B to go to the bottom of the page or press T to go to the top of the page. On the BlackBerry Storm, press the Menu key and choose Show Keyboard from the Browser menu. You can use the B and T keys on the virtual keyboard to go to the top or bottom of the page.

✦ To stop loading a page, press the Escape key.

✦ After a page fully loads, go back to the previous page by pressing the Escape key.

And don't forget the Browser menu (press the Menu key). It has some useful shortcuts, as shown in Figure 1-5.

**Figure 1-5:** The Browser menu has lots of good stuff.

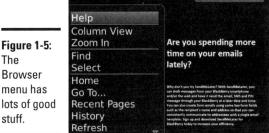 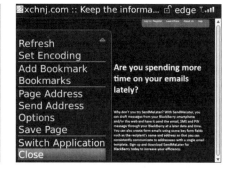

**Book III**
**Chapter 1**

Surfing the Web with a BlackBerry

Here are the Browser menu options:

✦ **Page View:** This is the default view and normally does not appear as a menu option. It shows up only if you are currently in Column view. This view allows you to see the page as you would normally see it on your PC's Internet browser. The compressed version of the Web page takes up the entire screen first. A menu option allows you to zoom in and out.

✦ **Column View:** Appears only if you are currently in Page view. With this view, the Web page is displayed vertically, meaning that a wide Web page wraps down and you can use the trackball, trackpad, or touch screen to scroll up and down the page.

✦ **Find:** Locates and highlights text within the current page. Like any other basic Find tool, choosing this option displays a prompt to enter the text you want to find. After the initial search, a Find Next menu appears for finding the next matching text.

✦ **Select:** Shows up only if the pointer is placed on text. This feature allows you to highlight text on the screen for copying.

✦ **Stop:** Shows up only if you're in the middle of requesting a page and allows you to cancel such request. This is the same as pressing the Escape key.

✦ **Copy:** This menu item appears if you have currently highlighted text. Selecting Copy copies the highlighted text into memory so that you can use it later for pasting somewhere else, such as in MemoPad.

✦ **Full Image:** This menu item appears only if you highlight an image and only a portion of the image is displayed on the screen.

✦ **Save Image:** Also appears if you highlight an image and allows you to save the image in the built-in memory or microSD card.

✦ **Home:** The shortcut to your home page. The default home page can vary from carrier to carrier, but to change it, follow these steps:

   1. *Open the Browser.*

   2. *Press the Menu key and choose Options.*

   2. *Go to Browser Configuration.*

   3. *Change the Home Page Address field.*

   4. *Press the Menu key and choose Save.*

✦ **Get Link:** This menu item appears if you have a currently highlighted link. Choosing this menu item opens that page of the link. *Hint:* The faster way to open a link is to press Enter.

✦ **Go To:** Allows you to open a Web page by entering the Web address and pressing Enter. As you enter more addresses, the ones you entered before are listed in the History portion of the screen and stored for possible future use so that you don't have to type them again. To find out how to clear that list, see the "Cache operations" section, later in this chapter.

✦ **Back:** Goes back to the previous page you viewed. This menu item displays only if you navigated to more than one Web page.

   You can achieve the same function by pressing the Escape key.

✦ **Forward:** If you've gone back at least one Web page in your browsing travels, use Forward to progress one page at a time.

✦ **Recent Pages:** Browser can track up to 20 pages of Web addresses you've visited, which you can view on the History screen. From there, you can jump to any of those Web pages by highlighting the history page and pressing the Enter key twice.

✦ **History:** Displays a list of the Web pages you visited and allows you to jump back quickly to those pages. It is grouped by date.

✦ **Refresh:** Updates the current page. This is helpful when you're viewing a page with data that changes frequently (such as stock quotes).

✦ **Set Encoding:** Specifies the encoding used in viewing a Web page. This is useful when viewing foreign languages that use different characters. Most BlackBerry users don't have to deal with this and probably don't know what type of encoding a particular language could display.

✦ **Add Bookmark:** Allows you to add a bookmark for this page. The process for adding a bookmark is discussed later in this chapter.

✦ **Bookmarks:** Takes you to the BlackBerry Bookmarks screen. Here you can choose one of your bookmarks to load that Web page.

When you open a Web page, indicators appear at the bottom of the screen, showing you the progress of your request. The left screen in Figure 1-6 shows that Browser is requesting a page. The right screen of Figure 1-6 shows that you've reached the page and that the page is still loading.

**Figure 1-6:**
Requesting a page (left) and then loading it (right).

You see either four or five icons in the upper-right corner of Browser. We explain them, from right to left:

✦ The **bars** show the strength of the network signal. (BlackBerry uses the same signal indicator for phone and e-mail.)

✦ Your **connection type** also appears. In Figure 11-6, *WiFi* means that the connection is using a Wi-Fi network.

✦ The **lock icon** indicates whether you're at a secure Web page. Figure 1-6 is showing a non-secure page. Whether a page is secure depends on the Web site you're visiting. If you're accessing your bank, you most likely see the secured icon (a closed lock). On the other hand, most pages don't need to be secure, so you see the unsecured icon (an open lock).

✦ The **rightmost arrow icon** appears when Browser is processing or receiving data.

If you lose patience waiting for a page to load and want to browse somewhere else, press the Escape key to stop the page from loading.

When you see a phone number or an e-mail address on a Web page, you can scroll to or touch that information to highlight it. When the information is highlighted, pressing the trackball, trackpad, or touch screen initiates a phone call or opens a new e-mail message (depending on which type of link you highlighted).

## Saving or sending a Web page address

Entering a Web address to view a page can get tedious. Fortunately, you can return to a page without typing the same address. While you're viewing a Web page, simply use the Browser menu (shown in Figure 1-7) to save or send that page's address.

Saving a Web page address is different from bookmarking it. Saving the Web page puts it in a message in your Messages screen, whereas bookmarking a Web page places it in your Bookmarks screen.

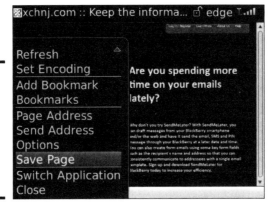

**Figure 1-7:** Use the Browser menu to save or send a Web page address.

You can save a Web page address in a couple of ways:

+ **Page Address:** This option allows you to view the Web address of the current page through a pop-up screen, which presents you with two options to act on.

  - *Copy Address:* Saves the page address on your BlackBerry clipboard and allows you to paste it somewhere else.

  - *Send Address:* Presents another screen so that you can choose whether to send the address by e-mail, MMS, PIN, or SMS. See Book III, Chapter 2 for more on sending e-mail messages and Book III, Chapter 3 for MMS, PIN, and SMS.

    For a more *direct* way of sending an address, select Send Address from the Browser menu while the Web page is displayed. If you know right away that you're going to send an address to someone, use the more direct method. It saves you a couple of clicks.

+ **Save Page:** Use this option to save the Web address of the current page to Messages. A message appears with the Browser globe icon to indicate that it's a Web link, as shown in Figure 1-8. Scrolling to that entry and pressing the trackball, trackpad or touch screen launches Browser and opens the page for your viewing pleasure. See Book III, Chapter 3 for more on using messages on the BlackBerry.

Saving a page to your message list has a different purpose than book-marking a page has. When you save a page to your message list, you can mark the page as unread, like an e-mail message, to remind yourself to check back later.

| | | |
|---|---|---|
| 7:15 PM | | edge ᵀ.ıll |
| | Default 3G Network | |
| | Sat, Jan 23, 2010 | |
| ● xchnj | | 7:15p |
| ◉ smrtcase | | 7:14p |
| 📝 Yosma Sarigumba | | 7:13p |
| Can you stop by | | |
| 💬 Dante Sarigumba (Mobile) | | 7:13p |
| Need you to check on the app | | |
| 📝 Dante Sarigumba | | 7:11p |
| Status | | |

**Figure 1-8:**
Save a Web
page link in
Messages.

*Note:* If you do not have network coverage when you try to access a Web page, you're prompted to save your request. When you do, your request is automatically saved in the message list. When you do have coverage later, you can open the same Web page from the message list, with the content loaded already!

Pressing a letter key while a menu appears selects the first menu item that starts with that letter. Pressing the same letter again selects the next menu item that starts with that letter.

## Saving Web images

Your BlackBerry smartphone allows you to view and save pictures or images from a Web page. You can save images of JPEG, PNG, GIF, and BMP formats. Any saved image is kept in the Pictures application, which enables you to view it later, even when you are out of coverage range. To save it from a Web page, highlight the image you want to save, press the Menu key, and select Save Image from the menu that appears.

To view the images you saved, do this:

**1. From the Home screen, select the Media icon and then select Pictures.**

The Pictures application opens.

**2. Scroll to and select the image you want to see.**

You can check out Book V, Chapter 3 for more information on saving and viewing images on your Blackberry.

**Book III
Chapter 1**

**Surfing the Web
with a BlackBerry**

## Changing your Home screen background

This is a neat trick. You can use an image that you have saved in your pictures list as the background on your Home screen. Here's how:

1. From the Home screen, select the Media icon and then select Pictures.

   The Pictures application opens.

2. Scroll to and select the image you want to set as your background.

3. Press the Menu key and then select Set as Home Screen Image.

# Bookmarking Your Favorite Sites

You don't have to memorize all the addresses of your favorite sites. BlackBerry Browser allows you to keep a list of sites you want to revisit. In other words, make a *bookmark* so that you can come back to a site quickly.

## Adding and using a bookmark

On the page you want to bookmark, select Add Bookmark from the Browser menu. Remember that the menu is always accessible by pressing the Menu key. Enter the name of the bookmark and then select the folder where you want to save the bookmark in the Add Bookmark dialog box (as shown in Figure 1-9). The default folder is BlackBerry Bookmarks, but you can save the bookmark in folders that you create. To see how to create a bookmark folder, skip ahead to the section "Adding a bookmark subfolder."

Auto Synchronize is a cool feature that you can enable for your bookmarks. It checks a bookmarked Web page at an interval you specify and lets you know if there have been any changes. It does that by displaying the bookmark in bold in your Bookmark screen. While creating a Bookmark, just highlight Auto Synchronize and change it from Never to an interval between 1 and 24 hours.

The next time you want to go to a bookmarked page, return to the Bookmarks screen. Press the Menu key and choose Bookmarks from the Browser menu. From this screen, you can find all the pages you bookmarked. Just highlight the name of the bookmark and press the Enter key to open that page.

**Figure 1-9:**
Specify the name and the folder in which to store the bookmark.

## Accessing Web pages offline

The Add Bookmark dialog box includes an Available Offline check box, which you might be wondering about. If that check box is selected, you not only save a page as a bookmark, but you also store the page itself so that you can view it even when you're out of your network coverage area (like when you're stuck deep in a mountain cave or underground on the subway). Even if you are in coverage, the next time you click the bookmark, that page comes up very fast. However, if the page changes often you will need to click the Menu button and choose Refresh to make sure you load the most recent version of the page.

We recommend that you create bookmarks to search engines (such as Google) available offline because the content of the initial search page is not likely to change from day to day. Although you won't be able to perform a Web search while you're out of coverage, caching a search engine page helps it load more quickly, even when you're in coverage, so you can get to submitting your search faster.

The Bookmarks screen appears by default when you open Browser.

## Modifying a bookmark

Changing a bookmark is a snap. Just follow these steps:

1. **Press the Menu key and choose Bookmarks from the Browser menu to go to the Bookmarks screen.**

2. **Highlight the name of the bookmark you want to modify, press the Menu key, and then select Edit Bookmark.**

3. **On the screen that follows, you can edit the existing name, the address the bookmark is pointing to, or both.**

4. **Select Accept to save your changes.**

## Organizing your bookmarks

Over time, the number of your bookmarks will grow. A tiny screen can make it tough to find a certain site. You can organize your bookmarks by using folders to help work around this problem. For example, you can group related sites in a folder, and each folder can have one or more folders inside it (subfolders). Having a folder hierarchy narrows your search and allows you to easily find a site.

For example, your sites might fall into these categories:

+ Reference

    NY Times

    Yahoo!

+ Fun

    Flickr

    The Onion

+ Shopping

    Etsy

    Gaiam

### Adding a bookmark subfolder

Unfortunately, you can add subfolders to only those folders that are already listed on the Bookmark page. That is, you can't create your own root folder. Your choices for adding your first subfolder are under WAP Bookmarks, BlackBerry Bookmarks, or WiFi Bookmarks (if your BlackBerry smartphone model supports Wi-Fi).

Suppose that you want to add a Reference subfolder within your BlackBerry Bookmarks folder. Here are the quick and easy steps:

1. **On the Bookmarks screen, highlight BlackBerry Bookmarks.**

    The initial folder you highlight is the parent of the new subfolder. So in this case, the BlackBerry Bookmarks folder will contain the Reference subfolder.

2. **Press the Menu key and then select Add Subfolder, as shown in Figure 1-10.**

    You see a dialog box where you can enter the name of the folder.

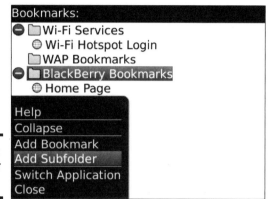

**Figure 1-10:**
Add a folder here.

*3.* **Enter the name of the folder (as shown in Figure 1-11) and use the trackball, trackpad, or touch screen to select OK.**

We named our folder *Reference.* The Reference folder appears on the Bookmarks screen (as shown in Figure 1-12), bearing a folder icon.

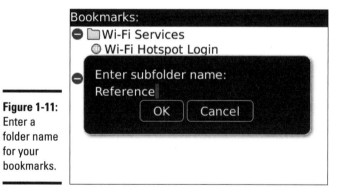

**Figure 1-11:**
Enter a folder name for your bookmarks.

**Figure 1-12:**
The Bookmarks screen showing the folder you just added.

### Renaming a bookmark folder

Renaming a bookmark folder you've created is as easy as editing a bookmark. Follow these steps:

1. **Go to the Bookmarks screen.**

2. **Highlight the name of the folder you want to change.**

3. **Press the Menu key.**

   A menu appears.

4. **Select Rename Folder.**

5. **Type the name of the folder.**

6. **Select OK to save your changes.**

### Moving your bookmarks

If you keep going astray looking for a bookmark that you think exists in a particular folder but is instead in another, move that bookmark where it belongs. Follow these steps:

1. **Highlight the bookmark, press the Menu key, and select Move Bookmark.**

2. **Use the trackball, trackpad, or touch screen to move the bookmark to the location in the list where you want it to appear.**

3. **After you find the correct location, press the trackball, trackpad, or touch screen.**

   Your bookmark is in its new home.

### Cleaning up your bookmarks

Cleaning up your bookmarks list can help you keep organized. On the Bookmarks screen, highlight the name of the bookmark you want to delete, press the Menu key, and select Delete bookmark from the menu that appears. You'll be asked to confirm that you want to delete the bookmark; choose Delete again and the bookmark is gone.

You can clean up bookmarks wholesale by deleting an entire folder. However, if you delete a folder, you delete the contents of that folder as well, so purge with caution.

# Exercising Options and Optimization Techniques

Browser works out of the box. But everyone has his or her own tastes, right? You can check out Browser Options for Browser features and attributes that you can customize. Here's how:

1. **Press the Menu key.**

2. **Select Options.**

   The Browser Options screen offers three main categories to choose from, as shown in Figure 1-13:

   - *Browser Configuration:* A place to toggle Browser features.

   - *General Properties:* Settings for the general look and feel of the Browser.

   - *Cache Operations:* Allows you to clear file caches used by Browser.

   If you feel speed-greedy after adjusting the options, see the sidebar "Speeding up browsing," later in this chapter.

**Figure 1-13:** The Browser Options screen.

## Configuring Browser

You can define browser-specific settings from the Browser Configuration screen, which you access from the Browser Options screen. The customization items you can choose (shown in Figure 1-14) are as follows:

✦ **Support JavaScript:** JavaScript is a scripting language used heavily to make dynamic Web pages. A Web page might not behave normally when this option is turned off. This option is off by default on the latest BlackBerry smartphones including Bold, Storm, Pearl Flip, and Curve 8900. The only downside to turning this on is that some Web pages may load a little more slowly — a small price to pay for greater compatibility.

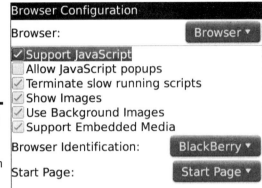

**Figure 1-14:**
The
Browser
Configuration
screen.

+ **Allow JavaScript Popups:** Most ad pages are launched as JavaScript pop-ups. So, deselecting this check box minimizes these ads. But be aware that some important pages are also displayed as JavaScript pop-ups. (Note that this option shows up only if you select the Support JavaScript check box.)

+ **Terminate Slow Running Scripts:** Sometimes you find Web pages with scripts that aren't written well. Keep this selected to keep Browser from hanging. (Note that this option shows up only if you select the Support JavaScript check box.)

+ **Show Images:** Controls the display of images, depending on the content mode of WML, HTML, or both. Think of *WML pages* as Web pages made just for mobile devices, such as the BlackBerry. We recommend leaving this selected for both.

Turn on and off the display of image placeholders if you opt to not display images.

+ **Use Background Images:** A Web page background image can make the page look pleasing, but if the image is big, it could take time to download it.

+ **Support Embedded Media:** Select this option to support media such as SVG (scalable vector graphics). Think of it as Flash for mobile devices such as the BlackBerry. SVG can be a still image or an animated one.

+ **Browser Identification:** Specifies which browser type your browser emulates. The default is BlackBerry, but Browser can also emulate these instead:

  • Microsoft Internet Explorer

  • Firefox

We don't see much difference in their behavior, so we recommend emulating the default BlackBerry mode.

✦ **Start Page:** Allows you to specify the default view your Browser uses when you open it. You can choose Start Page, Home Page, or Last Page Loaded. We recommend leaving this on Start Page because that view gives you access to enter a URL, search using the search bar, choose a recent bookmark, and view your recent history all in one screen.

✦ **Home Page Address:** Allows you to set the URL that is loaded as your home page. This page is loaded if you change your Start Page option to Home Page, or if you choose Home on the Browser menu.

## General Browser properties

The General Properties screen does something similar to the Browser Configuration screen (see the preceding section): It lets you customize some Browser behaviors. This screen, however, is geared more toward the features of the Browser content. As shown in Figure 1-15, you can do the following:

✦ Configure features

✦ Turn on features

✦ Turn off features

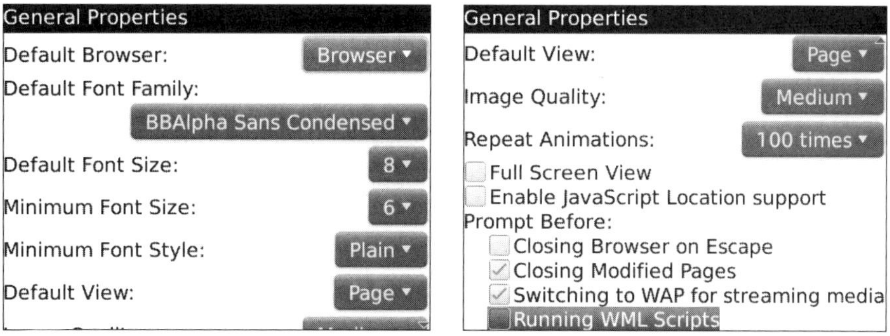

**Book III
Chapter 1**

Surfing the Web
with a BlackBerry

**Figure 1-15:**
The General
Properties
screen.

| General Properties | |
| --- | --- |
| Default Browser: | Browser ▾ |
| Default Font Family: | |
| | BBAlpha Sans Condensed ▾ |
| Default Font Size: | 8 ▾ |
| Minimum Font Size: | 6 ▾ |
| Minimum Font Style: | Plain ▾ |
| Default View: | Page ▾ |

| General Properties | |
| --- | --- |
| Default View: | Page ▾ |
| Image Quality: | Medium ▾ |
| Repeat Animations: | 100 times ▾ |
| ☐ Full Screen View | |
| ☐ Enable JavaScript Location support | |
| Prompt Before: | |
| ☐ Closing Browser on Escape | |
| ☑ Closing Modified Pages | |
| ☑ Switching to WAP for streaming media | |
| ☐ Running WML Scripts | |

From this screen, you use the Space key or touch screen to change the value of a field. You can configure the following features by selecting from these choices:

✦ **Default Browser:** If you have multiple browsers available, such as Internet Browser, BlackBerry Browser, and WiFi Browser, use this to specify which one you want to use when opening a Web link.

✦ **Default Font Family:** When a Web page doesn't specify the text font, Browser uses the one you select here.

✦ **Default Font Size:** When a Web page doesn't specify the text font size, Browser uses the one you select here. The smaller the size, the more text fits on the screen.

✦ **Minimum Font Size:** A Web page may specify a font size that is too small to be legible. Specifying a legible font size will override the Web page.

✦ **Minimum Font Style:** When Browser uses the minimum font size, you can choose what font to use. Some fonts are more legible, even in small size, than others. If you're not sure which one to use, leave the default untouched.

✦ **Default View:** You can toggle the default view between these two options:

  • *Column:* Wraps all Web page elements vertically, so you just scroll up and down by panning the page.

  • *Page:* Displays the page like you normally see in your PC's Internet browser. Pan the page to scroll left, right, up, and down.

✦ **Image Quality:** The higher the quality, the slower the page loads. The default quality is medium. You have three options: low, medium, and high.

✦ **Repeat Animations:** Sets the number of times an animation repeats. The default is 100, but you can change this setting to

  • Never

  • Once

  • 10 Times

  • 100 Times

  • As Many as the Image Specifies

✦ **Full Screen View:** Browser will try to display all Web pages in full screen view if this is selected. We haven't noticed a difference whether this is checked or not, so it is fine to leave it unchecked.

✦ **Enable JavaScript Location Support:** Web pages that have scripts that take advantage of your BlackBerry's location through GPS will work if you have this selected.

✦ **Prompt Before:** You can have BlackBerry Browser give you a second chance before you do the following things:

  • *Closing Browser on Escape:* You're notified right before you exit BlackBerry Browser.

  • *Closing Modified Pages:* You're notified right before you exit a modified Web page (for example, some type of online form you fill out).

  • *Running WML Scripts:* WML is a script that tells a wireless device how to display a page. It was popular years ago when resolutions of device screens were low, but very few Web sites use it now. We recommend leaving this field deselected because this type of scripting is old and benign.

## Cache operations

At any given time, your BlackBerry smartphone uses a few cache mechanisms. A *cache* (pronounced *cash*) temporarily stores information used by Browser so that the next time the info is needed, Browser doesn't have to go back to the source Web site. The cache can speed up displays when you want to view the Web page again and is also useful when you're suddenly out of network coverage. When you visit a site that uses *cookies,* Browser caches that cookie. (Think of a cookie as a piece of text that a Web site created and placed in your BlackBerry's memory to remember something about you, such as your username.)

Some Web sites *push* (send information to) Web pages to BlackBerry devices. These sites have a link on their home page that lets you set up push service on your BlackBerry. Once you subscribe, an icon will appear on the Home screen allowing you to quickly view the page. After the Web page is delivered to your BlackBerry, it becomes available even if you go out of the coverage area. If you subscribed to this service, your device will have Web pages stored in the cache. Also, the addresses of the pages that you visited (or your latest 20 in your history list) comprise a cache.

The Cache Operations screen, shown in Figure 1-16, allows you to manually clear your cache. To view the Cache Operations screen, follow these steps:

*1.* **From the Browser screen, press the Menu key.**

*2.* **Select Options.**

*3.* **Select Cache Operations.**

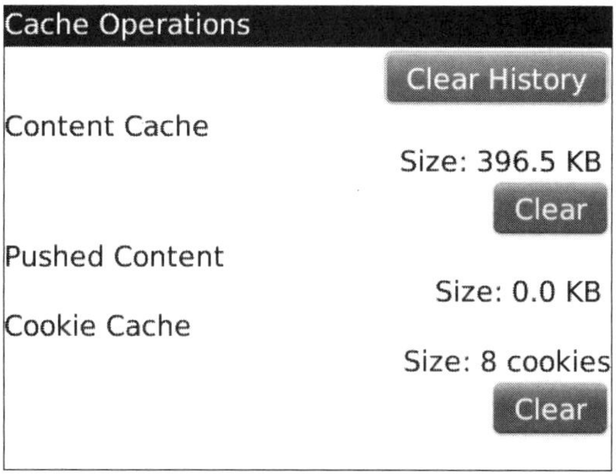

**Figure 1-16:**
The Cache
Operations
screen.

```
Cache Operations
                              [ Clear History ]
Content Cache
                              Size: 396.5 KB
                              [ Clear ]
Pushed Content
                              Size: 0.0 KB
Cookie Cache
                              Size: 8 cookies
                              [ Clear ]
```

# Speeding up browsing

On a wireless network, many factors can affect the speed with which Web pages display. If you find that browsing the Web is extremely slow, or you're just looking to squeeze every bit of speed out of your browsing experience, you can make your pages load faster in exchange for not using a few features. Here are some of the techniques you can use:

✔ Don't display images. You can see a big performance improvement by turning off the display of images. From the Browser menu, choose Options⇨Browser Configuration, scroll to Show Images, and change the value to No.

✔ Don't use the Wi-Fi browser. The Internet Browser **and** BlackBerry Browser use your carrier BIS **and** company BES respectively to encrypt, compress, and transcode Web content before delivering it to your BlackBerry smartphone. This means the pages display faster for you than they do for non-BlackBerry users. The Wi-Fi browser doesn't do this, so although your connection is faster, pages may still load slower. Changing your default browser to Internet Browser or BlackBerry Browser can improve performance. From the BlackBerry Home screen, choose Settings⇨Options⇨Advanced Options⇨Browser and change Default browser configuration to BlackBerry Browser or Internet Browser. Don't worry; these browsers still use your faster Wi-Fi connection to pass data between your BlackBerry and the BIS/BES server so that your pages will load faster than ever.

✔ Make sure that your BlackBerry is not low on or out of memory. When your BlackBerry's memory is very low, its performance degrades. The BlackBerry low-memory manager calls each application every now and then, telling each one to free resources.

*Hint #1:* Don't leave many e-mail messages unread. When the low-memory manager kicks in, Messages tries to delete old messages, but it can't delete those that are unread.

*Hint #2:* You can also clean up the BlackBerry event log to free needed space. Enter the letters LGLG while holding the Shift key. An event log opens, where you can clear events to free memory.

✔ Turn off other features. If you're mostly interested in viewing content, consider turning off features that pertain to how the content is processed, such as Support HTML Tables, Use Background Images, Support JavaScript, Allow JavaScript Popups, and Support Style Sheets. To turn off other browser features, open the Browser menu and choose Options⇨General Properties.

***Warning:*** We don't advise turning off features and performing an important task such as online banking. If you do, you might not be able to perform some of the actions on the page. For example, the Submit button might not work.

The size of the saved data in each type of cache is displayed on this screen. If the cache has content, you also see the Clear button, which you can use to clear (delete the content in) the specified cache type. This is true for all types of cache except for history, which has its own Clear History button. There are four types of cache:

+ **Content Cache:** Any offline content. You might want to clear this whenever you're running out of space on your BlackBerry and need to free some memory. Or maybe you're tired of viewing old content or tired of pressing the Refresh option.

+ **Pushed Content:** Any content that was pushed to your BlackBerry from Push Services subscriptions. You might want to clear this to free memory on your BlackBerry.

+ **Cookie Cache:** Any cookies stored on your BlackBerry. You might want to clear this for security's sake. Sometimes you don't want a Web site to remember you.

+ **History:** This is the list of sites you've visited by using the Go To function. You might want to clear this for the sake of security if you don't want other people knowing which Web sites you're visiting on your BlackBerry.

# Installing and Uninstalling Applications from the Web

You can download and install applications on your BlackBerry by using Browser — that is, if the application has a link that lets you download and install the files. Book VI introduces you to the wonderful world of BlackBerry applications. It describes how to search for applications, the different methods you can use to install and uninstall them, and even recommends applications for popular areas like news, entertainment, and social networking. In case you already have some applications in mind, the downloading and installing parts are easy. Follow these steps:

1. **Click the link from Browser.**

   It displays a simple prompt that looks like the screen shown in Figure 1-17.

2. **Click the Download button.**

   The download starts and displays a progress bar.

As long as you stay within network coverage while the download is progressing, your BlackBerry can finish the download *and* install the application for you. If you lose coverage, or if the download fails for some other reason, you will need to start the download again. If it finishes without any problems, you see a screen similar to Figure 1-18. Unless you changed your download folder, you can find the application in the Downloads folder located on your BlackBerry Home screen. See Book I, Chapter 6 for more information on changing your download folder.

**Figure 1-17:**
A typical page that lets you download an application on your BlackBerry.

**Figure 1-18:**
The download and installation are complete.

As with a desktop computer, the application might or might not work for a variety of reasons. For instance, the application may require you to install libraries, or the application may work on only a certain version of the BlackBerry handheld firmware. These issues can be prevented, depending on the sophistication of the site where the link is published. With most reputable sources, these issues are considered, and successful downloading and installation are a snap.

You can check your firmware version manually using the following steps:

*1.* **From the BlackBerry Home Screen, choose Settings⇨Options.**

*2.* **Select About.**

Your BlackBerry firmware version is displayed on the third line. You usually only need the first two digits, like 4.7 or 5.0.

Installing applications from nonreputable sources can cause your BlackBerry smartphone to become unstable. Before you download an application from the Web, be sure to read reviews about that particular application. Most of the time, other people who tried the software provide reviews or feedback. Don't be the first to write the bad review!

Your BES administrator can disable the feature in your BlackBerry to download and install an application. This is mostly the case for a company-issued device. If you have problems downloading and installing an application, check your company policy or contact the BlackBerry support person in your company.

If you download an application that turns out to be a dud, you need to uninstall it. See Book VI, Chapter 1 for more information on installing and uninstalling applications on your BlackBerry.

# Browser's Behavior in an Enterprise Environment

Getting a device from your employer has both a good and an ugly side:

✦ On the good side, your company foots the bill.

✦ On the bad side, your company foots the bill.

Because your company pays for the bill, the company has a say in what you can and can't do with your BlackBerry device. This is especially true with respect to browsing the Web.

Two scenarios come into play when it comes to your browser:

✦ Your browser might be running under your company's BlackBerry Enterprise Server (BES). With this setup, your BlackBerry Browser is connecting to the Internet through your company's Mobile Data Service (MDS). It's like using your desktop machine at work.

✦ Your browser is connected through a network service provider. Most of the time, this kind of browser is called by the company's name.

In most cases, your device fits in only one scenario, which is the case where your browser is connected through your company's BES server. Some lucky folks might have both. Whatever scenario you're in, the following sections describe the major differences between the two and indicate what you can expect.

## Using Browser on your company's BES

In an enterprise setup, your BlackBerry Browser is connected through your company's BES server. With this setup, the browser is actually named

*BlackBerry Browser.* BES is located inside your company's intranet. This setup allows the company to better manage the privileges and the functions you can use on your device.

For the BlackBerry Browser application, this setup allows your company to use its existing Internet infrastructure, including the company's firewall. Because you are within the company's network, the boundaries that your network administrator set up on your account apply to your BlackBerry as well. For example, when browsing the Web, your BlackBerry won't display any Web sites that are blocked by your company's proxy server. The good thing, though, is that you can also browse the company's intranet.

Know your company's Web-browsing policy. Most companies keep logs of the sites you view on your browser and might even have software to monitor usage. Also, your company might not allow downloading from the Web.

## Using your network provider's browser

Any new device coming from a network service provider can come with its own branded Web browser. It's the same BlackBerry Browser, but the behavior might differ in the following ways:

✦ **The name is different.**

✦ **The default home page usually points to the provider's Web site.** This isn't necessarily a bad thing. Most of the time, the network provider's Web site is full of links that you might not find on the BlackBerry Browser.

✦ **You can browse more sites.** You're not limited by your company's policy.

Most of the time, if your browser is through a BES, surfing the Web is much faster. This is not true in all cases, however, because the network bandwidth of your BES affects the speed.

## Setting the default browser

If you have more than one Web browser on your device, you have the option to set the default browser. This comes into play when you view a Web address by using a link outside the browser application. For example, when you view a contact with a Web page in Contacts, the Contacts menu contains a Get Link option. Clicking Get Link launches the default browser.

To set up the default browser, follow these steps:

1. **Go to the Home screen.**

2. **Select Settings⇨Options⇨Advanced Options⇨Browser.**

3. **Use the trackball, trackpad, or touch screen to change the value of the default browser configuration, as shown in Figure 1-19.**

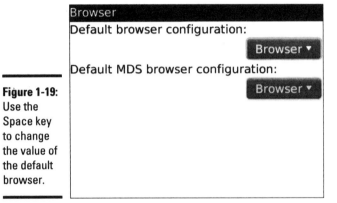

**Figure 1-19:**
Use the
Space key
to change
the value of
the default
browser.

# Chapter 2: Using E-Mail

## In This Chapter

✔ **Linking your e-mail accounts to your BlackBerry**

✔ **Adding your own e-mail signature**

✔ **Reconciling e-mails on your BlackBerry and PC**

✔ **Receiving, sending, and spell-checking e-mails**

✔ **Deleting and filtering your e-mails**

✔ **Searching your e-mail**

✔ **Saving messages**

*Y*our BlackBerry brings a fresh new face to the convenience and ease of use that you associate with e-mail. You can "hook" as many as ten e-mail accounts (from your work e-mail to personal e-mail from providers such as Yahoo! and AOL) to your BlackBerry. You can set up an e-mail signature, configure e-mail filters, and search for e-mails.

In this chapter, you find answers on how to use and manage the e-mail capabilities of your BlackBerry to their full potential. From setup to sorts, it's all covered here.

## Getting Up and Running with E-Mail

Regardless of your network service provider (Verizon, Rogers, Vodafone, or other), you can set up your BlackBerry to receive e-mail from at least one of your e-mail accounts. Thus, with whatever address you use to send and receive e-mail from your PC (Yahoo!, Gmail, and so on), you can hook up your BlackBerry to use that same e-mail address. Instead of checking your Gmail from your desktop, for example, you can now get it on your BlackBerry.

Most network service providers allow you to connect as many as ten e-mail accounts to your BlackBerry. This provides you with the convenience of one central point from which you can get all your e-mail, without having to log on to multiple e-mail accounts. Sweet!

In an enterprise environment — depending on your company policy — you might not be able to access the BlackBerry Internet Service site to link your personal e-mail accounts to your BlackBerry. If you work for a Fortune 500 company, most likely, you can't access BlackBerry Internet Service (BIS). However, you can still configure e-mail settings (such as the BlackBerry e-mail filter and BlackBerry e-mail reconciliation) to make your e-mail experience that much better. (See the upcoming section "Enabling Wireless Reconciliation.") If you're an enterprise user, skip to the parts where you see the Enterprise icon to configure your e-mail settings. If you haven't set up e-mail on your company-owned BlackBerry, see the upcoming section "Setting up e-mail in an enterprise environment."

## Using the BlackBerry Internet Service client

You can pull together all your e-mail accounts into one by using the BlackBerry Internet Service. The BIS client allows you to do the following:

+ **Manage multiple e-mail accounts.** As we mention earlier, you can combine as many as ten e-mail accounts onto your BlackBerry. See the next section for details.

+ **Use wireless e-mail reconciliation.** No more trying to match your BlackBerry e-mail against e-mail in your combined account(s). Just turn on wireless e-mail reconciliation, and you're good to go. For more on this, see the upcoming section, "Enabling Wireless Reconciliation."

+ **Create e-mail filters.** You can filter e-mails on your BlackBerry so that you get only those messages that you truly care about. See the "Filtering your e-mail" section, near the end of this chapter.

Think of the BlackBerry Internet Service client as an online e-mail account manager, but one that doesn't keep your e-mails. Instead, it routes the e-mails from your other accounts to your BlackBerry (because it's directly connected to your BlackBerry).

## Combining your e-mail accounts into one

To start herding e-mail accounts onto your BlackBerry, you must first run a setup program from the BIS client.

You can access the BIS client from your BlackBerry or from your desktop computer. For PC access, you need the URL specific to your network service. Contact your network service provider (such as Verizon or Telus) directly to get that information.

We cover accessing the BIS client from your BlackBerry, but you can do all these functions from a Web browser as well.

To get started with the BIS client from your BlackBerry, do the following:

1. **From the BlackBerry Home screen, select the Setup folder.**

2. **Select the Email Settings icon.**

   You're prompted with a login screen similar to Figure 2-1. If you haven't created your account, click the Create button to create your BIS account.

**BlackBerry® Internet Service**

**Existing Users**

If you want to access your email settings or you changed your device, log in.

**User name:**

mytestbb

**Password:**                                    Forgot Password?

********

☑ Remember me on this device.

**Log In**

**Close**

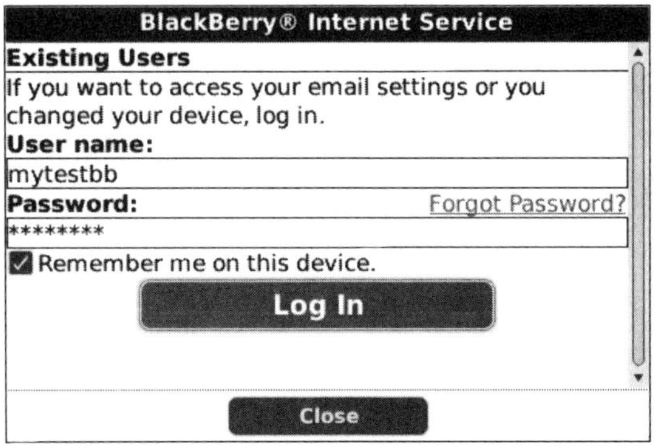

**Figure 2-1:**
The BIS client login screen.

After you're logged in, you see a list of e-mail accounts that have been set up. If you haven't set up any e-mail accounts yet, you see an Add button. We show you how to add accounts in the next section.

## Adding an e-mail account

You can set up your BIS account directly from your BlackBerry. (As we mention earlier in this chapter, you can have as many as ten e-mail accounts on your BlackBerry.) To add an e-mail account to your BlackBerry account, follow these steps on your BlackBerry:

1. **From your BlackBerry Home screen, select the Setup folder.**

2. **Select the Person E-mail Setup icon.**

   You're prompted with a login screen similar to that shown in Figure 2-1. If you haven't created your account, click the Create button to create your BIS account.

3. **After you log in, select Add.**

   A screen with different e-mail domains (Yahoo!, Google) appears, as shown in Figure 2-2.

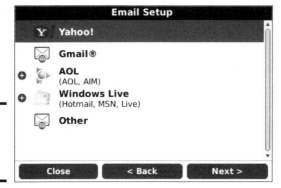

**Figure 2-2:**
Select
an e-mail
domain.

4. **Select an e-mail domain.**

5. **Enter the e-mail address and password, and then select Next.**

   If you entered your e-mail credentials correctly, you see the setup confirmation screen.

As we mention earlier in this chapter, you can also add an e-mail account to your BlackBerry from your desktop PC or Mac via a Web browser. Just follow these steps:

1. **From the BIS client, click Setup Account.**

   Contact your network carrier for the URL to access the BIS client on your PC or Mac Web browser.

2. **Enter the address and login credentials for that e-mail address.**

   • The *e-mail address* is the address from which you want to receive e-mail — for example, myid@yahoo.com.

   • The *account login* is the one you use to log in to this particular e-mail account.

   • The *password* is the one you use with the login.

3. **Click the Next button.**

   You're finished. It's that easy!

If ever in doubt, simply go to Google and enter *carrier-name BIS login,* and the first result returned will almost always be the BIS login page for your carrier.

## Setting up e-mail in an enterprise environment

This section is for you if your BlackBerry can't receive and send e-mail yet — such as when you first get your BlackBerry or you swap an old model for a new one.

If your e-mail function works properly on your BlackBerry, you can skip this section.

Follow these steps to activate your BlackBerry for enterprise use:

**1. From the Home screen, press the Menu key and select Enterprise Activation.**

The Enterprise Activation screen opens, with two fields for you to fill in:

- *Your corporate e-mail address:* For example, `myaccount@abc Company.com`

- *Your password:* From your IT department

**2. Type your corporate e-mail account along with the appropriate password.**

If you don't know this information, contact your corporate system administrator.

**3. Press the Menu key and select Activate.**

Your BlackBerry attempts to activate itself with your corporation.

Some corporations don't allow any employee-purchased BlackBerry smartphones to be activated with corporate e-mail. Check with your system administrator for corporate BlackBerry policies.

## Getting e-mail in an enterprise environment using Desktop Redirector

If you're a sole proprietor or a consultant who works in a corporation that runs Exchange or Lotus and you would like to get enterprise e-mails on your BlackBerry, this section is for you.

Typically, to get enterprise e-mail, your BlackBerry would have to be configured with the BlackBerry Enterprise Server (BES). Expect this if your employer hands you a BlackBerry. However, if you work for a company as a contractor, you probably won't be getting a BlackBerry from that company. When you want to get enterprise e-mail so that you don't fall behind, you need Desktop Redirector so that you can get company e-mail on your personal BlackBerry.

**Book III
Chapter 2**

**Using E-Mail**

To start using Desktop Redirector, you first need to install BlackBerry Desktop Manager; see Book IV, Chapter 1 details about how to do this. After you install Desktop Manager with Redirector, make sure that Redirector starts every time you boot your PC.

Depending on the corporate security policy, some corporations allow Desktop Redirector, and some do not. Before you start using Desktop Redirector, contact the IT department in the company you work for.

Here are just a few caveats when using Desktop Redirector:

✦ You can get enterprise e-mail as long as your PC is turned on and has an Internet connection.

✦ When someone sends you an attachment, you can't retrieve it from your BlackBerry. Unfortunately, that is the limitation for Desktop Redirector.

✦ When someone sends you a meeting notice, you can't accept or reject it.

## Configuring Your E-Mail Signature

By default, your e-mail signature is something like *Sent via My BlackBerry,* which can be cool the first week, showing off to people that you're à la mode with your BlackBerry. But sooner or later, you might not want people to know that you are out and about while answering e-mail. Or you might want something more personal.

Follow these steps to configure your e-mail signature by using the BIS client on your BlackBerry:

1. **From your BlackBerry Home screen, select the Setup folder.**

2. **Select the Email Setting icon.**

   You are prompted with a login screen similar to that shown earlier in Figure 2-1. If you haven't created your account, click the Create button to create your BIS account.

3. **Log in to the BIS client on the BlackBerry.**

   You see the BIS main screen.

4. **Select the e-mail account for which you want to set up an e-mail signature.**

5. **In the Signature field, type the text for your e-mail signature.**

6. **Select Save.**

# Enabling Wireless Reconciliation

With wireless reconciliation, you don't need to delete the same e-mail in two places (on your computer and on your BlackBerry). The two e-mail inboxes reconcile with each other — hence, the term *wireless reconciliation.* Convenient, huh?

## Enabling wireless e-mail synchronization

You can start wireless e-mail synchronization by configuring your BlackBerry. Follow these steps:

1.  **From the Home screen, press the Menu key and then select Messages.**

    The Messages application opens. You see the message list.

2.  **In the message list, press the Menu key and select Options.**

    The Options screen appears, with two option types: General Options and E-mail Reconciliation.

3.  **Select E-mail Reconciliation.**

    The E-mail Reconciliation screen opens, which has the following options:

    *   *Delete On:* Configures how BlackBerry handles your e-mail deletion

    *   *Wireless Reconciliation:* Turns on or off the wireless sync function

    *   *On Conflict:* Controls how BlackBerry handles any inconsistencies between e-mail on your BlackBerry and the BlackBerry Internet Service client

        With this option, you can choose who "wins": your BlackBerry or the BlackBerry Internet Service client.

4.  **Select Delete On and then select one of the following from the drop-down list:**

    *   *Handheld:* A delete on your BlackBerry takes effect only on your BlackBerry.

    *   *Mailbox & Handheld:* A delete on your BlackBerry takes effect on both your BlackBerry and your inbox on the BlackBerry Internet Service client.

    *   *Prompt:* This option prompts your BlackBerry to ask you at the time of deletion where the deletion takes effect.

5.  **Select Wireless Reconciliation and then select On from the drop-down list.**

6.  **Select On Conflict, and make a selection from the drop-down list.**

    If you choose Handheld Wins, the e-mail messages in your e-mail account will match the ones on the handheld.

**Book III
Chapter 2**

**Using E-Mail**

After your device is configured, you need to enable Synchronize Deleted Item on the BIS client. Follow these steps:

1. **From your BlackBerry Home screen, select the Setup folder.**

2. **Select the Email Setting icon.**

   You are prompted with a login screen similar to Figure 2-1, shown earlier in this chapter. If you haven't created your account, click the Create button to create your BIS account.

3. **Log in to the BIS client on the BlackBerry.**

   You see the BIS main screen.

4. **Select an e-mail account for which you want to enable Synchronize Deleted Item.**

5. **Select Synchronization Options to expand the options.**

6. **Make sure that the Deleted Items check box is selected.**

7. **Select Save.**

Unfortunately, some e-mail accounts might not work well with the e-mail reconciliation feature of the BlackBerry, so you might have to delete an e-mail twice.

## Permanently deleting e-mail from your BlackBerry

When deleting e-mail on your BlackBerry, the same message in that e-mail account is placed in the Deleted folder. You can set up your BlackBerry to permanently delete e-mail, but use this option with caution because *after that e-mail is gone, it's truly gone.*

To permanently delete e-mail on your BlackBerry Internet Service client from your BlackBerry, follow these steps:

1. **Open the Messages application.**

2. **In the message list, press the Menu key and select Options.**

3. **On the Options screen, select E-mail Reconciliation.**

4. **On the E-mail Reconciliation screen, press the Menu key and select Purge Deleted Items.**

   You see all your e-mail accounts.

5. **Choose the e-mail account from which you want to purge deleted items.**

   A screen appears, warning you that you are about to purge deleted e-mails on your Service client.

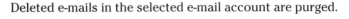

**6. Select Yes.**

Deleted e-mails in the selected e-mail account are purged.

Unfortunately, some e-mail accounts may not work with the Purge Deleted Items feature.

# Working with E-Mail

From the Messages application, you send and receive your e-mails and also configure wireless e-mail reconciliation with your e-mail account(s). See the preceding section for information on e-mail reconciliation.

To access Messages, press the Menu key from the Home screen and select Messages. The first thing you see after opening Messages is the message list. Your message list can contain e-mail, voice mail messages, missed phone call notices, Short Messaging Service (SMS) messages, and even saved Web pages.

## Receiving e-mails

Whether you're concerned about security or speed of delivery, with BlackBerry's up-to-date secured network, you're in good hands when receiving e-mail on your BlackBerry.

And whether you've aggregated accounts or just use the plain-vanilla BlackBerry e-mail account, you receive your e-mail the same way. When you receive an e-mail, your BlackBerry notifies you by displaying a numeral next to a mail icon (an envelope) at the top of the screen. This number represents how many new (unread) e-mails you have (see Figure 2-3). A red asterisk next to the envelope indicates that you have new mail and that you haven't opened the Messages application yet. Your BlackBerry can also notify you of new e-mail by vibration, by a sound alert, or both. You can customize this from the Profile application, as we detail in Book I, Chapter 3.

### Retrieving e-mail

Retrieving your e-mail is simple. Follow these steps:

**1. From the Home screen, press the Menu key and select Messages.**

**2. In the message list, scroll to any e-mail, and press the trackball or trackpad.**

You can tell whether an e-mail is unopened by the small, unopened envelope icon on the left side of the e-mail. A read e-mail bears an opened envelope icon, a sent e-mail has a check mark as its icon, and a document icon represents a draft e-mail.

3. **After you finish reading the message, press the Escape key to return to the message list.**

**Figure 2-3:**
You've got
333 e-mails!

### Sorting the message list

Your BlackBerry mail lists messages in order by the date and time they were received, but you can sort by different criteria. For example, to see only incoming e-mail, press Alt+I. (For more predefined hot keys, see the upcoming section "Reusing saved searches.")

Sorting and searching are closely related on your BlackBerry. In a sense, searching is really sorting your e-mail based on your search criteria. You can search your e-mail by the name of the sender or by keywords. Or you could run a search as broad as looking through all the e-mail that has been sent to you. See the later section "Searching through Messages Like a Pro" for more on searching and sorting.

### Saving a message to the saved folder

You can save any important e-mail in a folder so that you can find it without sorting through tons of e-mail. To do so, simply scroll to the e-mail you want to save, press the Menu key, and select Save from the menu. A pop-up message confirms that your e-mail has been saved. Your saved e-mail remains in the message list.

To retrieve or view a saved e-mail, follow these steps:

*1.* **Open the Messages application.**

*2.* **In the message list, press the Menu key and select View Saved Messages.**

You see the list of messages you saved.

*3.* **Select the message you want and then press the trackball or trackpad to open it.**

### Viewing attachments

Your BlackBerry is so versatile that you can view most e-mail attachments just like you can on a desktop PC. And we're talking sizable attachments, too, such as JPEGs (photos), Word docs, PowerPoint slides, and Excel spreadsheets. Table 2-1 shows a list of supported attachments viewable from your BlackBerry.

If you're using BlackBerry Desktop Redirector to get your e-mail onto your BlackBerry, you won't get attachments on your BlackBerry.

| Table 2-1 | BlackBerry-Supported Attachments |
|---|---|
| **Supported Attachment Extension** | **Description** |
| .bmp | BMP image file format |
| .doc, .docx | MS Word document |
| .dot | MS Word document template |
| .gif | GIF image file format |
| .htm | HTML Web page |
| .html | HTML Web page |
| .jpg | JPEG image file format |
| .pdf | Adobe PDF document |
| .png | PNG image file format |
| .ppt, .pptx | MS PowerPoint document |
| .tif | TIFF image file format |
| .txt | Text file |
| .wpd | Corel WordPerfect document |
| .xls, .xlsx | MS Excel document |
| .zip | Compressed file format |

**Book III
Chapter 2**

**Using E-Mail**

To tell whether an e-mail has an attachment, look for the standard paper-clip icon next to your e-mail in the message list.

You retrieve all the different types of attachments the same way. This makes retrieving attachments an easy task. To open an attachment, follow along:

1. **While reading an e-mail, press the Menu key and select Open Attachment.**

   You see a screen that contains the name of the file, a Table of Contents option, and a Full Contents option. For Word documents, you can see different headings in outline form in the Table of Contents option. For picture files, such as JPEGs, you can go straight to the Full Contents option to see the graphic.

   For all supported file types, you see Table of Contents and Full Contents as options. Depending on the file type, use your judgment on when you should use the Table of Contents option.

2. **Scroll to Full Contents, press the Menu key, and select Retrieve.**

   Your BlackBerry attempts to contact the BlackBerry Internet Service client to retrieve your attachment. This retrieves only part of your attachment. As you peruse a document, your BlackBerry retrieves more as you scroll through the attachment.

### Editing attachments

Your BlackBerry comes with Documents To Go, which means that out of the box, you not only can view, but also edit Word and PowerPoint documents. You can even save the documents to your BlackBerry and transfer them later to your PC.

As an example, imagine editing a Word document that you receive as an attachment to an e-mail, following these steps:

1. **Open the e-mail.**

2. **In the message list, open an e-mail with a Word document attached.**

   The e-mail opens for you to read. Notice the little paper clip, indicating that it has an attachment.

3. **Press the Menu key and select Open Attachment.**

   You're prompted with a pop-up that asks whether you want to view the Word document or Edit with Documents To Go.

4. **Select Edit with Documents To Go.**

   Here, you can view and edit a document.

5. **Press the Menu key and select Edit Mode.**

   In Edit mode, you can edit your document.

6. **When you're finished editing and viewing, you can either save the document on your BlackBerry or e-mail it:**

   - *To e-mail the edited document:* Press the Menu key and select Send via E-mail.

     You see an e-mail message with the Word document. Follow the steps described in the next section to send this e-mail attachment as you would any other e-mail.

   - *To save the document:* Press the Menu key and select Save.

     If you want to save the attachment to your BlackBerry, you have to navigate its folder structure. For documents, the default save location usually is the Documents folder.

## Sending e-mail

The first thing you probably want to do when you get your BlackBerry is to write an e-mail to let your friends know that you just got a BlackBerry. Follow these steps to send an e-mail message:

1. **Open the Messages application.**

2. **In the message list, press the Menu key, and select Compose E-mail.**

   You're prompted with a blank e-mail that you need to fill out, just as you would do on your PC.

3. **In the To field, type the recipient's name or e-mail address.**

   As you type, you see a list of contacts from your Contacts that match the name or address you're typing. You can make a selection from this list.

4. **Enter a subject in the Subject field, and type your message in the Body field.**

5. **When you're finished, press the Menu key and select Send.**

   Your message has wings.

### Forwarding e-mail

When you need to share an important e-mail with a colleague or a friend, you can forward that e-mail. Simply do the following:

1. **Open the e-mail.**

   For information on opening e-mail, see "Retrieving e-mail," previously in this chapter.

2. **Press the Menu key and select Forward.**

*3.* **Type the recipient's name or e-mail address in the To field and then add a message if needed.**

When you start typing your recipient's name, a drop-down list of your contacts appears, and you can choose the recipient from it.

*4.* **Press the Menu key and select Send.**

Your message is on its way to your recipient.

### Sending e-mail to more than one person

When you need to send an e-mail to more than one person, just keep adding recipient names as needed. You can also add recipient names to receive a Cc (carbon copy) or Bcc (blind carbon copy). Here's how:

*1.* **Open the e-mail.**

For information on opening e-mail, see the earlier section "Retrieving e-mail."

*2.* **Press the Menu key and select Compose E-mail.**

*3.* **Specify the To field for the e-mail recipient and press the Return key.**

Another To field is added automatically below the first. The Cc field works the same way.

*4.* **To add a Bcc recipient, press the Menu key and select Add Bcc.**

You see a Bcc field. You can specify a Bcc recipient the same way you do To and Cc recipients.

Whether you're composing a new e-mail, replying, or forwarding an e-mail, you add new Cc and Bcc fields the same way.

### Attaching a file to your e-mail

Many people are surprised that you can attach any document on your BlackBerry or on the microSD card. You can attach Word, Excel, and PowerPoint documents as well as pictures, music, and videos. To send an e-mail with a file attached, follow these steps:

*1.* **Open the Messages application.**

*2.* **In the message list, press the Menu key and select Compose E-mail.**

You're prompted with a blank e-mail that you can fill out just as you would on your PC. Enter the recipient's name in the To field and then enter the subject and body of the message.

*3.* **Press the Menu key and select Attach File.**

You're prompted with a list of your folders. Think of these as the folders on your PC.

4. **Navigate to the file of your choice, and press the trackball or trackpad.**

   After you select a file, you see the file in the e-mail message.

5. **When you're finished, press the Menu key and select Send.**

   Your message and attached file wing their way to the recipient.

## Spell-checking your outgoing messages

Whether you're composing an e-mail message or an SMS text message, you can always check your spelling with the built-in spell checker. Simply press the Menu key and then select Check Spelling. When your BlackBerry finds an error, the spell checker makes a suggestion, as shown in Figure 2-4. To skip the spell check for that word and go on to the next word, press the Escape key. If you want to skip spell-checking for an e-mail, simply press and hold the Escape key.

**Figure 2-4:**
The
BlackBerry
spell
checker in
action.

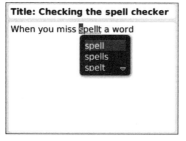

Your BlackBerry, just like Microsoft Word, underlines a misspelled word.

By default, the spell checker doesn't kick in before you send your message, but you can configure it to check spelling before you send an e-mail. Follow these steps:

1. **Open the Messages application.**
2. **Press the Menu key and select Options.**
3. **Select the Spell Check option.**
4. **Select the Spell Check E-mail before Sending check box.**
5. **Press the Menu key and select Save.**

The underline feature is a default setting called Check Spelling as You Type. To turn off this feature, disable the Spell Check option in Message Options.

## Deleting e-mail

If you want to really clean up your old e-mails but don't want to scroll through tons of messages, you can do the following:

1. **Open the Messages application.**

2. **From the message list, highlight a horizontal date mark, press the Menu key, and then choose Delete Prior.**

   The *date mark* is simply a horizontal bar with dates. Just as you can highlight e-mails in the message list, you can highlight the date mark. You are prompted to confirm the deletion.

   Before you take the plunge, remember that going ahead will *delete all the e-mails before the particular date mark.* You cannot retrieve deleted items from your BlackBerry.

3. **Select Delete to confirm your deletion.**

   All your e-mails before the date mark are history.

## Filtering your e-mail

Most of your e-mail messages aren't urgent. Instead of receiving them on your BlackBerry — and wasting both time and effort — filter them out. While in the BIS client, set up filters to make your BlackBerry mailbox receive only those e-mails that you care about. (Don't worry; you'll still receive them on your main computer.)

The following example creates a simple filter that treats work-related messages as urgent and forwards them to your BlackBerry:

1. **From your BlackBerry Home screen, select the Setup folder.**

2. **Select the Email Settings icon.**

   You are prompted with a login screen similar to Figure 2-1, shown earlier in this chapter. If you haven't created your account, click the Create button and create your BIS account.

3. **Log in to the BIS client on the BlackBerry.**

   You see the BIS main screen.

4. **Select the e-mail account for which you want to set up filters.**

5. **Press the Menu key and select Filters.**

   You see a list of filters, if any, and an Add Filter button, as shown in Figure 2-5.

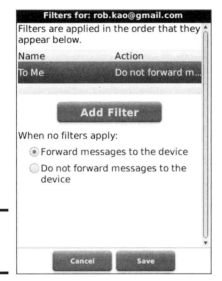

6. **Select the Add Filter button.**

   The Add Filter screen appears, as shown in Figure 2-6.

7. **Enter a filter name.**

   The filter name can be anything you like. In Figure 2-6, we entered **To Me**.

**Figure 2-6:**
Create a
filter for
your e-mail
here.

8. **From the Filter On drop-down list, choose the condition to place on the filter:**

   - *High-Priority Mail Email:* Select this option if the filter applies only to urgent e-mail.

   - *Subject:* When this option is selected, the Contains field is enabled and you can type text in it. Specify what keywords the filter will look for in the subject field, separating each entry with a semicolon (;).

   - *From Address:* When this option is selected, the Contains field is enabled and you can type text in it. Type a full address or part of an address. For example, you can type *rob@robkao.com* or just *kao.* Separate each entry with a semicolon (;).

   - *To Address:* This option is similar to the From Field Contains option.

   - *CC Address:* This option is similar to the From Field Contains option.

   This example selects From Address.

9. **Specify the text in the Contains field.**

   See details in the preceding step for what to enter in the Contains field. This example types the domain of your work e-mail address. For example, if your work e-mail address is `myName@XYZCo.com`, enter **XYZCo.com**.

10. **Select one of the following options for the Action field:**

   **Forward Messages to the Device:** You can select either or both of the following two check boxes:

   - *Header Only:* Choose this if you want only the header of the e-mails that meets the condition(s) you set in Steps 7–9 to be sent to you. (A *header* doesn't contain the message — just who sent it, the subject, and the time it was sent.) Choose this if you get automated alerts, for which receiving only the subject is sufficient.

   - *Level1 Notification:* Level1 notification is another way of saying *urgent e-mail.* A Level1 e-mail is bold in Messages.

   **Do Not Forward Message to the Device:** Any e-mail that meets the conditions you set in Steps 7–9 doesn't go to your BlackBerry.

11. **Confirm your filter by selecting the Save button.**

   You return to the Filter screen, where you can see your newly created filter in the list.

If you have a hard time setting the criteria for a filter, guesstimate and then check it by having a friend send you a test e-mail. If the test e-mail isn't filtered correctly, set the conditions until you get them right.

# Searching through Messages Like a Pro

Searching is a function you probably won't use every day, but when you run a search, you usually need the information fast. Take a few minutes here to familiarize yourself with general searching.

The BlackBerry Messages application provides three ways to search through your messages. Two of the three ways are specific, and one is a broad search:

✦ **By sender or recipient:** Specific. This method assumes that you already know the sender or recipient.

✦ **By subject:** Specific. This approach assumes that you already know the subject.

✦ **General search:** Broad. You don't have a specific assumption.

You can search through anything listed in the messages list. This means that you can search through SMS and voice mail as well as e-mail.

## Searching by sender or recipient

Search by sender or recipient when you're looking for a specific message from a specific person. For example, suppose that your brother constantly sends you e-mail (which means your message list has many entries from him). You're trying to locate a message he sent you approximately two weeks ago regarding a fishing-trip location. You scrolled down the message list, but you just can't seem to find that message. Or maybe you want to find a message you sent to Sue but can't lay your hands on it.

To find a message when you know the sender or recipient, follow these steps:

1. **Open the Messages application.**

2. **In the message list, highlight a message that you sent to or received from that particular person.**

   The choice you get in the next step depends on whether you highlighted a sent message or a received message.

3. **Press the Menu key and then select one of these options:**

   • *To search for a message* from *someone specific:* Because that certain someone sent you the message, choose Search Sender.

   • *To search for a message* to *someone specific:* Because you sent that certain someone the message, choose Search Recipient.

   Your search starts. Any results appear onscreen.

**Book III
Chapter 2**

**Using E-Mail**

## Searching by subject

Search by subject when you're looking for an e-mail titled by a specific subject that you already know. As is the case when running a search by sender or recipient, first scroll to an e-mail that bears the same subject you're searching for. Then follow these steps:

1. **Open the Messages application.**

2. **In the message list, highlight an e-mail titled by the specific subject you're searching for.**

3. **Press the Menu key and select Search Subject.**

   The search starts, and the results appear onscreen.

## Running a general search

A *general search* is a broad search from which you can perform keyword searches of your messages. To run a general search, follow these steps:

1. **Open the Messages application.**

2. **In the message list, press the Menu key and select Search.**

3. **In the Search screen that appears, fill in your search criteria (see Figure 2-7).**

**Figure 2-7:**
The Search screen in Messages.

The search criteria for a general search follow:

- *Name:* This is the name of the sender or recipient to search by.

- *In:* This is related to the Name criterion. Use this drop-down list to indicate where the name may appear, such as in the To or Cc field. From the drop-down list, your choices are From, To, Cc, Bcc, and any address field.

- *Subject:* This is where you type some or all keywords that appear in the subject.

- *Message:* Here, you enter keywords that appear in the message.

- *Service:* If you set up your BlackBerry to receive e-mail from more than one e-mail account, you can specify which e-mail account to search.

- *Folder:* This is the folder in which you want to perform the search. Generally, you should select All Folders.

- *Show:* This drop-down list specifies how the search result will appear — namely, whether you want to see only e-mails that you sent or e-mails that you received. From the drop-down list, your choices are Sent and Received, Received Only, Sent Only, Saved Only, Draft Only, and Unopened Only.

- *Type:* This drop-down list specifies the type of message that you're trying to search for: e-mail, SMS, or voice mail. From the drop-down list, your choices are All, E-mail, E-mail with Attachments, PIN, SMS, Phone, and Voice Mail.

From the Search screen shown in Figure 2-7, you can have multiple search criteria or just a single one. It's up to you.

4. **Press the Menu key and select Search to launch your search.**

   The search results appear onscreen.

You can narrow the search results by performing a second search on the initial results. For example, you can search by sender and then narrow those hits (results) by performing a second search by subject.

You can also search by sender or recipient when you're looking for a specific message from a specific person. Scroll to an e-mail that bears the specific sender or recipient. Press the Menu key and then select Search Sender or Search Recipient. If the e-mail that you highlighted is an incoming e-mail, you'll see Search Sender. If the e-mail is outgoing, you'll see Search Recipient.

## Saving search results

If you find yourself searching with the same criteria over and over, you may want to save the search and then reuse it. Here's how:

1. **Follow Steps 1–3 in the preceding section for an outgoing e-mail search.**

2. **Press the Menu key and select Save.**

   The Save Search screen appears, from which you can name your search and assign it a shortcut key (see Figure 2-8).

3. **In the Title field, enter a name.**

   The title is the name of your search, which appears on the Search Results screen.

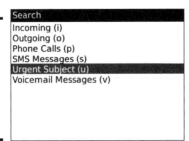

**Figure 2-8:**
Name your search, and assign it a shortcut key.

```
Save Search
Title: Urgent Subject|
Shortcut Key (Alt +):                        u
Name:
   In                     Any Address Field
Subject: Urgent
Message:
   Include Encrypted Messages:         Yes
Service:                       All Services
Folder:                        All Folders
Show:                 Sent and Received
Type:                                   All
```

4. **Scroll to the Shortcut Key field, press the trackball or trackpad, and select a letter from the drop-down list.**

   You can choose among ten letters.

5. **Confirm your saved search by pressing the Menu key and selecting Save.**

## Reusing saved searches

Right out of the box, your BlackBerry comes with five saved search results. Any new saved result will make your search that much more robust.

Follow these steps to see all saved search results:

1. **Open the Messages application.**

2. **In the message list, press the Menu key and select Search.**

3. **Press the Menu key and select Recall.**

   The Recall screen opens, and you can see the five preloaded search shortcuts, as well as any searches you saved, as shown in Figure 2-9.

**Figure 2-9:**
The Recall screen, showing default search hot keys.

```
Search
Incoming (i)
Outgoing (o)
Phone Calls (p)
SMS Messages (s)
Urgent Subject (u)
Voicemail Messages (v)
```

To reuse one of the saved search results, simply select a desired search from the list, press the Menu key, and select Search.

If you have multiple e-mail accounts set up, you can set up a search shortcut so that you view only one specific account. For example, say you have both your personal e-mail and your small-business e-mail accounts set up on your BlackBerry. In the Message application, you see e-mails from both, which can be overwhelming at times. From the general Search screen, set the Service drop-down list to the one you want, and follow the preceding steps to save the search and assign a shortcut key. The next time you want to see only a certain account, you can get to it in an instant!

## Follow Up Your E-Mail

Do you get lots of e-mails and sometimes find that you absentmindedly miss replying to an e-mail? No worries, with your BlackBerry, you can automatically add reminders to any e-mail that you want to follow up. This is similar to the follow-up flag in Microsoft Outlook. So the next time you get a flurry of e-mails in the morning, as you read through them, you can mark the ones you want to revisit later by using the Followup feature.

To add a follow-up flag, follow these steps:

1. **Open the Messages application.**

2. **Highlight the e-mail in need of a follow up, and press the Menu key.**

3. **Select Flag for Follow Up.**

   You see a red flag next to your message.

4. **While the flagged e-mail is still highlighted, press the Menu key again.**

5. **Select Flag Properties.**

   You can select the following:

   • *Request:* This is the type of follow up. You can choose from Call, Review, Forward, and more.

   • *Color:* Yup, you guessed it. This is the color of the flag.

   • *Status:* You can choose whether the status is completed or not.

   • *Due:* This is the due date for this follow up. When the due date arrives, you get a pop-up reminder, similar to a calendar reminder.

## Long Live E-Mail

No closet has unlimited space, and your BlackBerry e-mail storage has limits, too. You've likely pondered how long your e-mails are kept in your BlackBerry. (The default is 30 days. Pshew.) You can choose several options: from 15 days to forever (well, for as long as your BlackBerry has enough space for them).

Because any message you save is kept for as long as you want, a good way to make sure that you don't lose an important message is to save it.

To change how long your e-mails live on your BlackBerry, follow these steps:

1. **Open the Messages application.**

2. **Press the Menu key and select Options.**

3. **Select General Options.**

4. **Scroll to the Keep Messages option and then press the trackball or trackpad.**

5. **From the drop-down list that appears, choose the time frame that you want and then press the trackball or trackpad:**

   - *Forever:* If you choose Forever, you'll seldom need to worry about your e-mails being automatically deleted. On the downside, though, your BlackBerry will eventually run out of memory. At that point, you must manually delete some e-mails to free space to accept new e-mails.

   A good way to archive your e-mail is to back up your e-mails by using BlackBerry Desktop Manager. See Book IV, Chapter 4 for more on backing up your BlackBerry on your PC.

   - *Time option:* If you choose a time option, any message older than that time frame is automatically deleted from your BlackBerry the next time you reboot your BlackBerry. However, the message will be deleted only on your BlackBerry — even if you turn on e-mail reconciliation — because these deletions are only on the device.

6. **Confirm your changes by pressing the Menu key and then selecting Save.**

# Chapter 3: Messaging from Your BlackBerry

## In This Chapter

✔ Sending PIN-to-PIN messages

✔ Using SMS (text messaging)

✔ Using MMS (image and video messaging)

✔ Setting up and using IM applications

✔ Instant messaging other BlackBerry users

✔ Avoid common IM pitfalls

Your BlackBerry is primarily a communication tool, with e-mail messages and phone conversations as the major drivers. Being the social creatures that we are, however, we constantly come up with new ways to communicate — ever more ways to overcome the distance barrier, as it were. You shouldn't be surprised that the folks at RIM have moved beyond e-mail and phoning in their search for other ways to communicate. (And we're not talking semaphore or stringing together two tin cups.)

Not that we're dismissing the power of phones and e-mail. Both are wonderful technologies, but you might find yourself in a situation where other means of communication would be more appropriate. For instance, e-mail isn't the tool of choice for those who prefer instant messaging/chatting — most people would find e-mail slow and cumbersome. Nor is e-mail the best tool to use when you want to alert someone.

"What might be a better fit?" you ask. Read this chapter to find out. You can familiarize yourself with some less obvious ways you can use your BlackBerry to communicate — ways that might serve as the perfect fit for a special situation. You get the inside scoop on PIN-to-PIN messaging and text messaging (also known as short message service, or SMS). We also give you tips on how to turn your BlackBerry into a lean (and not so mean) instant messaging machine.

## Sending and Receiving PIN-to-PIN Messages

What happens when you use PIN-to-PIN messaging? First and foremost, get the acronym out of the way. *PIN* stands for *personal identification number* (familiar to anyone who's ever used an ATM) and refers to a system for uniquely identifying your device.

## A little bit of RIM history

Sometime during the last millennium, Research In Motion (RIM) wasn't even in the phone business. Before BlackBerry became all the rage with smartphone users, RIM was doing a tidy little business with its wireless e-mail. Back then, RIM's primitive wireless e-mail service was served by network service providers on radio bandwidth, namely the DataTAC and Mobitex networks. These were separate from a typical cellphone infrastructure's bandwidth. RIM devices at that time already had PIN-to-PIN messaging. This type of messaging is akin to a pager, where a message doesn't live in a mailbox but is sent directly to the BlackBerry with no delay. (No one wants a paging system that moves at turtle speed when you can get one that moves like a jack rabbit, right?)

Several interesting facts followed from RIM's initial decision. Of note, most cellphone users in New York City were left without service during the 9/11 disaster. As you can imagine, the entire cellphone infrastructure in New York and surrounding areas was overwhelmed when faced with too many people trying to use the bandwidth available. However, one communication device continued to work during that stressful time — RIM's PIN-to-PIN messaging kept the information flow going.

*PIN-to-PIN,* then, is another way of saying *one BlackBerry to another BlackBerry.*

As for the other details, they're straightforward. PIN-to-PIN messaging is based on the technology that underlies two-way pager systems. Unlike a standard e-mail, when you send a PIN-to-PIN message, the message doesn't venture outside RIM's infrastructure in search of an e-mail server and (eventually) an e-mail inbox. Instead, it stays solidly in the RIM world, where it is shunted through the recipient's network provider until it ends up in the recipient's BlackBerry.

Here's the neat part. According to RIM, the message isn't saved anywhere in this universe *except* on the one device that sends the PIN message and the other device that receives it. Compare that with an e-mail, which is saved in at least four separate locations (the mail client and e-mail servers of both sender and recipient), not to mention all the system's redundancies and backups employed by the server. Think of it this way: If you whisper a little secret in someone's ear, only you and that special someone know what was said. In a way, PIN-to-PIN messaging is the same thing, with one BlackBerry whispering to another BlackBerry. Now that's discreet.

If you tend to read the financial newspapers — especially the ones that cover corporate lawsuits extensively — you'll know that there's no such thing as privacy in e-mail. PIN-to-PIN messaging, in theory at least, is as good as the old Cone of Silence. Now, is such privacy really an advantage? You can argue both sides of the issue, depending on what you want to use PIN-to-PIN messaging for. Basically, if you like the idea that your communications can be kept discreet, PIN-to-PIN messaging has great curb appeal. If you don't care about privacy issues, though, you still might be impressed by the zippy nature of PIN-to-PIN messaging. (It really is the Ferrari of wireless communication — way faster than e-mail.)

The Cone of Silence in an enterprise environment has always been a thorny issue for companies that have strict regulatory requirements. As expected, RIM addressed this issue with a new feature in later operating systems, allowing BES administrators to "flip a switch," which forces BlackBerry smartphones to forward all PIN-to-PIN messages to the BES. A company can also install third-party applications on BlackBerry smartphones to report PIN-to-PIN messages.

## Getting a BlackBerry PIN

When you want to call somebody on the telephone, you can't get far without a telephone number. As you might expect, the same principle applies to PIN-to-PIN messaging: No PIN, no PIN-to-PIN messaging. In practical terms, this means that you need the individual PIN of any BlackBerry device owned by whomever you want to send a PIN message to. (You also need to find out your own PIN so that you can hand it out to those folks who want to PIN-to-PIN message you.) The cautious side of you might be thinking "Why on earth would I give my PIN to somebody?" This PIN is really not the same as your password. In fact, this PIN doesn't give anybody access to your BlackBerry or do anything to compromise security. It's simply an identification number; you treat it the same way as you treat your phone number.

RIM makes getting hold of a PIN easy. In fact, RIM even provides you with multiple paths to PIN enlightenment, as the following list makes clear:

✦ **From the Help screen:** You can find the PIN for any device right there on its Help screen. On most BlackBerry models, you can call up the Help screen by pressing Alt+Caps+H. This shortcut to the Help screen is not available on the BlackBerry Pearl Flip or BlackBerry Storm, but you can pull up the Help screen on the Storm by holding the Escape key and touch-pressing the top-left corner, top-right corner, top-left corner again, and then top-right corner again.

✦ **From the Message screen:** RIM also makes it easy for you to send your PIN from the Message screen with the help of a keyword. A *keyword* is a neat feature with which you type a preset word, and your BlackBerry replaces what you type with a bit of information specific to your device.

Sound wacky? It's actually easier than it sounds. To see what we mean, just compose a new message. (Book III, Chapter 2 gives you the basics on the whole e-mail message and messaging thing, if you need a refresher.) In the subject or body of your message, type **mypin** and add

**Book III**
**Chapter 3**

a space. As soon as you type the space, mypin is miraculously transformed into your PIN in the format pin: *your-pin-number*, as shown in Figure 3-1. Isn't that neat?

*mypin* isn't the only keyword that RIM predefines for you. *mynumber* and *myver* give you the phone number and operating system version, respectively, of your BlackBerry.

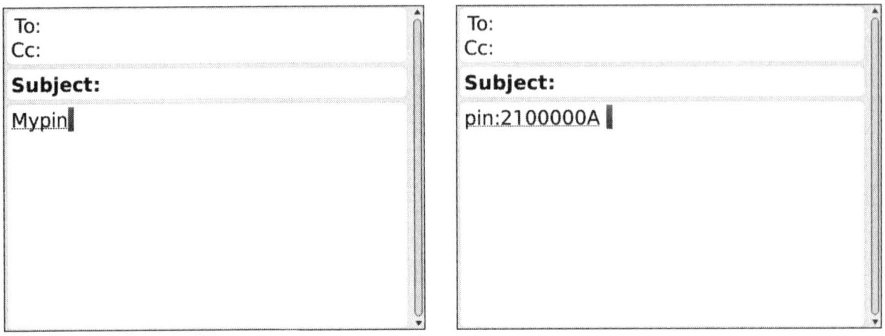

**Figure 3-1:**
Type a
keyword
(left), add a
space, and
the keyword
is translated
(right).

| To: | To: |
| Cc: | Cc: |
| **Subject:** | **Subject:** |
| Mypin | pin:2100000A |

✦ **From the Status screen:** You can also find your PIN on the Status screen. You display the Status screen by starting on your Home screen and choosing Settings⇨Options⇨Status. Figure 3-2 shows a typical Status screen.

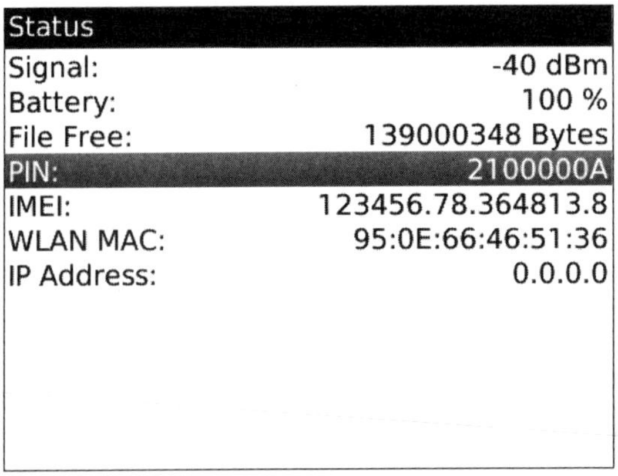

**Figure 3-2:**
Find your
PIN on
the Status
screen.

| Status | |
|---|---|
| Signal: | -40 dBm |
| Battery: | 100 % |
| File Free: | 139000348 Bytes |
| PIN: | 2100000A |
| IMEI: | 123456.78.364813.8 |
| WLAN MAC: | 95:0E:66:46:51:36 |
| IP Address: | 0.0.0.0 |

## Assigning PINs to names

So, you convince your BlackBerry-wielding buddies to go to the trouble of finding out their PINs and passing said PINs to you. Now the trick is finding a convenient place to store those PINs so that you can use them. Luckily for you, you have an obvious choice: the Contacts application. RIM, in its infinite wisdom, makes storing such info as easy as the following steps:

1. **From the BlackBerry Home screen, select Contacts.**

   Contacts opens.

2. **Highlight a contact name, press the Menu key, and then select Edit.**

   The Edit Contact screen for the contact name you selected makes an appearance.

3. **On the Edit Contact screen, scroll down to the PIN field (as shown in Figure 3-3).**

4. **Enter the PIN by pressing letters and numbers on the keyboard.**

5. **Press the Menu key and then select Save.**

   The edit you made for this contact is now saved.

```
Edit Contact                        [123]
 Mobile 2:                           ▲
 Pager:
 Work Fax:
 Home Fax:
 Other:
 PIN: A121212▉
┌─────────────────────────────────────
 Work Address
  Address 1:
  Address 2:
  City:
  State/Prov:
  Zip/Postal Code:                   ▼
```

**Figure 3-3:**
Add a
contact's
PIN info
here.

**Book III**
**Chapter 3**

**Messaging from Your BlackBerry**

It's that simple. Of course, it's even easier if you think ahead and enter the PIN information you have when you set up your initial contact info (by using the New Contact screen), but we understand that a PIN isn't usually the kind of info your casual acquaintances carry around with them.

If all this talk about New Contact screens and Edit Contact screens doesn't sound familiar, check out Book II, Chapter 1, which covers the Contacts application in more detail.

## Sending a PIN-to-PIN message

Sending a PIN-to-PIN message isn't much different from sending an e-mail. Here's how:

1. **From the BlackBerry Home screen, select Contacts.**

   Contacts opens.

2. **Highlight a contact name and then press the Menu key.**

   If a contact has a PIN, you see a menu item titled PIN *contact name*. Say, for example, you have a contact named Dante Sarigumba. When you highlight Dante Sarigumba in the list and then press the Menu key, the menu item PIN Dante Sarigumba appears as an option, as shown in Figure 3-4.

3. **Select PIN *contact name* from the menu.**

   The ever-familiar New Message screen, with the PIN of your buddy already entered as an address, makes an appearance.

4. **Treat the other e-mail-creation stuff — adding a subject line, entering the body of your message, and then sending the message — just as you would with a normal e-mail.**

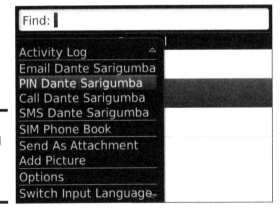

**Figure 3-4:**
Send a PIN message via your Contacts.

Alternatively, if you know the PIN, you can also type it directly. Here's how:

1. **From the BlackBerry Home screen, select Messages.**

   The Messages application opens.

2. **Press the Menu key and then select Compose PIN.**

   The ever-familiar New Message screen makes an appearance.

3. **In the To field, enter the PIN and then scroll down to the subject.**

   You just added a recipient in the To field.

4. **Add a subject line, enter the body of your message, and send the message just like you would a normal e-mail.**

Unlike e-mails, when you send a PIN-to-PIN message, you can tell whether your message reached its destination almost instantly. Viewing the message list, you see a letter D, which means *delivered,* on top of the check mark next to the PIN-to-PIN message you sent.

Because of the nature of PIN-to-PIN messaging (the conspicuous lack of a paper trail, as it were), RIM has set it up so that companies can disable PIN-to-PIN messaging on your BlackBerry device. (No paper trail can mean legal problems down the road — can you say *Sarbanes-Oxley?*) If your BlackBerry is from your employer and you don't see the PIN menu item allowing you to send PIN-to-PIN messages, you can safely assume that your employer has disabled it. Contact your BlackBerry Enterprise Server (BES) administrator to make sure. (See Book VIII, Chapter 1 for an explanation of BES.)

### Receiving a PIN-to-PIN message

Receiving a PIN-to-PIN message is not much different from receiving a standard e-mail. You get the same entry in your message list for the PIN-to-PIN message that you receive, and the same message screen appears when you open the message. By default, PIN-to-PIN messages are considered Level 1 Messages. This means they will show up in red in your message list, which makes it easy to tell them apart from regular e-mail messages. By default, your BlackBerry vibrates to alert you, but you can modify this behavior in Profiles. Because PIN-to-PIN messages are considered Level 1 Messages, you should change that setting to customize how you want to be alerted. (Check out Book I, Chapter 3 for more details on changing your profile.) When you reply to the message, the reply is a PIN-to-PIN message as well — that is, as long as your BlackBerry is set up to send PIN-to-PIN messages. Again, because this involves a BlackBerry-to-BlackBerry communication, you won't see a PIN message show up in your desktop's e-mail client's inbox.

**Book III
Chapter 3**

**Messaging from
Your BlackBerry**

# Keeping in Touch, the SMS/MMS Way

Short message service (also known as *SMS,* or simply *text messaging*) is so popular these days that you see it used in many TV shows, including the Fox Network's *American Idol,* which lets you vote for the show's contestants by using SMS. Moreover, SMS is an established technology (not a new and unproven thing, in other words) that has been popular for years in Europe and Asia, where the Global System for Mobile Communications (GSM) is the technology of choice among cellphone network providers. How short is short? The maximum size per message is about 160 characters. If you send more than that, it's broken down into multiple messages.

Multimedia Messaging Service (MMS) is the latest evolution of SMS. Rather than a simple text message, you can also send someone a contact, a picture, an appointment, an audio clip, or a video clip.

Whether you call it SMS, short Message service, or just plain text messaging, your BlackBerry makes it simple. But before we go over the details, we want to point out to all you BlackBerry Pearl and Pearl Flip owners that text messaging does pose a challenge for beginners. It's not that it's a difficult task; it's just that it takes time to adjust to typing using the limited keypad. See Book I, Chapter 2 if you're not familiar with the Pearl's keyboard.

# SMS is a smash!

In the early days of SMS, phone providers gave away their text messaging services for free. It caught on like wildfire, especially in developing countries where folks are a tad more cost conscious. In those countries, it didn't take long before lots of folks started carrying a cellphone just to avail themselves of the free text messaging. Adding to the appeal was the fact that many network service providers didn't even force you to pay for a plan; you could walk into any convenience store and buy a prepaid SIM card, no questions asked. By the way, a subscriber identity module, or SIM, is a tiny electronic chip found in the back of a GSM phone. This chip is your gateway to using the phone network.

In the United States, however, it took some time for SMS to catch up. For one thing, network service providers in the United States are divided between use of CDMA and GSM/GPRS technology. (See Book I, Chapter 4 for more on that divide.) GSM/GPRS phones have allowed SMS since 1991; CDMA phones did not until about 2000. (Competing innovations do have a downside, right?) The second reason is that most American homes were wired for regular phone service, so cellphones were not necessarily seen as something you just had to have. However, the convenience of using cellphones and SMS is making them more and more attractive, and nowadays almost everyone has one.

Also, you need to be aware of the trends and options for text messaging. There is a growing SMS subculture among teenagers and those who jumped onto the bandwagon early. These in-the-know folks use abbreviations that might be difficult for you to understand in the beginning, so don't dive in without your oxygen tank. A quick preparation goes a long way toward avoiding being labeled *uncool* when it comes to your SMS syntax. The upcoming sections help smooth your path a bit by filling you in on the basics of SMS-speak.

## Using shorthand for speedy replies

From the get-go, text messaging sprang up from the wild and crazy world of cellphone users. And those of you who've used a phone lately probably know that phone design revolves around entering phone numbers rather than entering the text to *War and Peace*. Time for a reality check: On a regular cellphone, three letters share a single key. As you can imagine, trying to bang out even a single paragraph can be a real pain.

Human ingenuity prevails. People have found ways to circumvent the fact that cellphones have such a limited number of keys at their disposal. One strategy that we highlight here involves using abbreviations that allow you to cut down on the amount of text you need to enter. These shorthand words have quickly become quite hip, especially among the 14–18-year-old set. Veteran text messagers (the hip ones, at least) can easily spot people who are new to SMS technology by how they don't use the correct lingo — or use such lingo incorrectly.

Proud owners of a BlackBerry Pearl or Pearl Flip who are beginners to SMS might be tempted to rely on BlackBerry's default SureType feature when they're text messaging. *SureType* is a technology used initially in BlackBerry 7100 series and carried on in the BlackBerry Pearl and Pearl Flip. SureType's purpose is to compensate for the limited keys at your disposal on this type of BlackBerry. The software is smart enough to predict what you want to type. It can also learn and become smarter as you use it. The technology is neat, making it possible for you to type words faster and more easily on the BlackBerry Pearl. However, you might find yourself in a situation in which you receive text from buddies who use cool, with-it, hip, shorthand lingo, and you respond with white-bread, uncool, complete words provided for you by SureType technology. To preserve (or create) your cool image, we recommend disabling SureType when you do any text messaging. (Book I, Chapter 2 fills you in on how to toggle SureType on and off.) That way, you can master shorthand without any interference from SureType and transform yourself into a text-messaging fool.

## *Awhfy?*

In text messaging, the challenge lies in using abbreviations to craft a sentence with as few letters as possible. Because text messaging has been around for a number of years, plenty of folks have risen to this challenge by coming up with a considerable pool of useful abbreviations. Don't feel that you have to rush out and memorize the whole shorthand dictionary at once, though. As with mastering a new language, start out with the most commonly used words, phrases, or sentences. Then when you become familiar with those, slowly gather in more and more terms. In time, the whole shorthand thing will be second nature.

And what are the most commonly used terms? Funny you should ask. Table 3-1 gives you our take on the most common abbreviations, which are enough to get you started. With these under your belt, you can at least follow the most important parts of an SMS conversation.

**Book III
Chapter 3**

**Messaging from
Your BlackBerry**

| Table 3-1 | SMS Shorthand and Its Meanings | | |
|-----------|----------------|-----------|---------|
| *Shorthand* | *Meaning* | *Shorthand* | *Meaning* |
| 2D4 | To die for | CUL8R | See you later |
| 2G4U | Too good for you | CUS | See you soon |
| 2L8 | Too late | F2F | Face to face |
| 4E | Forever | FC | Fingers crossed |
| 4YEO | For your eyes only | FCFS | First come, first served |
| A3 | Anytime, anywhere, anyplace | FOAF | Friend of a friend |

*(continued)*

### Table 3-1 *(continued)*

| Shorthand | Meaning | Shorthand | Meaning |
|---|---|---|---|
| AFAIK | As far as I know | FWIW | For what it's worth |
| ASAP | As soon as possible | GAL | Get a life |
| ASL | Age, sex, location | GG | Good game |
| ATM | At the moment | GR8 | Great |
| ATW | At the weekend | GSOH | Good sense of humor |
| AWHFY | Are we having fun yet? | H2CUS | Hope to see you soon |
| B4 | Before | IC | I see |
| BBFN | Bye-bye for now | IDK | I don't know |
| BBL | Be back later | IMHO | In my honest opinion |
| BBS | Be back soon | IMO | In my opinion |
| BCNU | Be seeing you | IOU | I owe you |
| BG | Big grin | IOW | In other words |
| BION | Believe it or not | KISS | Keep it simple, stupid |
| BOL | Best of luck | LOL | Laughing out loud |
| BOT | Back on topic | OIC | Oh, I see |
| BRB | Be right back | RUOK | Are you okay? |
| BRT | Be right there | W4U | Waiting for you |
| BTW | By the way | W8 | Wait |
| CMON | Come on | WTG | Way to go |
| CU | See you | TOM | Tomorrow |

## Showing some emotion

One aspect of written communication that has gotten a few folks in trouble now and then is that the very same words can mean different things to different people. A simple example is the phrase, "You're such a clueless individual." When you speak such a phrase (with the appropriate facial and hand gestures), your conversational partner knows right off the bat (we hope) that you're teasing and that it's all a bit of fun. Write that same phrase in a text message and, well, you might get a nasty reply in return — which you then have to respond to, which prompts another response, and soon enough you've just ended that seven-year friendship.

Because SMS is meant for quick and short messages akin to chatting — rather than long and drawn-out debates about what you actually *meant* when you said what you said — a quick-and-dirty system for characterizing what you've just written (I'm joking! I'm happy! I'm mad!) has sprung up. Known as *emoticons,* these cutesy typographical devices let you telegraph your meaning in sledgehammer-to-the-forehead fashion.

Yes, we're talking smileys here — those combinations of keyboard characters that when artfully combined, resemble a human face. The most popular example — one that you've probably encountered in e-mails from especially chirpy individuals — is the happy face, which you see used (usually at the end of a statement) to convey good intentions or imply a happy context, like this :).

Table 3-2 shows you the range of smiley choices. The trick to recognizing what each smiley conveys lies in your ability to view them sideways (hopefully without developing a crick in your neck). Just remember that smileys are supposed to be fun. They could be the one thing you need to make sure that your "gently teasing remark" isn't misconstrued as a hateful comment.

| **Table 3-2** | **Smileys and Their Meanings** | | |
|---|---|---|---|
| *Smiley* | *Meaning* | *Smiley* | *Meaning* |
| :) | Happy, smiling | :( | Sad, frown |
| :-) | Happy, smiling, with nose | :-( | Sad, frown, with nose |
| :D | Laughing | :-< | Super sad |
| :-D | Laughing, with nose | :'-( | Crying |
| :'-) | Tears due to laughter | :-0 | Yell, gasped |
| :-)8 | Smiling with bow tie | :-@ | Scream, what? |
| ;) | Winking | :-(o) | Shouting |
| ;-) | Winking, with nose | \|-0 | Yawn |
| 0:-) | I'm an angel (male) | :----( | Liar, long nose |
| 0*-) | I'm an angel (female) | %-( | Confused |
| 8-) | Cool, with sunglasses | :-\| | Determined |
| :-! | Foot in mouth | :-() | Talking |
| >-) | Evil grin | :-ozz | Bored |
| :-x | Kiss on the lips | @@ | Eyes |
| (((H))) | Hugs | %-) | Cross-eyed |
| @>--;-- | Rose | \|@@\| | Face |
| :b | Tongue out | #:-) | Hair is a mess |
| ;b | Tongue out with a wink | &:-) | Hair is curly |
| :-& | Tongue tied | $-) | Yuppie |
| -!- | Sleepy | :-($) | Put your money where your mouth is |
| <3 | For heart or love | <(^(oo)^)> | Pig |

## Sending a text message

After you have the shorthand stuff as well as the smileys under your belt, get your fingers pumped up and ready for action: It's message-sending time! Whether it's SMS or a richer (MMS) message, here's how it's done:

1. **From the BlackBerry Home screen, select Contacts.**

   Contacts opens.

2. **Highlight a contact for whom you have a cellphone number, press the Menu key, and select SMS (or MMS)** *contact name* **from the menu that appears.**

   The menu item for SMS or MMS is intelligent enough to display the name of the contact. For example, if you choose John Doe, the menu item reads SMS John Doe or MMS John Doe, as shown in Figure 3-5.

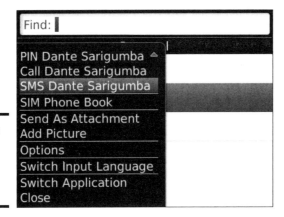

**Figure 3-5:**
Start
your text
message
here.

3. **If you chose MMS, press the Menu key and select the type of media file you would like to attach.**

   The File Explorer screen opens. Browse from your multimedia folders and select the file you want to send. See Book V, Chapter 1 if you're not sure how to use the BlackBerry File Explorer.

   When choosing MMS, this extra step allows you to choose the multimedia file. This is the only difference between SMS and MMS with regard to sending a message.

4. **Type your message.**

   Remember that shorthand business? You should start taking advantage of it the first chance you get. (Practice makes perfect.)

5. **Press the Menu key and select Send from the menu that appears.**

   Your message is sent on its merry way.

### Viewing a message you receive

If you have an incoming SMS or MMS message, you get a notification just like you do when you receive an e-mail. Also, like e-mail, the e-mail icon at the top of the Home screen indicates a new message. In fact, everything about viewing SMS and MMS messages is pretty much the same as reading an e-mail. So, if you have Book III, Chapter 2 loaded into your memory, you know how to read SMS messages. The basic run-through is as follows:

1. **Open Messages.**

2. **Scroll to the unread message and select it.**

   The message appears onscreen.

By default, SMS and MMS messages are displayed in the same list as your messages. This can make it difficult to distinguish them from regular e-mails. Luckily, RIM makes it possible for your text messages to stand out from the crowd by including a way for you to place them in their own separate list. Just follow these steps:

1. **From the BlackBerry Home screen, select Messages.**

2. **Press the Menu key and select Options.**

3. **Select General Options.**

4. **Scroll down to SMS and Email inboxes.**

5. **Press the trackball, trackpad, or touch screen and choose Separate.**

6. **Press the Menu key and select Save.**

   That's it! You see a new icon on your BlackBerry Home screen named SMS and MMS, and the message indicator for text messages changes as well. Whenever a new text message arrives, you can find it with this icon instead of in your message list.

You can customize how your BlackBerry notifies you when you receive an SMS message. Book I, Chapter 3 has the scoop on all the customization options for your BlackBerry, including options for SMS notification. (Look for the section about customizing your profile.)

## Always Online Using Instant Messaging

Real-time conversation with your friends or buddies over the Internet is easier with the advent of instant messaging (IM). This technology allows two or more people to send and receive messages quickly with the use of software that uses the Internet as the wire. It all started with pure text messages and evolved into a rich medium involving voice and even video conversation in real time.

**Book III
Chapter 3**

**Messaging from
Your BlackBerry**

## Messaging etiquette and a few words of caution

Here are some common-sense messaging rules as well as a few words of caution. Even if you're new to messaging, being a neophyte doesn't give you free license to act like a jerk. So remember, play nice and take the following pointers to heart:

- **Use smileys to avoid misunderstandings.** Read more about emoticons and smileys earlier in this chapter.

- **Do not ever forward chain letters.** We mean it. Never.

- **If you need to forward a message, check the entire message content first.** Make sure that nothing offends the recipient.

- **Some things in this world need to be said face to face, so don't even think of using messaging for it.** Ever try dumping your girl-friend or boyfriend over the phone? Guess what? Using messaging is a far worse idea.

- **Keep your tone gender neutral.** Some messages that are forwarded through e-mails are inappropriate to the opposite sex.

- **Capital letters are like shouting, so DON'T USE THEM.**

- **Know your recipient.** A newbie may not easily grasp smileys and shorthand at first, so act accordingly. (Read more about shorthand earlier in this chapter.)

- **Don't reply to any message when you're angry.** You can't unsend a sent message. It's better to be prudent than sorry.

- **Don't spread gossip or make personal remarks about other people.** Beware! Your messages can end up in the wrong hands and haunt you in the future.

- **Easy does it.** There is no documented evidence of the deleterious effects (physical or psychological) of too much text messaging. However, don't text-message as if you wanted to enter the books as the first recorded case of Instantmessagingitis. As your great-grandma would tell you, too much of anything is bad for you. It's easy to lose track of time doing IM.

- **Drive safely.** Tuck away your BlackBerry whenever you're in the driver's seat.

You may not find an instant messenger on your BlackBerry. That's because it's a service provider's prerogative to include IM on the BlackBerry it sells. (Most providers, however, do support it for most BlackBerry models.) Also, your company has the option of not including or allowing instant messaging. Here's the deal: You can add instant messengers to your BlackBerry even if they didn't come as a default. Now, for those who already have instant messengers installed, you're in luck. You can go on your merry IM way. However, for those of you lacking an IM application, don't despair. See the upcoming section, "Instant messaging on your BlackBerry," which shows you where to download instant messengers so you can install them on your BlackBerry.

## Chatting using IM rules

When America Online (AOL) came out with Instant Messenger (IM) in the mid-1990s, it was an overnight hit. What made it successful was that it could provide quick (instantaneous) responses to any messages you sent. In addition, the service introduced many simple (yet clever) functions that offer you a new way of communicating. For example, you can chat with multiple people at the same time. You can tell whether someone is trying to type a message to you. You can even tell whether your buddies are online, away from their computers, or simply too busy to be interrupted at the moment. IM adds a totally different slant on long-distance communication, opening a wide array of possibilities — possibilities that can be used for good (team collaboration) or ill (mindless gossip), depending on the situation.

As you might expect, IM is great for both personal and business applications. Whether you're maintaining friendships or busily working to create new ones, Instant Messenger is definitely one powerful tool to consider adding to your social-skills toolbox.

## Instant messaging on your BlackBerry

Most network providers include the most popular IM clients on their BlackBerry smartphones. The following clients should be preinstalled on your BlackBerry in the IM folder:

+ Google Talk

+ Yahoo! Messenger

+ Windows Live Messenger (from the giant Microsoft)

+ AOL Instant Messenger

+ ICQ Instant Messenger

There are several other popular IM services, like iChat and Facebook Chat, which RIM doesn't include a client for. If you want to access these on your BlackBerry, you'll need to install a third-party chat client like IM+. Check out Book VI, Chapter 5 for more details on all of these popular IM clients and services.

If you want to use an IM network that isn't already loaded, you can always check RIM's Web site to download the application: Go to www.mobile. blackberry.com, select Instant Messaging and click the list of IM application you want to download.

### IM basics: What you need

Assuming that you have the IM application available on your BlackBerry, you just need these two things to start using one of the standard four IM programs:

✦ User ID

✦ Password

Getting a user ID/password combo is a breeze. Just go to the appropriate registration Web page (listed next) for the IM application(s) you want to use. Use your desktop or laptop PC for signing up. It's easier and faster that way.

✦ **Google Talk:** www.google.com/accounts/NewAccount?service= talk

✦ **Yahoo! Messenger:** http://edit.yahoo.com/config/eval_ register?.src=pg&.done=http://messenger.yahoo.com

✦ **Windows Live Messenger:** http://download.live.com/messenger

✦ **AOL Instant Messenger (AIM):** https://reg.my.screenname.aol. com/_cqr/registration/initRegistration.psp

✦ **ICQ:** www.icq.com/register

Given the many IM network choices available, your friends are probably signed up on a bunch of different networks. You may end up having to sign up for multiple networks if you want to reach them all by using IM.

Several "all-in-one" instant messaging applications tie all of your instant messaging accounts into one easy-to-use interface. See Book VI, Chapter 5 for more information on instant messaging applications.

### Going online with IM

After you obtain the user ID/password combo for one (or more) of the IM services, you can use your BlackBerry to start chatting with your buddies by following these steps:

*1.* **From the BlackBerry Home screen, go to the IM Folder and select the IM application of your choice.**

   To illustrate how to do this, we use Google Talk. An application-specific logon screen shows up for you to sign on, similar to the one shown in Figure 3-6. It's straightforward, with the standard screen name or ID line and password line.

*2.* **Enter your screen name/ID and password.**

*3.* **(Optional) If you want, select the Remember Password check box. Also if you want, select the Automatically Sign Me In check box.**

**Google Talk**

ta k

Username:

|

Password:

**Sign In**

☐ Remember password
☐ Automatically sign me in

Need an account?
Go to google.com/accounts on your computer. ▾

**Figure 3-6:**
Logon
screen for
Google Talk.

When the Remember Password check box is enabled, it allows you to keep the ID/password information preentered the next time you come back to this screen. (Um, that is, you don't have to type this stuff every time you want to IM.) We recommend that you select this check box to save time, but you should also set your handheld password to Enabled so that security is not compromised. Refer to Book I, Chapter 3 if you need a refresher on how to enable passwords on your BlackBerry.

The Automatically Sign Me In check box allows you to toggle the feature of signing in automatically when your BlackBerry is powered up. This is helpful if you have a habit of turning off your BlackBerry periodically.

4. **Press the Menu Key and select Sign In.**

   At this point, IM tries to log you on. This can take a few seconds, during which time the screen reads Sending Request to Google, or whichever service you're using. After you're logged on, a simple listing of your contacts, or buddies, appears on the screen.

5. **Highlight the person you'd like to chat with, and then press the Menu key.**

   A menu appears, listing various things you can do. Features could differ a little bit for each IM application, but for Google Talk, here's a sample of what you can do:

   - *Start Chat:* Opens a chat with the selected contact.

   - *Send File:* Sends a file to the selected contact.

   - *Add a Friend:* Lets you search for friends who are using the IM service and send an invitation to add you to their buddy list.

   - *Rename:* Allows you to rename a contact in your buddy list.

   - *Remove:* Removes a contact from your buddy list.

   - *Block:* Prevents a contact on your buddy list from chatting with you.

6. **Select the action you'd like.**

## Adding a contact/buddy

Before you can start chatting with your buddies, you need to know their user IDs as well. Consult the following table for information on how to obtain user IDs in some of the most popular IM applications:

| Provider | Where You Get Someone's User ID |
|---|---|
| AIM | From your friend or by searching AOL's directory |
| ICQ | Your friend's e-mail or the ICQ Global Directory |
| Google Talk | The text before the @ sign in his or her Google e-mail address |
| Windows Live | E-mail address or Windows Live ID |
| Yahoo! | The text before the @ sign in his or her Yahoo! e-mail address |

Luckily for you, you don't need to search around for IDs every time you want to IM someone. You can store IDs as part of a contact list. Follow these steps:

1. **Starting within the IM application of your choice, press the Menu key.**

2. **Select Add a Friend, as shown in Figure 3-7.**

   Depending on the IM client, this may show slightly different wording, such as Add a Contact or Add a Buddy.

   The Add a Friend screen appears.

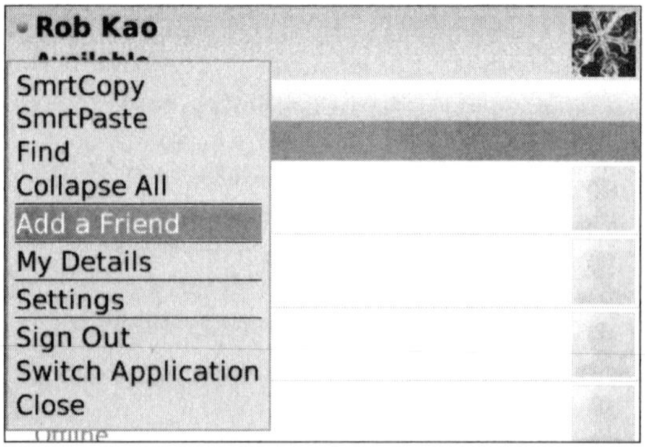

**Figure 3-7:**
The Google
Talk menu.

Rob Kao

SmrtCopy
SmrtPaste
Find
Collapse All
Add a Friend
My Details
Settings
Sign Out
Switch Application
Close

3. **Enter the user ID of your contact on the Add a Friend screen.**

*4.* **Press the trackball, trackpad, or touch screen.**

IM is smart enough to figure out whether this contact is a valid user ID. If the ID is valid, the application adds the ID to your list of contacts. The buddy goes either to the Online or Offline section of your list, depending on whether your buddy is logged on. You'll be warned if the ID you entered is not valid.

## Doing the chat thing

Suppose you want to start a conversation with one of your contacts (a safe assumption on our part, we think). By sending a message within the IM application, you're initiating a conversation. Here are the details on how to do it:

*1.* **Log on to the IM application of your choice.**

*2.* **Select the person you want to contact.**

An online chat screen shows up. The top portion lists old messages sent to and received from this contact. You type your message at the bottom part of the screen.

*3.* **Type your message.**

*4.* **Press the Enter key.**

Your user ID and the message you just sent show up in the topmost (history) section of the chat screen. When you get a message, it's added to the history section so that both sides of your conversation stay in view.

## Sending your smile

You can quickly add emoticons to your message (without having to remember all the character equivalents in Table 3-1). Follow these steps:

*1.* **While you're typing your message, press the Menu key.**

A menu appears.

*2.* **Select Show Symbols.**

The step is the same no matter which IM client you are using.

All the icons appear, as shown in Figure 3-8.

*3.* **Select the emoticon you want.**

The emoticon is added to your message.

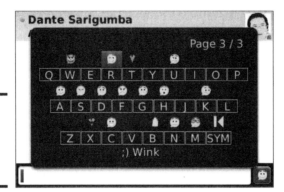

**Figure 3-8:**
You can
choose
from many
smileys.

## Taking control of your IM app

If you use IM frequently — and you tend to chat with many contacts all at
the same time — your BlackBerry's physical limitations may cramp your
IM style. No matter whether you use AIM, Yahoo! Messenger, ICQ, Windows
Live Messenger, or BlackBerry Messenger, it's still slower to type words on
the tiny keypad than it is to type on your PC.

Do you just give up on the dream of IMing on-the-go? Not necessarily. The
following sections show how you can power up your BlackBerry IM
technique.

### Limiting the number of people you instant message

If you can't keep up with all your buddies, your best bet is to limit your expo-
sure. Take a whack at your IM contact list so that only your true buddies
remain as contacts whom you want to IM from your BlackBerry. Trimming
your list is easy — to delete a contact from your IM application, press the
Menu key and use the Delete option from the menu.

Deleting a contact or buddy from an IM application on your BlackBerry also
deletes it from the desktop or laptop computer version of the app. That's
because the list of contacts is maintained at a central location — an IM
server, to be precise — and not on your BlackBerry.

A simple work-around here is to set up two accounts of your favorite IM
application — one for your BlackBerry and one for your desktop. By using
these accounts separately, you can limit the number of contacts you have on
your BlackBerry and still maintain a full-blown list of contacts on your
desktop.

### Giving your hands a break

Cut down your typing time. Don't forget the shorthand described previously. It's widely used in IM as well as texting, so refer to Table 3-1 whenever you can so that you can quickly respond. Before you know it, you'll have the abbreviations memorized and be using them with ease. The emoticons also can make your conversation more interesting. Always take them out of your toolbox. Refer to Table 3-2 for the list of the most common ones.

### Avoiding costly mistakes

SMS messages are short messages designed for cellphones. IM is a step up, evolving from the Internet, where bandwidth is no longer a concern. It provides a better real-time conversation experience across distances. These two technologies evolved in parallel. As more people use IM, it becomes apparent that this technology has a place in handheld devices, where mobility is an advantage. Some of the IM programs developed and used in the BlackBerry in the past use SMS behind the scenes. And because your BlackBerry can connect to the Internet, other programs use the Internet directly. These differences can affect your monthly bill as well as your messaging experience. Read on.

*If you don't have unlimited SMS but have an unlimited data plan,* be careful with any third-party IM software. Make sure that when you use it, you're using the Internet instead of SMS. If it requires you to use SMS, you'll incur charges for every message sent and received. As of this writing, most network providers charge 20 cents for every SMS message. If you're a heavy IM user, 20 cents adds up quickly and will be a nasty surprise on your monthly bill.

If you want to make sure that you won't have text-messaging fees for using IM, check out Book VI, Chapter 5 for more information on instant messaging clients that use your BlackBerry data connection.

**Book III
Chapter 3**

**Messaging from
Your BlackBerry**

# Chapter 4: Using BlackBerry Messenger

*I*n Book III, Chapter 3, you find a slew of ways to send messages on your BlackBerry. In this chapter, you get the scoop on another way to send messages, using a special application known and loved by BlackBerry.

RIM has entered the IM (instant message) horse race in the form of a spirited filly named (you guessed it) BlackBerry Messenger. This application is based on the PIN-to-PIN messaging technology (refer to Book III, Chapter 3), which means that it is mucho fast and quite reliable.

However, with BlackBerry Messenger, you can chat with only those buddies who have a BlackBerry and also have PIN-to-PIN messaging enabled. The application supports IM features common to many other applications, such as group chatting and the capability to monitor the availability of other IM buddies.

## Accessing BlackBerry Messenger

You access BlackBerry Messenger from the Home screen, as shown in Figure 4-1. If you don't see it on the Home screen, check the IM folder because some BlackBerry models place it there. The first time you run BlackBerry Messenger, a welcome screen asks you to enter your display name. This display name is the one you want other people see on their BlackBerry Messenger when you send them a message.

The next time you open BlackBerry Messenger, you see a contacts list, as shown on the left in Figure 4-2. (Okay, the picture here displays some contacts, but your list should be empty; we show you how to populate the list in a minute.)

**Figure 4-1:**
Launch
BlackBerry
Messenger
here.

**Figure 4-2:**
The
BlackBerry
Messenger
contacts list
(left) and
menu (right).

Pressing the Menu key lets you do the following, as shown on the right side
of Figure 4-2:

+ **Broadcast Message:** Send a message to multiple contacts in your
  BlackBerry Messenger. The messages appear as conversations in the
  recipients' BlackBerry Messenger.

+ **Start Chat:** Initiate a conversation with the currently highlighted
  contact.

✦ **Invite to Conference:** Initiate a group conversation. See the later section "Starting a group conversation" for details.

✦ **Forward to Messenger Contact:** Send the currently highlighted contact information to your other BlackBerry Messenger contacts.

✦ **Invite Contact:** Add a new contact to BlackBerry Messenger. See the next section.

✦ **Add Category:** Create custom groupings within your BlackBerry Messenger.

✦ **Contact Profile:** Display a screen showing the information of the currently highlighted contact.

✦ **Delete Contact:** Delete the currently highlighted contact.

✦ **Move Contact:** Delete the currently highlighted contact.

The menu has more items. Although not shown on the right of Figure 4-2, scroll down the menu to see the following:

✦ **My Profile:** Customize your personal information and control how you want others to see you from their BlackBerry Messenger Contacts list. You can set the following (see Figure 4-3):

  • Change your picture. Simply select the default image, navigate to your picture, scroll or slide the picture to center your face in the square, press the Menu key and select Crop and Save.

  • Change your display name.

  • Allow others to see the title of the song you're currently listening to.

  • Allow others to see that you're currently using the phone.

  • Enter a personal message that others could see.

  • Set your time zone.

  • Allow others to see your location and time zone information.

  • Display your barcode.

✦ **Options:** Customize the behavior of your BlackBerry Messenger.

✦ **Back Up Contacts List:** Save your list of BlackBerry Messenger contacts to the file system. The location defaults to the media card, but the screen that follows after you select this option allows you to select a different folder and enter a different name for the backup file.

✦ **Restore Contacts List:** Restore your list of BlackBerry Messenger contacts from the file you created from the Back Up Contacts List.

✦ **Delete Backup Files:** Delete a backup file. If you have multiple backups, this option allows you to choose which file to delete.

✦ **Add New Group:** Create custom groupings for your contacts.

This option is helpful if you have a lot of contacts in BlackBerry Messenger. Select this menu item and an Add Group screen appears so you can enter a group name.

✦ **Scan a Group Barcode:** Allows you to add members to the currently opened group by scanning your friend's BlackBerry bar code.

**Figure 4-3:**
Set your personal information (left) and the rest of the My Profile screen (right).

## Adding a Contact

With no one in your contacts list, BlackBerry Messenger is a pretty useless item. Your first order of business is to add a contact to your list — someone you know who:

✦ Has a BlackBerry

✦ Is entered in your Contacts

✦ Has PIN-to-PIN messaging enabled

✦ Has a copy of BlackBerry Messenger installed on his or her device

If you know someone who fits these criteria, you can add that person to your list by doing the following:

*1.* **In BlackBerry Messenger, press the Menu key.**

*2.* **Select Add a Contact.**

The Invite Contact screen appears, listing actions related to adding a contact, as shown on the left side of Figure 4-4.

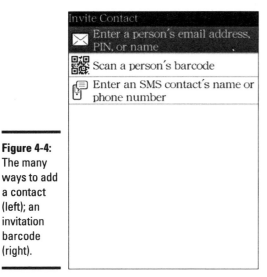

**Figure 4-4:**
The many
ways to add
a contact
(left); an
invitation
barcode
(right).

3. **To scan your friend's BlackBerry barcode, using your friend's BlackBerry, select the Scan a Person's barcode option.**

   A barcode image similar to Figure 4-4 appears on your friend's BlackBerry. A Camera application appears for you to capture the barcode. Once captured, the contact information is immediately added to your BlackBerry Messenger contacts list.

4. **If instead you want to enter the contact's e-mail address or BlackBerry PIN:**

   a. *Select Enter a person's email address, PIN, or name.*

      The Invite Contact screen appears.

   b. *Start typing the name of the contact, and when a list of the possible contacts appears, select the name you want to add.*

      BlackBerry sends the request to the potential contact with the message shown in Figure 4-5. You can edit this message.

   c. *Type your message.*

   d. *Select OK, and then select OK again in the screen that follows.*

      The application sends your request. As long as the person hasn't responded to your request, his or her name appears as part of the Pending group, as shown in Figure 4-6. When your contact responds positively to your request, the name goes to the official contacts list.

**Book III
Chapter 4**

**Using BlackBerry
Messenger**

**Figure 4-5:**
Potential
contacts
are asked
before being
added.

**Figure 4-6:**
To-be-
approved
contacts
are in the
Pending
group.

# Having Conversations

You come to this point because what you really want to do is chat with your friends with BlackBerry Messenger. The following sections give you a quick rundown on starting individual or group conversations. And we throw in some steps on sharing files and saving your conversation history.

## Starting a conversation

You can easily start a conversation with any of your contacts. Follow these steps:

*1.* **On the BlackBerry Messenger main menu, select the name in your contacts list.**

A traditional chat interface opens, with a list of old messages at the top and a text box for typing messages at the bottom.

2. **Type your message.**

3. **Press the Enter key.**

   Any messages you send (as well as any responses you get) are appended to the history list at the top.

## *Starting a group conversation*

You can also invite others to your BlackBerry Messenger conversation. Follow these steps:

1. **During a conversation, press the Menu key.**

   The BlackBerry Messenger main menu appears. This time, an Invite option has been added.

2. **Select Invite to Group Chat.**

   The Select Contacts screen opens, listing your BlackBerry Messenger contacts who are currently available (see Figure 4-7).

3. **Invite people to the chat by selecting the corresponding check boxes.**

   You can choose any number of people.

4. **Select OK.**

   You're back to the preceding conversation screen, but this time the history list shows the contacts you added to the conversation. The newly selected contact(s) can now join the conversation.

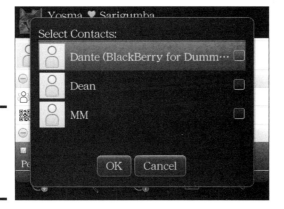

**Figure 4-7:**
See your
available
contacts
here.

**Book III**
**Chapter 4**

Using BlackBerry
Messenger

TIP

You can set a subject on your message. This is especially useful for group conversations:

1. **Press the Menu key while you're in the conversation screen, and select Set Subject.**

   On the screen that follows, the cursor is in the subject line, waiting for you to enter a subject (see Figure 4-8)

2. **On the screen that follows, enter the subject, and select OK.**

   The conversation screen is updated with the subject.

**Figure 4-8:** Add a subject to your conversation here.

You can make your name appear snazzy by adding symbols, such as DanteJ and YosmaC. (See Figure 4-9.)

1. **On the BlackBerry Messenger screen, press the Menu key.**

2. **Select Edit My Info.**

3. **Press the Menu key and select Add Smiley to choose the symbol you want.**

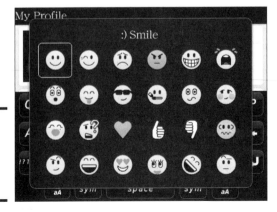

**Figure 4-9:** Add symbols to your name here.

## Sending a file

BlackBerry Messenger, like any other instant messaging application, can send files. While on a conversation, press the Menu key and you see several menu items allowing you to share a file:

✦ **Send Picture:** Displays a screen similar to Figure 4-10, left, where you can either launch the Camera application to take a picture to send or select the picture file you want to send (you can navigate to a specific folder if necessary). The default location for picture files is the Media Card pictures folder. Just select Camera to launch the Camera application.

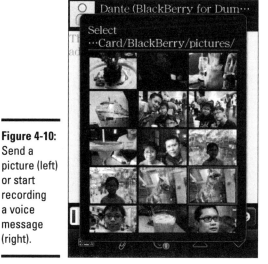

**Figure 4-10:** Send a picture (left) or start recording a voice message (right).

✦ **Send Voice Note:** Displays the Voice Note screen (refer to Figure 4-10, right), where you can record the voice message you want to send. When you're ready to record, follow these steps:

 a. *Select Start.*

 The BlackBerry is ready to record your voice message. A screen indicating recording appears with a Stop button, as shown to the left in Figure 4-11.

 b. *When you have finished speaking your message, select Stop.*

 A screen similar to the right of Figure 4-11 appears. You can play the message to review what you said, send the message if you're satisfied, or cancel sending a voice note.

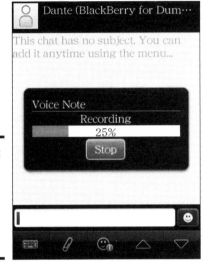

**Figure 4-11:** Record (left) and send (right) your voice message here.

 c. *Select Send.*

  You see a prompt asking you to add a description of your voice message. Selecting No sends the message right away.

 d. *Select Yes.*

  An Add File Description screen appears up.

 e. *Enter a description and then select OK.*

  A request to transfer the file is sent and your friend needs to accept it in his or her BlackBerry Messenger for the transmission to begin.

✦ **Send File:** Displays a screen that lists which file types you can send. Aside from giving you a different way of sending pictures and voice messages, the screen also lists the following types:

 • *File:* Sends any type of file. It shows an Explorer-type screen so you can navigate to the file you want to send. The default folders shown are Media Card and Device Memory.

 • *BlackBerry Contact:* Sends a vCard. When selected, it displays Contacts, allowing you to select the contact you want to send.

 • *Messenger Contact:* Lets you choose from your list of BlackBerry Messenger contacts and send the info as a file.

## Saving the conversation history

While you are in the conversation screen, you can save your chat history in two ways. Both methods are accessible by a simple press of the Menu key:

✦ **Copy History:** Copies the existing chat history to the Clipboard. Then you just need to paste it to the application where you want it saved, such as Calendar or MemoPad.

✦ **Email History:** Displays the Compose Message screen, shown in Figure 4-12, with the Subject field prepopulated with `Chat with <Contact> on <Date>` and the body of the message prepopulated with the chat history.

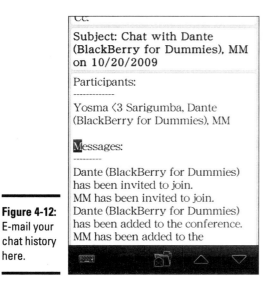

**Figure 4-12:** E-mail your chat history here.

# Broadcasting a Message

Do you feel the need to start a conversation on the same subject with several people? You can start with a group chat, but what if you want to get a personal opinion from each individual, something that they're not comfortable saying in front of the crowd? The best way to do this is to broadcast a message to multiple recipients, as follows:

*1.* **From the BlackBerry Messenger screen, press the Menu key and select Broadcast Message.**

The Broadcast Message screen appears, as shown in Figure 4-13, allowing you to enter your message and select the recipients.

**Figure 4-13:**
Broadcast
a message
here.

2. **Enter your message.**

3. **Select the recipients.**

4. **Select Send.**

# Book IV

# Desktop Operations

# Contents at a Glance

# Chapter 1: Introducing BlackBerry Desktop Manager

## In This Chapter

✔ Choosing the correct version of BlackBerry Desktop Manager

✔ Downloading and installing BlackBerry Desktop Manager

✔ Running Desktop Manager for the first time

The centerpiece of your desktop activities on the BlackBerry is *BlackBerry Desktop Manager,* which is a suite of programs that include the following:

✦ **Application Loader:** Installs BlackBerry applications and updates the BlackBerry handheld software. (In Book IV, Chapter 6, we tell you how to use Application Loader.)

✦ **Backup and Restore:** Backs up your BlackBerry data and settings. (Check out Book IV, Chapter 4 for details.)

✦ **Synchronize:** Synchronizes BlackBerry data to your PC (the topic of this chapter and discussed further in Book IV, Chapter 2).

✦ **Media Manager:** Uploads media files to your BlackBerry from your PC and vice versa. (See Book V, Chapter 1 for instructions on using this program.)

✦ **Device Switch Wizard:** Helps you transfer data from your existing mobile device to your BlackBerry. (See Book IV, Chapter 5 for details.)

This software allows you to synchronize your *personal information manager,* or PIM, data (see Book IV, Chapter 2) and upload or download media files between your PC and BlackBerry. (See Book V Chapter 1.) It also helps you upgrade your BlackBerry handheld software (see Book IV, Chapter 6) and back up or restore your device data. (See Book IV, Chapter 4.)

In this chapter, we introduce BlackBerry Desktop Manager. We show you how to choose, download, and install the correct version.

## Downloading BlackBerry Desktop Manager

Although the software is on the CD that comes with your BlackBerry smartphone, you are usually better off downloading the latest version from BlackBerry.

If you're a corporate user, check with your BlackBerry system administrator before downloading Desktop Manager. Your company may have restrictions in place that prevent you from downloading it, or may have an approved version that you should download.

Follow these steps to download the most up-to-date version of BlackBerry Desktop Manager:

*1.* **On your computer, go to www.blackberry.com.**

The BlackBerry Web page opens.

*2.* **Select your country from the drop-down list, and then click Go.**

If you would like BlackBerry.com to remember your country choice, select the Remember My Selection check box before clicking Go.

*3.* **Click the Software link near the top of the screen.**

*4.* **Click the BlackBerry Desktop Software link on the left side of the screen.**

*5.* **Click the Download for PC or Download for Mac button, depending on which you have.**

*6.* **Click the Download button.**

*7.* **Click Save to save the download to your desktop.**

You can also choose to open the file instead, but we recommend that you save the installation file to your desktop in case you need it again.

The BlackBerry Desktop Manager application for a PC is 260MB in size. That is a very large file. Depending on your Internet connection, it can take from 5 minutes to over an hour to download. For a Mac, the file is only 27MB and takes several minutes.

## Installing BlackBerry Desktop Manager and Desktop Redirector

After you finish downloading the setup file, it's time to install Desktop Manager. Double-click the file you saved to your desktop. When you start the setup, the installation wizard automatically runs.

*1.* **Choose your language.**

*2.* **Click Next on the welcome screen to choose your country.**

*3.* **Choose the I Accept the Terms in the License Agreement option and click Next.**

4. **On the Destination Folder, you may have two items that you can change. After making your selection, click Next.**

   - *Installation Folder:* You have the ability to change where the files will install. We recommend that you don't change this folder, as all of the troubleshooting documents refer to this location if you do have a problem.

   - *Install This Application For:* Depending on your computer's operating system, you may see the options to install the application for anyone or only for me. If you are the only person that uses the computer, it is safe to choose either one. If multiple people share a computer and log in with different usernames, they may not need to use this BlackBerry Desktop software — so you can choose Only for Me.

5. **The Setup Type screen allows you to either choose a default or custom setup. Choose one of the following and click Next.**

   - *Default:* This will have all the features except the Certificate Synchronization tool, which may be used by a corporation to sign and cipher their messages. We recommend that you select this option.

   - *Custom:* Choosing this option allows you install the Certificate Synchronization tool and not install the BlackBerry Automatic Update feature. You will also have a chance to change the default location as you did on the Destination Folder screen.

6. **You can choose to install these two add-on applications to your BlackBerry Desktop Software in the Media Options screen:**

   - *BlackBerry Media Sync:* If you have iTunes or Windows Media installed, this will allow you to synchronize music, including playlists, to your BlackBerry smartphone.

   - *Roxio Media Manager:* This allows you to manage your music and all of your other multimedia files, including pictures and videos. To find out how to manage your files with Roxio Media Manager, visit Book V Chapter 1.

   Integration Options allows you to choose which mail accounts you want to set up your BlackBerry with.

7. **Choose the type of account you'd like to integrate with and click Next.**

   - *Integrate with a personal account.* These types of accounts include Yahoo! Mail, Gmail.com, or some other account that your company hasn't set up for you.

   - *Integrate with a work email account.* Choose this option only if you intend to set up your BlackBerry in a work environment that has a BlackBerry Enterprise Server. In most cases, your IT department will assist in setting up your corporate BlackBerry.

**Book IV
Chapter 1**

**Introducing
BlackBerry Desktop
Manager**

When you (or your IT representative) select this option, it also adds a component called Desktop Redirector. (After it's installed, you can find Desktop Redirector by choosing Start➪Programs➪Desktop Manager➪BlackBerry and clicking Desktop Redirector.)

If you have a BlackBerry that is not connected to a BlackBerry Enterprise Server, but you still want to receive your corporate e-mail that you receive in Outlook or Lotus Notes on your desktop, you can. When Desktop Redirector is open and the BlackBerry is connected to the computer, a message that you receive on your desktop is also sent to your BlackBerry. To do this, make sure that you select the Redirect Messages Using the BlackBerry Desktop Redirector radio button on the installation screen, as shown in Figure 1-1.

For devices on a BlackBerry Enterprise Server, Desktop Redirector is not needed because all messages synchronize over the air.

**8. Choose your Installation Options and click Next.**

- *Start BlackBerry Desktop Manager Automatically Each Time the Computer Starts:* You don't have to worry about clicking the BlackBerry Desktop icon to start the program every time you turn on your computer. When you select this option, every time you connect your BlackBerry to the computer, Desktop Manager opens and begins synchronizing your information.

- *Create a shortcut for BlackBerry Desktop Manager on the Windows desktop:* From your desktop, you can always choose Start➪Programs➪BlackBerry and click BlackBerry Desktop Manager to open the application. However, if you find yourself doing this a lot, you may want to just have an icon on your Desktop for easy access.

- *Check for software updates:* We recommend selecting this option because it looks for updates for the Desktop Manager for bug fixes and also updates your device handheld code.

**Figure 1-1:**
Configure the BDM installation to include Desktop Redirector.

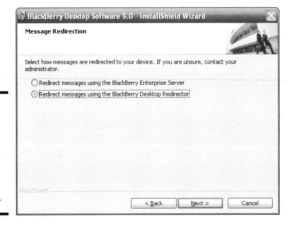

In most Windows installations, you find the shortcut to launch Desktop Manager through your computer's Start menu. Follow these steps to launch it:

1. **Connect your BlackBerry to your computer using the USB cable that came with your device.**

2. **On your computer, choose Start➪All Programs➪BlackBerry➪Desktop Manager.**

If your BlackBerry has a microSD card in it when you connect it to your computer, your BlackBerry screen displays a prompt for enabling mass storage mode. If you answer Yes to the prompt and type your BlackBerry password when prompted, your BlackBerry behaves like a flash drive. A drive letter is added to the drive list in Windows Explorer, allowing you to treat the microSD card as a normal flash drive.

Desktop Manager launches, with its opening screen; see Figure 1-2.

If you didn't install a shortcut on your desktop (PC) or your Dock (Mac) when going though the setup, it's not a problem because it's easy to add one. Follow these instructions:

✦ **PC:** Choose Start➪Programs➪BlackBerry, and then right-click BlackBerry Desktop Manager and choose Send to a Select Desktop.

✦ **Mac:** To add an icon to your Dock, go to the Applications icon on your dock and find BlackBerry Manager. Click and drag it to your Dock and release.

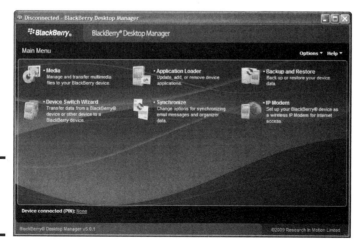

**Figure 1-2:**
BlackBerry Desktop Manager.

**Book IV Chapter 1**

Introducing BlackBerry Desktop Manager

Your Desktop Manager installation may vary depending on your wireless carrier and options that your BlackBerry administrator sets (if you are on a BES). You should see at least the following four icons, or applications:

+ Application Loader (See Book IV, Chapter 6.)

+ Backup and Restore (See Book IV, Chapter 4.)

+ Media (See Book V, Chapter 1.)

+ Synchronize (See Book IV, Chapter 2.)

If you're running Desktop Manager for the first time, the program does these things:

+ **Tries to make the initial configuration on your machine.** If your BlackBerry will be activated on a BES but hasn't been activated yet, this includes security encryption setup and asks you to randomly move your mouse to generate security encryption keys.

+ **Checks what applications you have on your device and what required applications need to be installed.** If it can't find a required application on your device, it prompts you to install it. Of course, you have the option to cancel and install it later.

+ **Looks at the settings you have for your Synchronize software.** If auto-synchronization is turned on, BDM attempts to run synchronization for your PIM. We discuss this further Book IV, Chapter 2.

## Connecting BlackBerry Desktop Manager to Your BlackBerry

You establish a connection between your BlackBerry and Desktop Manager through the USB cable that comes with your device. Plug your device into your desktop. After Desktop Manager is running, it tries to find a BlackBerry on the type of connection specified. The default connection is USB, so you shouldn't need to configure anything.

If you have more than one BlackBerry, distinguish between your smart-phones in Desktop Manager with your BlackBerry PINs. Book IV, Chapter 5 shows you how to do this.

Follow these steps to connect your BlackBerry to Desktop Manager:

1. **Connect your device to your computer using the USB cable.**

2. **Open Desktop Manager.**

   When Desktop Manager is running, it tries to find a BlackBerry on a USB connection.

3. **If your device has a password, you're prompted for the password.**

4. **Enter the password.**

   You see Connected as the screen heading. If for some reason, you see Disconnected and no password prompt, one of the following is happening:

   • Desktop Manager can't find the device being connected via the USB cable.

   • The connection setting isn't set to use USB.

5. **Choose Options⇨Connection Options at the right side of the screen.**

   The screen shown in Figure 1-3 appears. Make sure that the connection setting uses USB.

6. **In the Connection Type drop-down list, select the USB connection with your BlackBerry PIN.**

**Figure 1-3:**
Possible
connection
types
to your
BlackBerry.

# Chapter 2: Synchronizing Data through Desktop Manager

## In This Chapter

✔ **Configuring synchronization**

✔ **Mapping fields**

✔ **Confirming changes and resolving conflicts**

*W*hat better way to keep your BlackBerry updated than to synchronize it with your desktop application's data?

Arguably, most of the data you need to synchronize is from your *personal information manager (PIM)* applications: memos, calendar, contacts, and tasks. The Synchronize option within BlackBerry Desktop Manager allows you to synchronize your PIM data between your PC and BlackBerry.

In this chapter, we introduce the Synchronize application. We show you how to manually and automatically synchronize your BlackBerry with your desktop computer. We also offer tips about which options you might want to use.

If you're using a corporate BlackBerry that's connected to a BlackBerry Enterprise Server (BES), skip this chapter. BlackBerry smartphones running under BES synchronize over the air (OTA) or wirelessly.

## Setting Up Synchronize

Synchronize is the program in BlackBerry Desktop Manager that allows you to synchronize your data between your desktop computer and your BlackBerry. You'll find the Synchronize icon on the BlackBerry Desktop Manager screen. To launch Synchronize, simply click the Synchronize icon. A screen like the one shown in Figure 2-1 appears.

The Synchronize screen is divided into two sections, which you can navigate through by using the links on the left:

♦ **Synchronize:** The default view. It allows you to manually trigger synchronization. (Refer to Figure 2-1.) See the "Using on-demand synchronization" and "Automatic synchronization" sections, later in this chapter, for more details on when you use this screen.

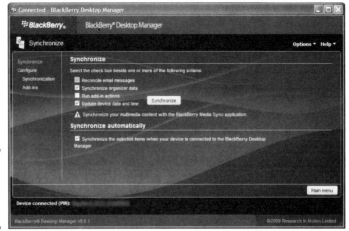

**Figure 2-1:**
The
Synchronize
screen.

✦ **Configuration:** The place where you can set up configuration and rules for reconciling data. Under the Configuration link are two subsections, Synchronization and Add-Ins, which further organize the interface. (See Figure 2-2.) The first thing you need to work with is the Synchronization Configuration screen. The following section helps you do that. You'll probably never need the Add-Ins screen, and we don't cover it in this book.

**Figure 2-2:**
The
Synchro-
nization
Config-
uration
screen.

## *Configuring PIM synchronization*

If you use any of the following popular PIM applications, you can sync the info contained in them to your BlackBerry: ACT!, ASCII Text File Converter, Lotus Notes, Lotus Organizer, Microsoft Outlook, Microsoft Outlook Express, and Microsoft Schedule.

Clicking the Synchronization button in the Synchronization Configuration section displays the screen shown in Figure 2-3. You can see that the names listed correspond to the BlackBerry applications, except for Contacts, which goes by the name Address Book. This is the entry point of the entire synchronization configuration for PIM applications. Selecting the application on this screen allows you to pair a PIM handheld application (such as Mail) to a desktop application (such as Outlook or Lotus Notes).

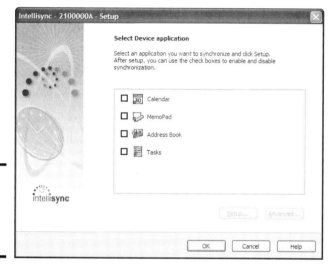

**Figure 2-3:**
The PIM config-
uration
screen.

Here are the four types of application data that can be synchronized to your Blackberry:

✦ **Calendar:** Synchronizes your appointments and events.

✦ **MemoPad:** Synchronizes any notes or text.

✦ **Address Book:** Synchronizes any contact information.

✦ **Tasks:** Synchronizes your to-do list.

Follow these steps to set up your device's synchronization:

*1.* **Open Desktop Manager and connect your BlackBerry smartphone.**

*2.* **Click the Synchronize icon.**

*3.* **Choose Configuration⟹Synchronization on the left.**

In the Synchronization Configuration section is a Configure Synchronization Settings for My Desktop Program label.

*4.* **Click the Synchronization button.**

The PIM configuration screen is displayed. (Refer to Figure 2-3.)

5. **Select the check box next to an application data type (such as Calendar, MemoPad, Address Book, or Tasks) that you want to synchronize. Click the Setup button.**

   For example, we selected the Calendar application data type. The Setup screen appears, as shown in Figure 2-4.

6. **Select a PIM application to retrieve application data from by highlighting your desired application.**

   Desktop Manager pulls your selected application data from the application selected on this screen.

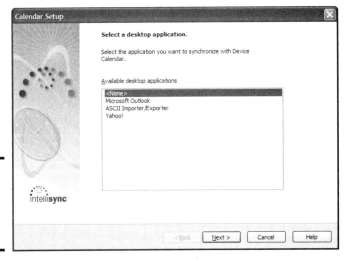

**Figure 2-4:** Choose the desktop application here.

7. **Click the Next button.**

   The Synchronization Options screen appears. (See Figure 2-5.)

8. **On the Synchronization Options screen, select which direction the synchronization will follow.**

   Here are the three available synchronization options:

   • *Two Way Sync:* Allows you to synchronize changes both in your BlackBerry smartphone and in your desktop application.

   • *One Way Sync from Device:* Synchronizes only the changes made on your BlackBerry to your desktop application. Changes to your desktop application aren't reflected in your Blackberry.

   • *One Way Sync to Device:* Synchronizes only changes made in your desktop application to your BlackBerry. Any changes made in your BlackBerry aren't reflected in your desktop application.

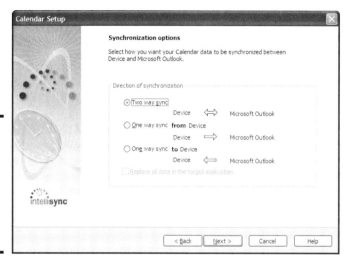

**Figure 2-5:**
Decide
which
direction
synchro-
nization
follows
here.

9. **Click the Next button.**

   The Options screen opens for the PIM application you selected in Step 6.
   Figure 2-6 shows the Microsoft Outlook Options screen.

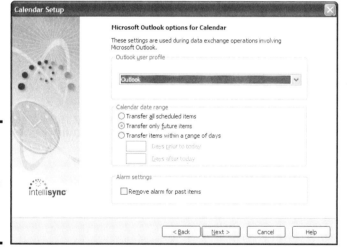

**Figure 2-6:**
Specific
application
settings can
be selected
on this
screen.

10. **Select your profile and choose a date range.**

   Make sure you select the correct user profile in the User Profile drop-
   down list. This is particularly pertinent in cases where you have multiple user profiles in your computer. Choosing the wrong one may result
   in you putting the wrong data into your BlackBerry. If you do choose

the wrong profile, you will need to delete all the information on your BlackBerry before choosing the correct profile. If you do not delete all the information, the wrong data will synchronize back to your computer.

You can also control the amount of data that is reconciled or synchronized. For example, you can specify whether to transfer all Calendar items, transfer just a set of appointments in the future, or transfer items within a range of dates you enter.

Select the Remove Alarm for Past Items check box if you don't want to keep the alarm setting for events that have already occurred.

*11.* **Click the Next button, and then click the Finish button.**

These steps will need to be repeated to set up all PIM items.

## *Mapping fields for synchronization*

Specific bits of information, or attributes, such as names, phone numbers, and addresses are *fields*. All the fields in your BlackBerry applications need to match up to fields in your PIM applications so that the data transfers correctly. For instance, the value of a home phone number field in Contacts needs to be mapped to the corresponding field in Outlook or Lotus Notes. For all four PIM applications, Synchronize is intelligent enough to know which fields in your BlackBerry smartphone correspond to the fields in your PIM application.

However, not all fields on the desktop side exist on the handheld (and vice versa). For example, a Nick Name field doesn't exist in the BlackBerry Contacts but is available in the Exchange (Outlook) Address Book. In some instances, Synchronize provides an alternate field and lets you decide whether to map it.

If you ever find the need to change the default mapping, you can. The interface is the same for all PIM applications. To illustrate how to map and unmap fields, we use Contacts.

Follow these steps to map the fields for Contacts:

*1.* **From Desktop Manager, click the Synchronize link.**

The Synchronize screen appears.

*2.* **Choose Configuration⇨Synchronization on the left side.**

*3.* **Click the Synchronization button next to Configure Synchronization Settings for My Desktop Program.**

The PIM configuration screen appears. (Refer to Figure 2-3.)

*4.* **Select the Address Book check box.**

The Advanced button is enabled.

**5.** **Click the Advanced button.**

The Advanced screen opens, as shown in Figure 2-7.

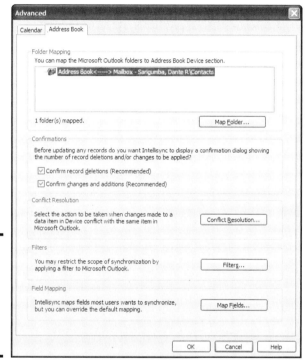

**Figure 2-7:**
The
Advanced
screen for
Address
Book
(Contacts).

**6.** **Click the Map Fields button.**

The Map Fields screen for the Address Book/Contacts application appears; see Figure 2-8. The field names for each application are in right and left columns.

**7.** **Drag the field names in the Device column up and down the column until they're aligned with the field names in the PIM column. To map or unmap, click the arrow icons.**

If you're not careful, you can inadvertently deselect a mapping such as Job Title, and suddenly your titles aren't in sync. Double-check your mapping before you click the OK button. If you think you made a mistake, click the Cancel button to save yourself from having to restore settings. If you do not select a mapping for First Name, Last Name, and Company Name, you will not be able to save your changes, as these are required fields.

**8.** **Click the OK button to save your changes.**

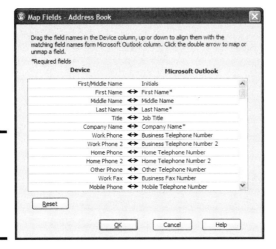

**Figure 2-8:**
The Map
Fields
screen for
Address
Book.

## Confirming record changes

Let's face the facts: Doing a desktop synchronization is not a very interesting task, and few people perform it on a regular basis.

You can tell Synchronize to prompt you for any changes it's trying to make (or perhaps undo) on either side of the wire. The Advanced screen comes in here. Follow these steps:

*1.* **From Desktop Manager, click the Synchronize link.**

The Synchronize screen appears.

*2.* **Choose Configuration➪Synchronization on the left side.**

*3.* **Click the Synchronization button next to Configure Synchronization Settings for My Desktop Program.**

The PIM configuration screen appears; refer to Figure 2-3.

*4.* **Select the PIM application.**

For example, if you want to check the Address Book, select that application from the list.

*5.* **Click the Advanced button.**

The Advanced screen for Address Book screen appears; refer to Figure 2-7.

*6.* **Choose how you want to confirm changes:**

- Confirm Record Deletions (Recommended)
- Confirm Changes and Additions (Recommended)

Regardless of whether you select the first option, Synchronize displays a prompt if it detects that it's about to delete *all* records.

7. **Click OK.**

## Resolving update conflicts

Synchronize needs to know how you want to handle any conflicts between your BlackBerry and your desktop application. A conflict normally happens when the same record is updated in the BlackBerry and also in Outlook. For instance, you change Jane Doe's mobile number in BlackBerry and also change her mobile number in Outlook. Where you resolve these conflicts is the same for all PIM applications. Again, for illustration, we use Address Book as an example:

1. **From Desktop Manager, click the Synchronize link.**

   The Synchronize screen appears.

2. **Choose Configuration⇨Synchronization on the left side.**

3. **Click the Synchronization button next to Configure Synchronization Settings for My Desktop Program.**

   The PIM configuration screen appears; refer to Figure 2-3.

4. **Select the PIM application.**

   If you want Address Book, select that application from the list.

5. **Click the Advanced button.**

   The Advanced screen for Address Book appears; refer to Figure 2-7.

6. **Click the Conflict Resolution button.**

   The Conflict Resolution screen, shown in Figure 2-9, appears.

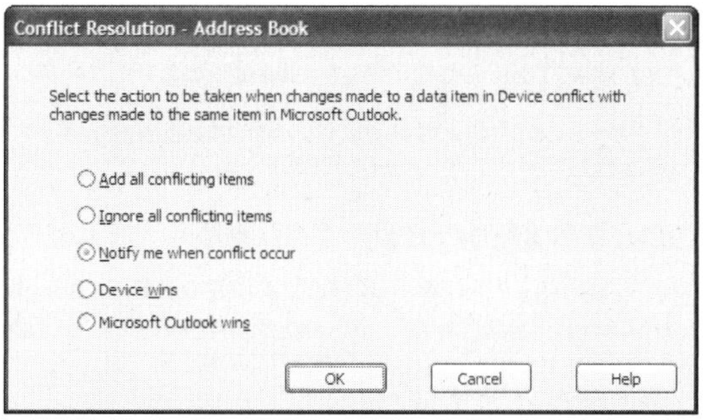

**Figure 2-9:**
Manage
conflicts
here.

> **Conflict Resolution - Address Book**
>
> Select the action to be taken when changes made to a data item in Device conflict with changes made to the same item in Microsoft Outlook.
>
> ○ Add all conflicting items
> ○ Ignore all conflicting items
> ● Notify me when conflict occur
> ○ Device wins
> ○ Microsoft Outlook wins
>
> [ OK ]   [ Cancel ]   [ Help ]

You can tell Synchronize to handle conflicts in a few ways. Here are the options shown in the Conflict Resolution - Address Book dialog box:

- *Add All Conflicting Items:* When a conflict happens, this option tells the application to add a new record to the BlackBerry for the changes on the desktop and add a new record to the desktop for the changes on the BlackBerry.

- *Ignore All Conflicting Items:* Ignores the changes and keeps the data the same on both sides.

- *Notify Me When Conflicts Occur:* This option is the safest. Synchronize displays the details of the conflict and lets you resolve it.

- *Device Wins:* Unless you're sure this is the case, you shouldn't choose this option. It tells Synchronize to disregard the changes in the desktop and use handheld changes every time it encounters a conflict.

- *Microsoft Outlook Wins:* If you're not using MS Outlook, this option is based on your application. This option tells Synchronize to always discard changes on the handheld and use the desktop application change whenever it encounters a conflict. Again, we don't recommend this option because there's no telling on which side you made the good update.

7. **Select the option you want.**

8. **Click OK to save the settings.**

Every time you synchronize and there is a conflict, this screen will appear and you will need to follow these steps again.

# Ready, Set, Synchronize!

Are you ready to synchronize? You can synchronize one of two ways:

+ Manually (by clicking the Synchronize Now icon)

+ Automatically (by choosing How Often on the calendar)

## Using on-demand synchronization

*On-demand synchronization* is a feature that lets you run synchronization manually. Remember that even if you set up automatic synchronization, actual synchronization doesn't happen right away. So, if you made updates to your appointments in Outlook while your BlackBerry was connected to your PC, this feature allows you to be sure that your updates made it to your BlackBerry smartphone before you head out the door.

Without delay, here are the steps:

**1. From Desktop Manager, click the Synchronize link.**

The Synchronize screen appears; refer to Figure 2-1. The following four check boxes let you be selective:

- *Reconcile Messages:* This option synchronizes your e-mails between Outlook and your smartphone.
- *Synchronize Organizer Data:* Includes memos, calendar, contacts, and tasks.
- *Run Add-In Actions:* This option comes into play when you have third-party applications that require data synchronization between your PC and your BlackBerry.
- *Update Device Date and Time:* You need this only if you want both the PC and BlackBerry to have the same time. This ensures that you're reminded of your appointments at the same time for both Outlook and your BlackBerry.

**2. Select the check boxes for the data you want to synchronize.**

**3. Click the Synchronize button.**

Synchronize starts running the synchronization, and you see a progress screen. If you set up prompts for conflicts or changes and Synchronize encounters one, a screen appears so that you can resolve it. When finished, the progress screen disappears and the Synchronize screen reappears.

If you've turned on automatic synchronization (see the next section), the items you select in Step 2 automatically sync every time you connect your BlackBerry to your PC.

**4. Click the Close button.**

## Automatic synchronization

How many times do you think you reconfigure your Synchronize setup? Rarely, right? After you have it configured, that's it. And if you're like us, the reason you open BlackBerry Desktop Manager is because you want to run Synchronize. So, opening Synchronize and clicking the Synchronize button is somewhat annoying.

To make Synchronize run automatically every time you connect your BlackBerry smartphone to your PC, select the last check box on the Synchronize screen (refer to Figure 2-1): Synchronize the Selected Items When Your Device Is Connected to the BlackBerry Desktop Manager.

You might be asking, "What items will autosynchronization sync?" Good question. Synchronize automatically syncs the items you selected in the top portion of the Synchronize screen. (Refer to Figure 2-1.) Note that if you make a change, selecting or deselecting an item on the Synchronize screen, only the selected items will be automatically synced the next time you connect your BlackBerry smartphone to your PC.

# Chapter 3: Synchronizing Data through the Mac

## In This Chapter

✔ **Installing BlackBerry Desktop Manager for Mac**

✔ **Preparing your desktop for PIM synchronization**

✔ **Using manual and automatic synchronization**

*W*hat better way to keep your BlackBerry updated than to synchronize it with your desktop application's data? Arguably, most of the data you need to synchronize is from your personal information manager (PIM) applications: notes, appointments, contacts, and tasks.

If you're a Mac user, good news! The folks at Research In Motion have finally rolled out a Mac version of BlackBerry Desktop Manager (see Book IV, Chapter 1). You are no longer left to use PocketMac, which is way behind on features and capabilities compared to its Windows cousin BlackBerry Desktop Manager. In this chapter, we show you how to use BlackBerry Desktop Manager on your Mac.

If you're using a corporate BlackBerry that's running under BlackBerry Enterprise Server, you can skip this chapter. BlackBerry smartphones running under BlackBerry Enterprise Server synchronize over the air (OTA), via serial bypass, or wirelessly.

## Installing BlackBerry Desktop Manager

The focus of your Mac activities — such as data synchronization and data backup on the BlackBerry — is *BlackBerry Desktop Manager.*

BlackBerry Desktop Manager for Mac has been a highly anticipated software program for a good reason: Folks in the Mac world have been waiting for a replacement of PocketMac, which is limited in functionality. BlackBerry Desktop Manager for Mac is fairly new, and if it doesn't come with your BlackBerry packaging, no worries; you can download an installation program from RIM's Web site.

We assume that you're not holding a CD of BlackBerry Desktop Manager for Mac, so in this section, we include the download steps. Here is a quick run-down on how to go about installing the application:

1. **On your computer, go to www.blackberry.com.**

   The BlackBerry Web page opens.

2. **Select your country from the drop-down list, and then click Go.**

   If you would like BlackBerry.com to remember your country choice, select the Remember My Selection check box before clicking Go.

3. **Click the Software link near the top of the screen.**

4. **Click the BlackBerry Desktop Software link on the left side of the screen.**

5. **Click the Download for Mac button, depending on which you have.**

6. **Click the Download button.**

7. **Click Save to save the download to your computer.**

   The BlackBerry Desktop Manager application for a Mac is 27MB and can take several minutes to download, based on your internet connection.

   After the file downloads, the screen in Figure 3-1 appears.

**Figure 3-1:**
Contents
of the
installation
file.

8. **Double-click BlackBerry Desktop Manager.mpkg.**

9. **Click Continue on the warning screen.**

   Another warning message appears. This time, the message tells you that PocketMac or the Missing Sync software will no longer be able to connect to your BlackBerry. Hey, you're probably installing the BlackBerry Desktop Manager because you're itching to replace PocketMac or Missing Sync, so no worries there.

*10.* **Click Install Anyway.**

An installation welcome screen appears.

*11.* **Click Continue.**

Another prompt appears, this time asking you to agree to the license agreement.

*12.* **Click Agree.**

A screen appears, allowing you to choose the location of the install. The default is your Mac's hard drive.

*13.* **Click Continue.**

You're prompted for your Mac password.

*14.* **Enter your Mac password and click OK.**

Before the installation starts, a screen appears, telling you that when the installation finishes, you need to restart your Mac.

*15.* **Click Continue Installation.**

At this point, the installation kicks in. This may take a minute or two. When the installation is complete, you see a message to restart the Mac.

*16.* **Click Restart.**

# Opening BlackBerry Desktop Manager

BlackBerry Desktop Manager doesn't automaticaly appear on your Mac's Dock (the bottom bar with application icons in Mac OS 10.5 or later) or on your Mac's desktop after your installation. You need to use the Finder app to locate it.

If you have Mac OS 10.5 or later, follow these instructions to find the BlackBerry Desktop Manager installation:

*1.* **Click the Finder application (the leftmost icon) on your Mac's Dock.**

The Finder is launched with a screen displaying your Mac file system. The top-right corner of the Finder screen has an empty text box, which you can use for entering search criteria.

*2.* **In the search text box (top-right corner) of the Finder screen, type** BlackBerry Desktop.

The listing on the screen is the result of your search, and BlackBerry Desktop Manager should be at the top of the list.

If you have Mac OS prior to 10.5, follow these instructions to find the BlackBerry Desktop Manager installation:

1. **Press ⌘+F to launch the Finder utility.**

   The Finder utility of the Finder application appears.

2. **In the Find text box of the Finder screen, type** BlackBerry Desktop.

   The listing on the screen is the result of your search, and BlackBerry Desktop Manager should be at the top of the list.

When you've found BlackBerry Desktop Manager, do the following to open it:

1. **Connect your BlackBerry to your Mac using the USB cable.**

2. **Click BlackBerry Desktop Manager in the Finder screen.**

   The screen in Figure 3-2 appears, displaying the BlackBerry connected to your Mac. In your case, it displays your BlackBerry's PIN.

**Figure 3-2:** BlackBerry Desktop Manager for Mac main screen.

To add an icon to your Dock, go to the Applications icon and find **BlackBerry Manager**. Click and drag it to your Dock and release. It's now on your dock so that you can access it quickly.

# Setting Synchronization Options

You need to set synchronization options probably only once, so you want to make sure you get them right. After all, the data between your Mac and your BlackBerry smartphone should be synced the way you want it.

## Device Options

At the bottom center of the BlackBerry Desktop Manager screen (refer to Figure 3-2 in the preceding section) is a Device Options button. Click this button, and you get the screen shown in Figure 3-3.

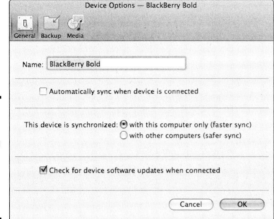

**Figure 3-3:** Decide whether you want your BlackBerry to sync only on this Mac.

An important option you need to set is This Device Is Synchronized. If you sync your BlackBerry with other machines or even if you have Google Sync for Calendars and Contacts, we advise that you choose With Other Computers (Safer Sync). This ensures that the automatic sync option is disabled. One side effect of automatic sync is creating duplicate contacts on your BlackBerry and your other desktop machines. However, if you sync BlackBerry only with this Mac, select With This Computer Only (Faster Sync).

The listing on the left (under the Summary heading) contains links for navigating to the option screens. We describe each option in the following sections.

**Book IV Chapter 3**

Synchronizing Data through the Mac

# Calendar

Clicking the Calendar link on the BlackBerry Desktop Manager screen displays the screen shown in Figure 3-4. You can configure how your appointments are synced as follows:

✦ **Sync Calendar:** Include Calendar in the sync by selecting Two Way, or skip it by choosing Do Not Sync.

A quick way to know that you set Calendar for Two Way sync is seeing two circular arrow icons next to the Calendar link on the left side of the BlackBerry Desktop Manager screen. This is true for Contacts, Notes, Tasks, and Music as well.

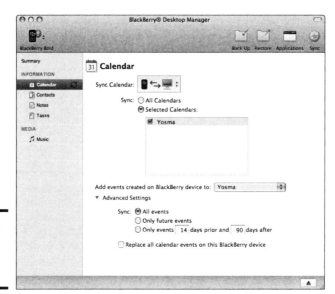

**Figure 3-4:**
Configure
Calendar
sync here.

✦ **Sync:** This section applies only if you have multiple calendar applications on your Mac. You can include all calendars or choose one from the list. Your BlackBerry handles the display of multiple calendar applications. The BlackBerry Calendar app uses different colors to indicate which calendar an appointment belongs to.

✦ **Add Events Created on BlackBerry Device To:** The default Calendar on your BlackBerry does not tie directly to any Mac applications. Appointments you create in your BlackBerry by default will not be synced to any Mac applications. Setting this option tells BlackBerry Desktop Manager to sync those appointments or events to a particular Mac application.

✦ **(Advanced Settings) Sync:** This setting allows you to limit the amount of appointments or events to sync on your BlackBerry. After all, past events just occupy valuable space on your smartphone with no purpose but a record. Here, you can control which ones your BlackBerry carries. Choose All Events (the default), Only Future Events, or Only Events [n] Days Prior and [n] Days After. The last option allows you to have a range of dates relative to the current day. The default is 14 days in the past and 90 days after.

✦ **(Advanced Settings) Replace All Calendar Events on This BlackBerry Device:** Keep this deselected unless you want a fresh start and want to copy to your BlackBerry appointments or events from your Mac.

## Contacts

The Contacts link on the BlackBerry Desktop Manager screen displays the screen shown in Figure 3-5. Here, you can do the following:

✦ **Sync Contacts:** Choose to include Contacts in the sync by selecting Two Way. Otherwise, select Do Not Sync.

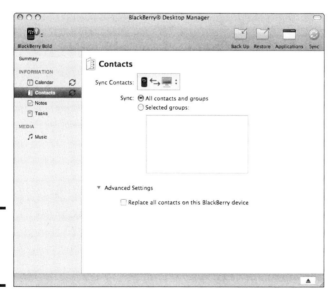

**Figure 3-5:** Configure Contacts sync here.

Book IV Chapter 3

Synchronizing Data through the Mac

✦ **Sync:** Choose here to include all contacts and groups on syncing or click Selected Groups to sync only the groups you want.

✦ **(Advanced Settings) Replace All Contacts on This BlackBerry Device:** Keep this deselected unless you want a fresh start and want to copy all contacts from your Mac to your smartphone.

## Notes

The Notes link on the BlackBerry Desktop Manager screen displays the screen shown in Figure 3-6. This screen is where you configure how you want Notes or MemoPad items on your BlackBerry synced.

+ **Sync Notes:** Lets you configure whether or not you want Notes included in the sync. Choose Two Way or Do Not Sync (the default).

+ **Sync Account:** If you have multiple note-keeping programs in your Mac, this setting allows you to choose which one you want to tie the sync into. The default is Apple Mail Notes.

+ **(Advanced Settings) Replace All Notes on This BlackBerry Device:** If you select this option, the Mac side becomes your master. Keep this deselected unless you want a fresh start and want to copy to your BlackBerry all notes from your Mac.

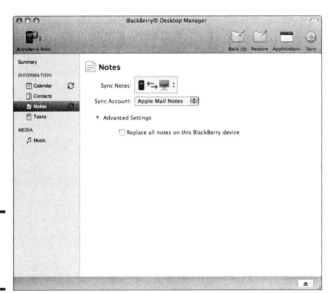

**Figure 3-6:**
Configure
Notes sync
here.

## Tasks

Clicking the Tasks link on the BlackBerry Desktop Manager screen displays the screen shown in Figure 3-7. Note that the screen is similar to Calendar because most of the Tasks items have associated dates and are essentially tied to your Calendar. You can configure how your tasks are synced as follows:

+ **Sync Tasks:** Choose to include Tasks in the sync (Two Way) or not (Do Not Sync).

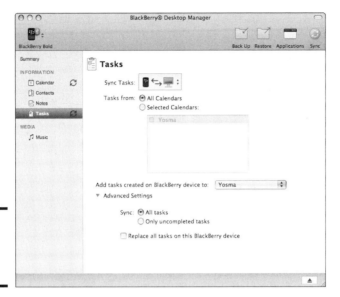

**Figure 3-7:**
Configure
the Tasks
sync here.

+ **Tasks From:** This section applies only if you have multiple calendar applications on your Mac. You can include all calendars or choose one of those listed. Your BlackBerry handles the display of multiple calendar applications. The BlackBerry Calendar app uses different colors to indicate which calendar an appointment belongs to.

+ **Add Tasks Created on BlackBerry Device To:** The default Calendar on your BlackBerry does not tie into any Mac application. Tasks you created in your BlackBerry will not be synced to any of your Mac applications. Setting this option tells BlackBerry Desktop Manager to sync those tasks to a particular Mac application.

+ **(Advanced Settings) Sync:** This setting allows you to limit the tasks to sync on your BlackBerry so that tasks that are completed don't simply occupy valuable space on your smartphone with no real purpose. Choose either All tasks (the default) or Only Uncompleted Tasks.

+ **(Advanced Settings) Replace All Tasks on This BlackBerry Device:** Keep this deselected unless you want a fresh start, deleting current calendar items in your smartphone and copying all appointments or events from your Mac.

## Music

You can easily sync your iTunes playlists to the BlackBerry by using the Music link on the BlackBerry Desktop Manager screen. And this is the place to configure how you want it synced, as shown in Figure 3-8.

**Book IV**
**Chapter 3**

**Synchronizing Data**
**through the Mac**

**Figure 3-8:**
Configure
Music sync
here.

Here, you can choose the following:

✦ **Sync Music:** Choose to include Music in the sync or not. Just below this check box, you have options to sync All Songs and Playlists (the default) or Selected Playlists. If you have a big music library in iTunes, we recommend that you choose Selected Playlists and choose only the music you want to carry with you, up to the capacity of your media card.

✦ **Add Random Music to Free Space:** If you want BlackBerry Desktop Manager to sync random songs from iTunes that are not included in your playlist, select this check box. After the sync, you can find the songs in the Random Music playlist in the Music application of your BlackBerry.

✦ **Memory Settings:** Displays the location where music is stored in your BlackBerry, which is the media card.

Some useful information on this screen is related to the memory space of your smartphone so that you have some idea how much memory your playlist occupies and how much free space remains on your device.

# Deleting All Music Files on Your BlackBerry

You might wonder why we bother to include a section on deleting music files. What does deleting music files have to do with data synchronization? With BlackBerry Desktop Manager, you have the option to sync your BlackBerry with iTunes playlists, including album art. If this is the first time you've installed and used BlackBerry Desktop Manager on your Mac, it's likely that you have the same music files in iTunes that are already in your BlackBerry. Any existing files in your BlackBerry will not be recognized by the BlackBerry Desktop Manager, and if you decide not to delete these existing files, you'll end up having duplicates after the sync. Therefore, it makes sense to start fresh, and clear your BlackBerry of whatever music files it has.

Establishing sync from iTunes to BlackBerry is a one-way direction only, from iTunes to the BlackBerry. This means that if you decide to sync your music to iTunes, iTunes will be the master location of your music files. This also means that any music you add to your BlackBerry (outside of syncing from iTunes) will not be recognized by your BlackBerry Desktop Manager and will not be copied to iTunes.

Be aware that doing this one-time delete will wipe all music files from your BlackBerry media card. This operation is unrecoverable, and if you have any music files that aren't saved anywhere else other than on your BlackBerry, you will lose them.

To do a one-time delete of all music files on your BlackBerry:

1. **Connect your BlackBerry to your Mac.**
2. **On your Mac, click BlackBerry Desktop Manager on the Dock or on the desktop.**

   If you can't find BlackBerry Desktop Manager, see the earlier section, "Opening BlackBerry Desktop Manager."
3. **Click the Device Options button.**
4. **Click the Media icon.**
5. **Click Delete.**

   A confirmation prompt asks whether you really want to delete your music files on the device.
6. **Click OK.**

## Doing a Manual Sync

Ready to sync? You've chosen the sync configuration you want. If you haven't, make sure to read the sections preceding this one.

To do a manual sync, click the Sync button, which is located in the top-right corner of the BlackBerry Desktop Manager screen. If this is your first attempt running the sync, you'll see the prompt shown in Figure 3-9. BlackBerry Desktop Manager needs to establish the latest copy of your data, and to do that, it needs to know how you want to proceed:

✦ **Merge Data:** Click this button if you want BlackBerry Desktop Manager to merge your Mac to your BlackBerry data. Merging is basically combining two sets of data with no duplicate checking. If you have synced your Mac before, using a different type of software such as PocketMac, you will end up with duplicates.

✦ **Replace Device Data:** Click this button if you want to have a fresh copy of Mac data on your smartphone. After the sync, your BlackBerry data will be the same as what you have on your Mac.

If you entered information, say an appointment in your BlackBerry, and have not done the same in your Mac, you will lose that appointment item if you choose Replace Device Data.

**Figure 3-9:**
The options you get when you do a manual sync for the first time.

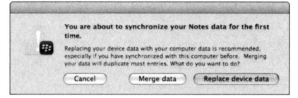

# Configuring an Automatic Sync

It's annoying to click the Sync button every time you want to sync. Simply follow these steps, and every time you connect the BlackBerry to the Mac, the sync will occur automatically:

*1.* **Connect the BlackBerry to the Mac.**

*2.* **On the Mac, click BlackBerry Desktop Manager on the Dock.**

*3.* **Click the Device Options button.**

*4.* **Select Automatically Sync When Device Is Connected.**

*5.* **Click OK.**

# Chapter 4: Protecting Your Information

## In This Chapter

✔ **Performing a full backup of your BlackBerry data**

✔ **Restoring from backups**

✔ **Selecting what data to back up**

✔ **Backing up and restoring wirelessly**

*I*magine that you left your beautiful BlackBerry in the back of a cab. You've lost your BlackBerry for good. Okay, not good. What happens to all your information? How are you going to replace all those contacts? What about security?

One thing that you *don't* need to worry about is information security — *if* you set up a security password on your BlackBerry. With security password protection on your BlackBerry, anyone who finds it has only ten chances to enter the correct password; after those ten chances are up, it's self-destruction time. Although it isn't as smoky as *Mission Impossible,* your BlackBerry does erase all its information, thwarting your would-be data thief.

BlackBerry passwords do not stop a thief from using the device. It merely will stop them from having any of your information. When the thief types the password wrong, on the final attempt the data on the device is wiped. This doesn't include contacts stored on your SIM card. The thief will be prompted to set up a new password and then can get into your BlackBerry and use the phone service, text messages, Web browsing, and set up a personal e-mail account. We recommend that you contact your mobile carrier if you lose your BlackBerry so that they can suspend your service until you get a replacement.

If you haven't set up a password for your BlackBerry, do it *now!* For information on how to do so, refer to Book I, Chapter 3.

Now, if you've lost your BlackBerry, how do you get back all the information that was on it? If you're like us and store important information on your smartphone, this chapter is for you. Vital information such as clients' and friends' contact information, notes from phone calls with clients — and, of course, those precious e-mail messages — shouldn't be taken lightly. Backing up this information is a reliable way to protect it from being lost forever.

If your BlackBerry is not connected to a BlackBerry Enterprise Server, BlackBerry Desktop Manager is the only way to back up and restore information to and from your computer. However, SmrtGuard (pronounced *smart guard)* has come up with a wireless backup and restore service for those of you BES-less users that aren't in the habit of plugging your BlackBerry into your computer. If this is you, go to the end of this chapter for an overview of how SmrtGuard's backup and restore solution can give you peace of mind when it comes to protecting your BlackBerry data. You can also flip to Book VI, Chapter 4 for more information on the SmrtGuard application.

# Accessing Backup and Restore

Backup and Restore is a BlackBerry Desktop Manager application. It allows you to back up all the sensitive data on your BlackBerry, including contacts, e-mails, memos, tasks, calendar items, personal preferences and options, and more.

For most users, your e-mails are already stored in accounts such as Gmail or Yahoo! Mail. But you can still back up the e-mails on your BlackBerry just in case.

To back up information on your BlackBerry, follow these steps:

*1.* **Open BlackBerry Desktop Manager on your computer by choosing Start➪All Programs➪BlackBerry➪Desktop Manager.**

   If you haven't already installed Desktop Manager on your computer, see Book IV, Chapter 1.

*2.* **Connect your BlackBerry to your computer with the USB cable that came with your BlackBerry.**

   If everything is set up right, a pop-up window on your computer asks you to type your BlackBerry security password.

*3.* **Type your password and click OK.**

   The BlackBerry connects to the computer.

*4.* **Click the Backup and Restore icon on the BlackBerry Desktop Manager screen.**

   The Backup and Restore screen opens; see Figure 4-1. You're ready to back up data from or restore information to your BlackBerry.

**Figure 4-1:**
The Backup and Restore screen.

# Backing Up, BlackBerry Style

We all know that backing up your data provides tremendous peace of mind. So do the folks at RIM, which is why backing up your information is quite easy. Both the manual and autopilot feature back up all your data, including your custom settings such as font sizes and ring tones. You can back up your BlackBerry manually or by autopilot.

Using the advanced option under Backup and Restore will allow you to selectively choose which items you would like to back up or restore on your BlackBerry.

## Backing up your BlackBerry manually

To back up your BlackBerry on demand, follow these steps:

*1.* **Open BlackBerry Desktop Manager, connect your BlackBerry, and enter your security password.**

*2.* **From the Desktop Manager screen, click the Backup and Restore icon.**

The Backup and Restore screen appears; refer to Figure 4-1.

*3.* **Click the Backup button.**

The dialog box shown in Figure 4-2 appears, so you can name the backup file and figure out where on your computer you want to save it.

*4.* **Name your backup file and choose a place to save it.**

Make sure you remember where you save this file because you are going to need it if or when you perform a backup.

**Book IV Chapter 4**

**Protecting Your Information**

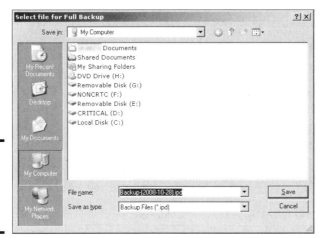

**Figure 4-2:**
Decide
where to
save your
backup file.

5. **Click the Save button.**

    BlackBerry Desktop Manager starts backing up your BlackBerry information to your computer. Figure 4-3 shows the backup progress in the Transfer in Progress window.

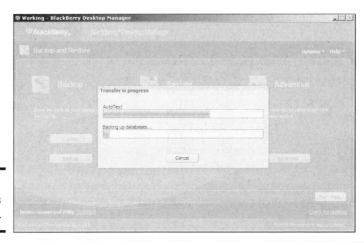

**Figure 4-3:**
A backup is
in progress.

Don't unplug your BlackBerry from the computer until the backup is finished! Depending on how much information you have on your BlackBerry, the backup might take ten minutes to finish.

6. **When the Transfer in Progress window disappears, unplug the BlackBerry from the computer.**

## Setting up automatic backups

What's better than backing up your information once? Remembering to back up regularly! What's better than backing up regularly? You guessed it — running backups automatically. After you schedule automated BlackBerry backups, you can really have peace of mind when it comes to preventing information loss.

We recommend that you connect your BlackBerry to your computer every few weeks to perform a backup. You never know when you might lose your BlackBerry.

Follow these steps to set up an autobackup:

*1.* **From the BlackBerry Desktop Manager screen, click the Backup and Restore icon.**

The Backup and Restore screen appears.

*2.* **Click the Options button.**

The Backup Options screen appears. This screen includes the option to schedule automatic backups. (See Figure 4-4.)

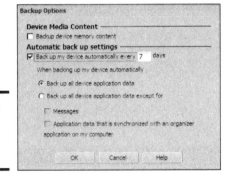

**Figure 4-4:** Set auto-backups here.

*3.* **Select the Back Up My Device Automatically Every check box.**

This lets you make more decisions (check boxes and options become active), such as how often you want Desktop Manager to back up your BlackBerry.

*4.* **In the Days field, select a number of days between 1 and 99.**

This interval sets how often your BlackBerry is backed up. For example, if you enter 14, your BlackBerry is backed up every 14 days.

*5.* **Select the Back Up All Device Application Data option.**

This option backs up all the data on your BlackBerry each time auto-backup runs.

Although you can exclude e-mail messages and information such as contacts, tasks, and memos, we recommend that you back up *everything* each time.

6. **Click the OK button.**

Now you can go on with your life without worrying about when to back up.

Make sure you remember where you save this file because you are going to need it if or when you perform a backup.

To run a backup, your BlackBerry must be connected to your computer. Make sure that you plug your BlackBerry into your computer once in a while so that autobackup has a chance to back up your information.

# Restoring Your Data from Backup Information

We hope that you never have to read this section more than once. A *full restore* means bringing back all your information from a backup. It probably means that you've lost important information that you hope to find in the backup you created on your computer.

Using the advanced option under Backup and Restore will allow you to selectively choose which items you would like to back up or restore on your BlackBerry.

The steps to fully restore information from your backup are simple:

1. **Open BlackBerry Desktop Manager, connect your BlackBerry to the USB cable, and enter your security password.**

2. **From the BlackBerry Desktop Manager screen, click the Backup and Restore icon.**

   The Backup and Restore screen appears.

3. **Click the Restore button.**

   An Open File dialog box asks where the backup file is on your computer.

4. **Choose a backup file and click the Open button.**

   A Warning window appears when you're about to do a full restore (see Figure 4-5), alerting you that you're about to overwrite existing information.

5. **Click the Yes button to go ahead with the full restore.**

   A progress bar appears.

It might take a while for the full restore to finish. Don't unplug your BlackBerry from your computer during this time!

6. **When the progress bar disappears, unplug your BlackBerry from the computer.**

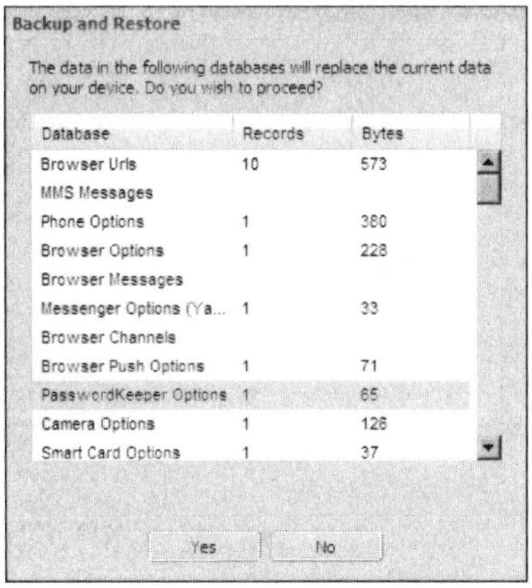

**Figure 4-5:**
Be careful when overwriting existing info.

Backup and Restore

The data in the following databases will replace the current data on your device. Do you wish to proceed?

| Database | Records | Bytes |
|----------|---------|-------|
| Browser Urls | 10 | 573 |
| MMS Messages | | |
| Phone Options | 1 | 380 |
| Browser Options | 1 | 228 |
| Browser Messages | | |
| Messenger Options (Ya... | 1 | 33 |
| Browser Channels | | |
| Browser Push Options | 1 | 71 |
| PasswordKeeper Options | 1 | 65 |
| Camera Options | 1 | 126 |
| Smart Card Options | 1 | 37 |

Yes    No

## Protecting Your Data, Your Way

A certain burger joint and BlackBerry both say that you can have it *your way* with their products. Just like you can get your burger with or without all the extras (such as pickles and onions), you can choose to not back up and restore things that you know you won't need.

For example, say you've accidentally deleted all your Internet bookmarks and now you want them back. *Don't* restore all the information from your last backup. That could be more than 90 days ago (depending on how often your autobackup runs, if at all). You might unintentionally overwrite other information, such as e-mail or new contacts. You want to restore bookmarks only.

If you lose something in particular, or want something specific back on your BlackBerry, use the selective backup and restore function in BlackBerry Desktop Manager to restore only what you need. The same goes with backing up. If you're a big e-mail user, back up *just* your e-mails and nothing else.

In the following sections, we use the term *databases*. Don't worry; this isn't as technical as you think. Think of a database as an information category on the BlackBerry. For example, saying, "backing up your Browser bookmarks database" is just a fancy way of saying, "backing up all the Browser bookmarks on your BlackBerry."

We'll start with a selective backup, and then describe a selective restore.

**Book IV
Chapter 4**

**Protecting Your
Information**

## Backing up specific information

To back up specific information, follow these steps:

1. **Open BlackBerry Desktop Manager, connect your BlackBerry to the USB cable, and enter your security password when prompted.**

2. **From the BlackBerry Desktop Manager screen, click the Backup and Restore icon.**

   The Backup and Restore screen appears.

3. **Click the Advanced button.**

   The advanced Backup and Restore screen appears, as shown in Figure 4-6. The right side of the screen shows different information categories, or *databases.*

4. **In the Device Databases list on the right, you can select multiple databases by Ctrl+clicking each database(s) you want to back up.**

   Have you ever had a really high Brick Breaker score that you were afraid of losing if your device was lost, stolen, or had its data erased? Well, you can save your Brick Breaker score for all time if you back up the RMS database!

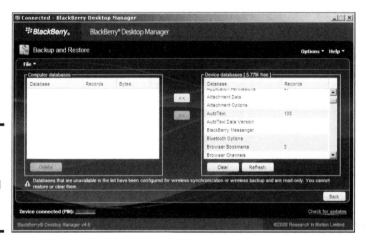

**Figure 4-6:**
The advanced Backup and Restore screen.

Worried about losing the pictures you've downloaded to your device if it's lost, stolen, or erased? Never fear; you can save those pictures by backing up the Content Store database!

5. **Click the left-pointing (backup) arrow.**

A progress bar moves while your BlackBerry is backing up. This step merely transfers the databases onto your computer; it doesn't save them. When the backup transfer is finished, you can see the databases on the left side of the window.

6. **Choose File⇨Save As.**

   A file chooser appears.

7. **Name your file and specify where you want to save it on your computer.**

   Make sure you remember where you save this file because you are going to need it when you perform a backup.

   Your selective backup is saved on your computer. Make sure to name it something specific so that you know exactly what is in the backup.

You need to manually save a selective backup file on your computer, even if you already set up an automatic backup. A selective backup doesn't automatically save your backup on your computer.

## *Restoring certain information*

When you're restoring selectively, you must already have a backup file to restore from. Although this might sound obvious, the point we're making is that you can selectively restore from any backup — auto or manual, full or partial.

For example, say you have autobackup running every other day and you want to restore only your e-mail messages from two days ago. You don't need to do a full restore; that would overwrite the new contact you put in Contacts yesterday. Rather, you can use the selective restore method and get only your e-mail messages.

---

## Looking at backup BlackBerry files

Whether you use the one-button-push backup method or you manually back up your data, the files are saved on your computer as .ipd files. You may be asking, "Can I read these backup files without a BlackBerry?" The answer is yes! With a third-party product called ABC Amber BlackBerry Converter, you can view any .ipd file. However, it can't read encrypted .ipd files. What's the point? Suppose you lost your BlackBerry but need to read an old e-mail or get contacts from your backup files. This tool allows you to convert anything in your backup file (e-mails, SMS messages, PIM messages, calendar entries, contacts, and so on) into PDF or Word documents. For more information, or to try ABC Amber BlackBerry Converter for free, go to www.processtext.com/abcblackberry.html.

To restore your way, follow these steps:

*1.* **Open BlackBerry Desktop Manager, connect your BlackBerry to the USB cable, and enter your security password if prompted.**

*2.* **From the BlackBerry Desktop Manager screen, click the Backup and Restore icon.**

The Backup and Restore screen appears.

*3.* **Click the Advanced button.**

The advanced Backup and Restore screen appears; refer to Figure 4-6. The right side of the screen shows your different information categories, or *databases.*

*4.* **Choose File⇨Open.**

A window opens so that you can choose which backup file you want to restore from.

A BlackBerry backup file has an .ipd suffix.

*5.* **Select a backup file.**

*6.* **Click the Open button.**

The different information categories, or databases, appear on the left side of the screen. You are now ready for a selective restore.

*7.* **Select the database(s) you want to restore.**

You can select multiple databases by Ctrl+clicking the databases you want.

*8.* **Click the right-pointing (Restore) arrow.**

You see a warning window asking whether you want to replace all the information with the data you're restoring. Refer to Figure 4-5.

If your BlackBerry has the same categories as the ones you're restoring (which is likely), you'll overwrite *any* information you have on your BlackBerry.

You can confidently move on to Step 9 (clicking the Yes button) if you know the database you're restoring has the information you're looking for.

*9.* **Click the Yes button.**

A progress bar appears while the selected databases are being restored. When the progress bar window disappears, the information categories that you've selected are restored on your BlackBerry.

## Selectively clearing BlackBerry information

You can also delete information on your BlackBerry from BlackBerry Desktop Manager. When would you use *selective deletion?* Suppose you want to clear only your phone logs from your BlackBerry. One way is to tediously select one phone log at a time and press Delete, repeating until all phone logs are gone. However, you could delete a database from Desktop Manager instead. When you delete a database, you are deleting directly off the BlackBerry smartphone.

To selectively delete databases on your BlackBerry, follow these steps:

*1.* **Open BlackBerry Desktop Manager, connect your BlackBerry to the USB cable, and enter your security password if prompted.**

*2.* **From the BlackBerry Desktop Manager on your computer, click the Backup and Restore icon.**

   The Backup and Restore screen appears.

*3.* **Click the Advanced button.**

   The advanced Backup and Restore screen appears; refer to Figure 4-6. The right side of the screen shows your BlackBerry's different databases.

*4.* **Ctrl+click the database(s) you want to delete.**

   The database is highlighted.

*5.* **Click the Clear button on the right side of the screen.**

   A warning window asks you to confirm your deletion.

*6.* **Click the Yes button.**

   A progress bar shows the deletion. When the progress bar disappears, the information categories you selected are cleared from your BlackBerry.

# *Backing Up and Restoring Wirelessly*

Yes, that's right! It's possible for your BlackBerry to back up and restore data without being connected to a BlackBerry Enterprise Server. However, you do have to pay a little bit for this service.

A company called SmrtGuard, `www.smrtguard.com`, offers a BlackBerry application that can wirelessly back up your data. In addition to its backup and restore capabilities, SmrtGuard also has features to help you locate, recover, and destroy data on your device.

SmrtGuard has a BlackBerry tracking feature that helps you determine whether you simply misplaced your device or whether your device was stolen. If you determine your device was stolen, you can send a signal to have your data destroyed via the SmrtGuard Dashboard on the Web site.

See Book VI, Chapter 4 for more information on this application.

**Book IV**
**Chapter 4**

**Protecting Your Information**

# Chapter 5: Transferring Data to a New BlackBerry

## In This Chapter

✔ Switching from an old BlackBerry to a new BlackBerry

✔ Switching from a PDA to a BlackBerry

*W*ouldn't it be nice if you could just make one device's data available to another? That's the future. But right now, RIM wants to make switching devices as painless as possible. That's why an application called Device Switch Wizard is part of the suite of applications in BlackBerry Desktop Manager.

In this chapter, we cover how to move data from one device to another. First, we show you how to transfer from an old BlackBerry to a new one. Then it's on to how to switch from another PDA device (such as Palm or Windows Mobile) to a BlackBerry. Read on to master the art of the switch.

## Switching to a New BlackBerry

Switching from an older BlackBerry to your new BlackBerry is no big deal. When you want to transfer application data (e-mails and contacts, for example) to your new BlackBerry, BlackBerry Desktop Manager's Device Switch Wizard backs up your old BlackBerry and loads that backup to your new device. If you don't have BlackBerry Desktop Manager installed, you can visit Book IV, Chapter 1 to guide you through the process of downloading and installing the software.

The following steps help you transition from your old BlackBerry smartphone to your new one with ease:

**1. On your PC, choose Start⇨All Programs⇨BlackBerry⇨Desktop Manager.**

The Desktop Manager screen opens, where you can find Device Switch Wizard, as shown in Figure 5-1.

**Figure 5-1:**
Launch
Device
Switch
Wizard
here.

**2. Click the Device Switch Wizard icon.**

The Device Switch Wizard screen (shown in Figure 5-2) lets you choose whether to switch from BlackBerry to BlackBerry or from non-BlackBerry to BlackBerry. The BlackBerry to BlackBerry section tells you to connect your current (old) BlackBerry to your PC.

**Figure 5-2:**
You can
switch
from one
BlackBerry
to another,
or switch
from a non-
BlackBerry
to a
BlackBerry.

**3. Connect your old BlackBerry to your PC with the USB cable.**

If you have a password on your BlackBerry smartphone, you will be prompted to enter this on your computer.

4. **Click the Start button below Switch BlackBerry Devices.**

The next screen will have your old device's PIN displayed, as shown in Figure 5-3.

Your BlackBerry PIN isn't a password, but your BlackBerry smartphone identifier. You can find the PIN by choosing Options⟶Status on your BlackBerry.

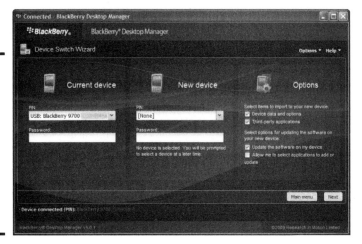

**Figure 5-3:** Verify that your old BlackBerry is connected to the PC and decide what data to include here.

5. **In the Options area, decide whether to include user data and third-party applications, and then and click the Next button.**

If you want all the data, leave the screen untouched; this backs up all your personal information such as e-mail, contacts, notes, and tasks, as well as customized settings including fonts, auto text, and tones. *Third-party applications* are all the programs you installed — the ones that didn't come with the device originally.

A status screen appears, showing the progress of the backup operation. When the backup is finished, the next screen prompts you for the PIN of your new BlackBerry. At this point, all your data is on your computer and ready to be put onto your new device. It is safe to disconnect your old device and plug in the new one.

6. **Connect your new BlackBerry to your PC with the USB cable.**

The next screen, shown in Figure 5-4, lets you verify that your BlackBerry is connected properly with the PIN displayed. It also asks you for the password.

Book IV
Chapter 5

Transferring Data to a New BlackBerry

**Figure 5-4:**
Type your
device
password
here.

7. **Enter the password of your BlackBerry and click the OK button.**

   A screen similar to Figure 5-5 tells you what will be restored to the new device. Nothing has been done to your new BlackBerry yet, and this is your last chance to cancel the process.

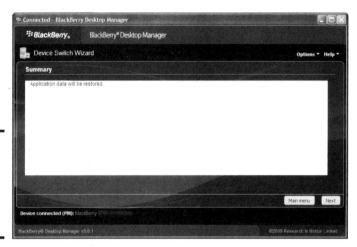

**Figure 5-5:**
Confirming
the loading
of data to
your new
BlackBerry.

8. **Click the Next button.**

   A progress screen shows you the loading process.

9. **When the Success screen appears, click the Close button.**

   At this point, your new BlackBerry is all set up, with your data, settings, and e-mail working.

## Switching from a Non-BlackBerry Device

Device Switch Wizard supports two types of non-BlackBerry devices:

✦ Palm

✦ Microsoft Windows Mobile

This doesn't mean that you can't import your old data if you have another device. The Device Switch Wizard just makes it simpler for these two types of devices. Check out Book IV, Chapter 2 for synchronization options to your Desktop PIM application if your old device is neither Palm nor Microsoft Windows Mobile.

## Palm device requirements

Your equipment has to meet three prerequisites for Device Switch Wizard to import data from Palm to BlackBerry:

✦ Your PC must be running Windows 2000 or later.

✦ Palm Desktop software 4.0.1 or later must be installed on your desktop.

✦ The Palm Desktop software installed is synchronizing properly with the Palm device.

You can check your Palm user guide for more details about your Palm device and on synchronizing it to your PC. You can also download the user guide from www.palm.com/us/support. Navigate to this page by selecting the Palm model you have and the wireless network provider it is running on.

## Windows Mobile device requirements

You need the following things for the wizard to work properly with a Windows Mobile device:

✦ Your PC must be running Windows 2000 or later.

✦ Microsoft ActiveSync versions must be installed on your PC.

✦ The Mobile device must run one of the following operating systems:

   • Microsoft Windows Mobile 2000, 2002, 2003, 2003 Second Edition, 2005/5.0, or 6.1 for Pocket PC

   • Microsoft Windows Mobile SmartPhone software 2002, 2003, 2003 Second Edition, 6.1, or 6.5

## Running the wizard

Before you run the wizard, make sure that all the requirements for your device are in place. We also recommend hot-syncing or synchronizing your Palm or Windows Mobile device; this ensures that the data you're sending to your BlackBerry is current. Palm Desktop software as well as Microsoft ActiveSync should come with help information on how to hot-sync.

Although the following steps migrate Windows Mobile data into the BlackBerry, the steps are similar for Palm as well. We indicate at what point the steps vary. Do the following to get your other device's data migrated to your new BlackBerry:

**Book IV
Chapter 5**

**Transferring Data to a New BlackBerry**

1. **Connect both the Windows Mobile device and the BlackBerry to your desktop computer.**

2. **On your PC, choose Start➪All Programs➪BlackBerry➪Desktop Manager.**

   The Desktop Manager screen appears; refer to Figure 5-1.

3. **Click the Device Switch Wizard icon.**

   The Device Switch Wizard screen appears.

4. **Choose Switch from Another Device to a BlackBerry Device.**

   The welcome screen, as shown in Figure 5-6, describes what the tool can do.

**Figure 5-6:**
Migrating
data from
a non-
BlackBerry
device.

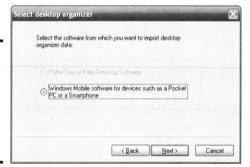

Welcome to the Migration Wizard

This tool enables you to import desktop organizer data from a Palm®/Treo™ device or a Windows® Mobile™-based device to your BlackBerry® device. You can import contacts, calendar entries, tasks, and memos.

**Note:** To import desktop organizer data, your computer must be running Microsoft® Windows 2000 or later.

Verify that both the Palm/Treo device or Windows Mobile device and your BlackBerry device are connected to your computer.

Click **Next** to begin.

Next >        Cancel

5. **Click the Next button.**

   A screen prompts you to decide whether you're migrating from Palm or Windows Mobile, as shown in Figure 5-7. The wizard is intelligent enough to enable the option associated to the connected device, which in the figure is the Windows Mobile device.

**Figure 5-7:**
The wizard
has already
selected
which
device to
port.

Select desktop organizer

Select the software from which you want to import desktop organizer data:

○ Palm/Treo or Palm Desktop Software

◉ Windows Mobile software for devices such as a Pocket PC or a Smartphone

< Back        Next >        Cancel

6. **Click the Next button.**

   Hot-syncing of the Windows Mobile device kicks in at this point. You see a series of progress screens that appear for each of the application's data, such as Calendar, Contacts, and MemoPad. A sample for the Calendar progress screen is shown in Figure 5-8. The screen indicates the number of records to sync and will be empty if you already performed a hot-sync prior to running the wizard; otherwise, it will take some time depending on how much data it has to sync between the device and the desktop software.

**Figure 5-8:**
A message showing hot-syncing on your device.

7. **Click the OK button.**

   A progress screen appears. Before the data is applied to your BlackBerry, the wizard prompts you about the change, as shown in Figure 5-9. Click the following buttons on this screen to either confirm or reject the change:

   • *Details:* See the records the wizard is trying to update.

   • *Accept:* Migrate the date.

   • *Reject:* Ignore this data and continue.

   • *Cancel:* Change your mind and cancel the whole operation.

**Figure 5-9:**
Confirm the importing of data here.

8. **Click the Accept or Reject button on any confirmation screens that appear.**

   The wizard migrates all the data you accepted. Obviously, the wizard skips everything you rejected. When the migration process is finished, a success screen appears.

9. **Click the Finish button.**

   At this point, your new BlackBerry is all set up for you to use with everything that you selected to transfer. You are free to unplug your BlackBerry and start using it.

# Chapter 6: Using Application Loader for Downloading Success

## In This Chapter

✔ Getting started with Application Loader

✔ Installing a BlackBerry application with BlackBerry Desktop Manager

✔ Uninstalling an application with BlackBerry Desktop Manager

✔ Upgrading your operating system

*T*hink of your BlackBerry as a mini-laptop where you can run preinstalled applications as well as install new applications. You can even upgrade your BlackBerry's operating system. (Yup, that's right; your BlackBerry has an operating system.)

We start this chapter by introducing Application Loader, which allows you to load applications (who'd have guessed?) onto your BlackBerry. Then we show you how to install and uninstall an application from your BlackBerry. Finally, we show you how to use Application Loader to upgrade your operating system.

You don't have to use Application Loader to get the goods onto your BlackBerry. You can install applications other ways as well:

✦ **Wirelessly through an over-the-air (OTA) download:** See Book VI, Chapter 1 for more on wireless application installation.

✦ **Through a BlackBerry Enterprise Server (BES) install (if your BlackBerry is employer provided):** In this case, you have no control over the installation process. Your company's BlackBerry system administrator controls which applications are on your BlackBerry. See Book VIII, Chapter 3 for more information on wireless application installation through a BES.

✦ **Through your PC via Microsoft Installer:** Some application installations automate the steps. All you need to do is connect your BlackBerry to the PC and click the installation file. Applications that install using the Microsoft Installer have the file extension `.msi`.

## Accessing Application Loader

In this chapter, as with other chapters in Book IV, you work closely with your computer and your BlackBerry. On your PC, you use an application called BlackBerry Desktop Manager — see Book VI, Chapter 1 for more information. Application Loader is part of BlackBerry Desktop Manager.

With the introduction of over-the-air application installations and device operating system upgrades on the newer BlackBerry smartphones, application Loader is not that widely used. However, if you have an older operating system, prior to 4.7, you will need to use Application Loader to upgrade or install a new operating system. Additionally, some BlackBerry application sites do not offer over-the-air installation of their software. In that case, the only way to install the application is with the use of Application Loader.

After installing BlackBerry Desktop Manager on your PC, do the following to access Application Loader:

*1.* **On your PC, choose Start⇨All Programs⇨BlackBerry⇨Desktop Manager.**

BlackBerry Desktop Manager opens.

*2.* **Connect your BlackBerry to your PC via your USB cable.**

If the connection is successful, you see the password dialog box, as shown in Figure 6-1. If not, see whether the USB cable is connected properly to both your PC and your BlackBerry and then try again. If all else fails, contact the technical support of your service provider.

**Figure 6-1:**
The password dialog box on your PC.

Device Password Required

Device: USB-PIN:
Please enter your device password (1/10).

Password:

Cancel

*3.* **Enter your password.**

Your BlackBerry-to-PC connection is complete.

*4.* **On your PC, click the Application Loader icon in BlackBerry Desktop Manager.**

The Application Loader screen opens, as shown in Figure 6-2. At this point, you're ready to use the Application Loader.

If your handheld isn't connected properly, your device's PIN won't show up in the Application Loader screen. Connect your BlackBerry to the USB cable and connect the USB cable to the PC.

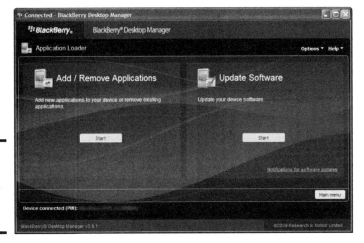

**Figure 6-2:**
The
Application
Loader
screen.

# Installing an Application

When you find an application — whether it's online or from a CD —, you can download it to your BlackBerry smartphone via Application Loader. In this section, we install iSkoot for Skype for BlackBerry. Skype is a tool that acts like a phone with video- and file-sharing abilities. Versions of Skype are free, so you can talk with someone around the world at no cost. *iSkoot* is a free application that allows you to use Skype through the Web. You can download this application at `www.download.com/iSkoot-for-Skype-BlackBerry-/3000-7242_4-10797721.html`.

No matter what application you're installing from your PC to your BlackBerry, the steps are the same. You can use the following steps as a guide to installing the application of your choice:

*1.* **Download the application to your PC, and then extract it.**

The steps for downloading the application vary depending on the application vendor. Check the application vendor's Web site for download instructions.

Make sure you remember where you save this file because you are going to need it when you install it on your BlackBerry.

2. **Locate the application's ALX file.**

You can usually find a file with the `.alx` extension in the folder where you extracted the application on your PC.

The ALX file doesn't get installed on your BlackBerry; rather, it tells Application Loader where the actual application files are located on your PC.

3. **Connect your BlackBerry to your PC using the USB cable.**

A screen prompting you to enter your BlackBerry password appears. Refer to Figure 6-1.

4. **Enter your password.**

After entering your password, the Application Loader screen indicates that your device is connected.

5. **Click the Application Loader icon in BlackBerry Desktop Manager.**

The Application Loader screen appears. Refer to Figure 6-2.

6. **Click the Start button below Add/Remove Applications.**

The list of applications available for installation appears.

7. **Click the Browse button and locate and select the ALX file you want to install.**

You return to the Application Loader screen, where your application — for example, iSkoot — is one of the applications in the list, as shown in Figure 6-3.

**Figure 6-3:**
Your
application
is added to
the list of
available
applications
and can be
installed
on your
BlackBerry.

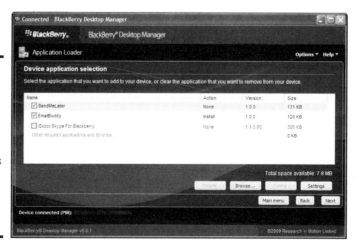

8. **Select the application you want to install and click the Next button.**

   We're installing iSkoot Skype for BlackBerry, so that's what we selected.

   A summary screen appears, listing only the applications that will be installed or upgraded.

9. **Click the Finish button.**

   The installation process starts, and a progress window appears. When the progress window disappears — and if all went well — the application is on your BlackBerry. You can find it in the Applications folder on your BlackBerry home screen.

If you get an invalid signature error after clicking the Finish button, and

✦ **You didn't get your BlackBerry from your employer.** Something is probably wrong with the application. You need to contact the software vendor.

✦ **You got your BlackBerry from your employer.** You don't have permission to install applications on your BlackBerry. The IT department rules the school, and you will need to contact them to request installation.

# Uninstalling an Application

You can uninstall an application in two ways:

✦ **Using Application Loader:** We talk about this method in this section.

✦ **Using your BlackBerry handheld:** Turn to Book VI, Chapter 1, where we discuss this method.

As we did in the preceding section on installing, we use iSkoot as an example and assume that you already installed the iSkoot application. Of course, you can follow the same steps for uninstalling other applications.

The steps to uninstalling a BlackBerry application are similar to installing:

1. **Open BlackBerry Desktop Manager on your PC.**

2. **Connect your BlackBerry to your PC using the USB cable.**

   A screen prompting you to enter your BlackBerry password appears. (Refer to Figure 6-1.)

3. **Enter your password.**

   After entering your password, the Application Loader screen indicates that your device is connected.

**Book IV
Chapter 6**

Using Application
Loader for Down-
loading Success

4. **Click the Application Loader icon.**

5. **Click the Start button below Add/Remove Applications.**

   The screen listing of applications appears; refer to Figure 6-3.

6. **Scroll to the application you want to delete and deselect its check box.**

   The Action column for the application changes from None to Remove. For example, when we deselect the iSkoot Skype for BlackBerry check box, the Action column for iSkoot indicates Remove.

7. **Click the Next button.**

   A summary screen that lists the action of Application Loader appears. It indicates that iSkoot is to be removed from your BlackBerry.

8. **Click the Finish button.**

   The uninstall process starts, and a progress window appears. When the progress window disappears, you have uninstalled the application from your BlackBerry.

# Upgrading Your BlackBerry OS

Having the latest BlackBerry OS, also called the BlackBerry device software, allows you to have the latest and greatest features available. New operating systems also include bug fixes from prior ones, so having the latest is always best. You can get the latest OS from two sources:

+ **Your network service provider**

+ **Your BlackBerry system administrator (if you're using your BlackBerry through a BES)**

The handheld software may differ from provider to provider, so get it from the service provider's Web site. You can find links to the latest software downloads at `http://na.blackberry.com/eng/services/devices/`.

You should upgrade your BlackBerry device software only when you have to or are told to by your network service provider or your corporate BlackBerry system administrator. There is always the risk that some third-party applications that you installed may stop working. We advise you to consult the third-party vendor prior to upgrading to ensure they support the upgrade.

To address the risk of information loss, RIM designed Application Loader to automatically perform a backup for you as part of the OS upgrade. However, our experience has taught us to perform a full backup ourselves, before starting the upgrade, as an extra precaution. The Application Loader backup isn't always complete, in our experience. See Book IV, Chapter 4 for more on backing up data manually.

After you finish a manual backup of your BlackBerry, you can start the upgrade process by doing the following:

1. **Enter your BlackBerry password (if you have set one) into BlackBerry Desktop Manager on your PC.**

2. **Click the Application Loader icon on the Desktop Manager screen.**

   The Application Loader screen appears. Refer to Figure 6-2.

3. **Click the Start button below Add/Remove Applications.**

   A list of software appears, as shown in Figure 6-4.

**Figure 6-4:**
Application updates that are available.

4. **Select the check box next to the OS portion.**

   This appears as BlackBerry 5.0 System Software shown in Figure 6-4.

   The upgrade option is listed only if the handheld software installed on your desktop computer is newer than the version installed on your BlackBerry. If the upgrade option doesn't appear in the list, the device software you installed on the desktop machine is the same as the one installed on your device, or a prior version compared to the one installed on your device.

   You also need to back up your device in case something goes wrong with the upgrade, which you do in the next step.

5. **Click the Options button.**

   The Options screen appears, as shown in Figure 6-5. This is where you decide whether you want to back up your BlackBerry content before upgrading your handheld software.

**Book IV
Chapter 6**

**Using Application
Loader for Down-
loading Success**

**Figure 6-5:**
Command
Application
Loader to
back up the
device here
(Options
screen).

6. **Select Back Up Device Automatically During the Installation Process and click the OK button.**

   You're back to the previous screen, shown earlier in Figure 6-4.

7. **Click the Next button.**

   A summary page confirms your actions — a final chance for you to either proceed with the OS upgrade or not.

8. **Click the Finish button.**

   The BlackBerry OS upgrade starts, complete with a progress window that shows a series of steps and a progress bar. The entire process takes ten minutes or more, depending on your PC model, how much data needs to be backed up, and the BlackBerry device software version you're upgrading to.

   At times during the BlackBerry device software upgrade, your BlackBerry display goes on and off. Don't worry; this is normal.

   When the progress window disappears, the handheld software upgrade is complete. You will also receive a message stating that it is now okay to disconnect your BlackBerry smartphone from your PC.

# Book V

# Music, Photos, Videos, and TV

"You should see the detail in this Topo map of the area. It's like we're standing right there."

# Contents at a Glance

# Chapter 1: Managing Media Files through Your Computer

## In This Chapter

✔ **Using Roxio Media Manager**

✔ **Synchronizing media files with Media Sync**

✔ **Managing media files with Windows Explorer**

The ways of finding media files are evolving. Ten years back, who would have thought that you could buy music stored in a tiny card or download music with an "all you can eat" monthly subscription?

Someday, you'll wake up with a technology that doesn't require you to constantly copy media files to your handheld music player. But for now, enjoying music while on the move means managing these files.

The modern BlackBerry offers a multitude of options for getting music — or other files — from your PC onto your BlackBerry. These options range from Roxio Media Manager, which is part of Desktop Manager to the Media Sync application that can synchronize with iTunes to the good old standby of drag, drop, copy, and paste through Windows Explorer. In this chapter, we go over all of the options so you can choose the one (or two, or three, or all) that work best for you. In the next chapter, we show you how to play all that music on your BlackBerry smartphone.

## Greeting Roxio Media Manager for BlackBerry Smartphones

Heard of Roxio? Roxio is known for its CD-ripping software. (*Ripping* is a process that converts music files in CD format to other popular compressed formats.) RIM licensed a portion of Roxio RecordNow and packaged it with BlackBerry Desktop Manager. Even though it's not the whole suite of Roxio software, that's still good news for you because you can now avail yourself of fantastic features, such as

✦ Ripping CDs

✦ Converting files to get the best playback on your BlackBerry

✦ Managing music files

✦ Syncing media files to your device

In the following sections, we familiarize you with the Roxio Media Manager interface and then show you how to copy a media file into your BlackBerry.

## Accessing Media Manager

You can get to Roxio Media Manager through BlackBerry Desktop Manager, which Book IV, Chapter 1 describes in detail. Go to Desktop Manager on your PC by following these steps:

*1.* **Click the Windows Start button.**

*2.* **Choose All Programs⇨BlackBerry⇨Desktop Manager.**

BlackBerry Desktop Manager appears, as shown in Figure 1-1.

**Figure 1-1:** Access Media Manager here.

*3.* **Click the Media icon.**

A screen appears, showing Media Manager and BlackBerry Media Sync sections. Each section has a Start button.

*4.* **Click the Start button in the Media Manager section.**

The initial Media Manager screen is well organized and gives you the following options:

• Manage Pictures

• Manage Music

- Manage Videos
- View Connected Devices

**5. Click one of the options.**

When you see the Media Manager screen shown in Figure 1-2, it may look intimidating. But it's really easy to use; it has the same interface as Windows Explorer:

✦ The left side is where you navigate to your folders and files.

✦ The right side displays the files in the folder currently selected on the left.

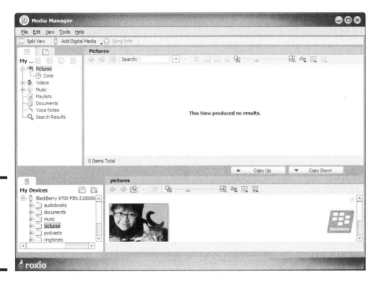

**Figure 1-2:** View your media files on this screen.

The top section looks the same as the bottom section except that the top represents your desktop, and the bottom represents the BlackBerry; they're named My Media and My Devices respectively. You can move or copy files easily. When you're copying, for example, one section can be the source and the other section can be the destination. By simply dragging the files between the two sections, you can copy on the same screen. Nice and simple, right?

Do you have an old version of Media Manager? No problem. Point your desktop Internet browser to http://na.blackberry.com/eng/services/desktop for directions on downloading the latest version of BlackBerry Desktop Manager for free and installing it on your PC. Installing the software from this link will update the entire BlackBerry Desktop Manager. Media Manager is packaged inside BlackBerry Desktop Manager.

## Importing media files to Media Manager

Want a quick and easy way to import media files? Follow these steps:

*1.* **Use Windows Explorer to find the media files you want.**

*2.* **Drag and drop the files into Media Manager.**

You can drag and drop files to the folder in the left part of the screen, where the folder tree appears, or the right part, where the files are listed. Just make sure that when you're doing the latter, the current folder in the tree view is the folder where you want the media files to be imported.

Without using Windows Explorer, you can also use Media Manager to locate the files you want. The trick is to change the upper-left-side view in one of the sections to Folders. If you look closely at the upper-left section, you see two tabs. The first tab, My Media, is the default view. The Folders tab has the icon of — guess what? — a folder.

Click the Folder icon. You see a tree view, but this time it looks exactly as you see it in Windows Explorer, as shown in Figure 1-3. The files can be on your local hard drive or in a network folder accessible by your desktop computer.

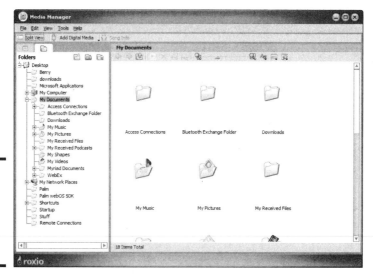

**Figure 1-3:** Navigate to your desktop media files here.

## Adding a media file to your BlackBerry

After you're familiar with the Media Manager, get those files copied to your BlackBerry. Here's the rundown:

*1.* **Connect the BlackBerry to your desktop computer with the USB cable that came with the device.**

Enter your device password if prompted.

*2.* **On the Media Manager screen, drag and drop your media files from the My Media view to any folder in My Devices.**

You can drag and drop an entire album. After dropping a media file, you're prompted to convert the file into a format that's viewable by your BlackBerry, as shown in Figure 1-4.

**Figure 1-4:**
Choose to
convert your
media files
for optimum
playback.

*3.* **Choose an option:**

- *Convert for Optimal Playback:* This is the safest bet and the default.

- *Copy with No Conversion:* Copies the file faster. The file is copied to your BlackBerry as is, but it might not play in your BlackBerry.

- *Advanced Conversion Options:* From here, another screen lets you downgrade the quality to minimize the file size. It also allows you to crop video so that the entire screen is filled, instead of showing dark margins.

*4.* **Click OK to begin the transfer.**

---

## Other features of Media Manager

Spend some time exploring Media Manager. It has interesting features that you might find useful. Here's a quick rundown of what you can do with Media Manager:

✔ Import media files.

✔ E-mail media files.

✔ Enhance photos and apply special effects to photos by using PhotoSuite.

✔ Set song info, such as title, artist, album, genre, year, or an image, to show as track art when playing the song.

✔ Record audio.

✔ Customize photo printing.

# Synchronizing with iTunes Using BlackBerry Media Sync

If you have an iPod, you're probably using iTunes and maintaining a play-list and perhaps a subscription to podcasts or videocasts. Podcast files are downloaded to iTunes using RSS. To clarify the jargon, RSS is short for *Really Simple Syndication,* a kind of digital file publish-subscribe mechanism. This is the mechanism iTunes uses to receive audio and video recordings, which most people refer to as podcasts and videocasts. Would you like to sync your BlackBerry with iTunes? Wouldn't we all?

### Installing BlackBerry Media Sync

*BlackBerry Media Sync* is an application that you can use to sync your BlackBerry to iTunes or Windows Media Player on your Windows computer. Although integrated into the BlackBerry Desktop Manager, BlackBerry Media Sync is actually a separate application that does not come preinstalled with the BlackBerry Desktop Manager.

Because you will be downloading the installation software for Media Manager, make sure that your desktop computer is connected to the Internet.

To install BlackBerry Media Sync, follow these steps:

1. **On your PC, click the Windows Start button.**

2. **Choose All Programs⇨BlackBerry⇨Desktop Manager.**

   BlackBerry Desktop Manager appears; refer to Figure 1-1, earlier in the chapter.

3. **On the Desktop Manager screen, click the BlackBerry Media Sync icon.**

   BlackBerry Media Sync screen appears. The screen is divided into two parts:

   • On the left is Media Manager.

   • On the right is BlackBerry Media Sync.

   On each side is an Install button.

4. **Click the Install button located on the BlackBerry Media Sync portion of the screen.**

   A prompt appears, asking whether you want to download and install BlackBerry Media Sync.

5. **Click Yes for the prompt.**

   Desktop Manager starts downloading the installation file for BlackBerry Media Sync. During this process, you'll see a status screen displaying the progress of the download. This might take a couple of minutes, depending on the speed of your Internet connection.

   After the download, the installation follows. When the installation completes, you'll get a screen similar to Figure 1-5. On this screen, you'll see that BlackBerry Media Sync now has a Launch button.

**Figure 1-5:**
Launch
BlackBerry
Media Sync
here.

## Synchronizing your BlackBerry to iTunes

After you have the BlackBerry Media Sync installed, synchronizing your BlackBerry to iTunes is simple. Just follow these quick and easy steps:

1. **Click the Windows Start button.**

2. **Choose All Programs⇨BlackBerry⇨Desktop Manager.**

   BlackBerry Desktop Manager appears. (Refer to Figure 1-1.)

3. **When BlackBerry Desktop Manager appears, click the Media icon.**

   A screen appears, showing Media Manager and BlackBerry Media Sync sections. Each section has a Launch button.

4. **In the BlackBerry Media Sync section, click the Launch button.**

   A prompt appears allowing you to choose which Media application you want your BlackBerry synchronized to. You have two options here: iTunes and Windows Media Player.

5. **Select the iTunes check box and then click OK.**

   The BlackBerry Media Sync screen appears. This screen allows you to pick and choose what type of media you want to synchronize. The screen has two tabs:

   • Music (shown in Figure 1-6)

   • Pictures

   The contents of the tab are the details of what you're about to sync.

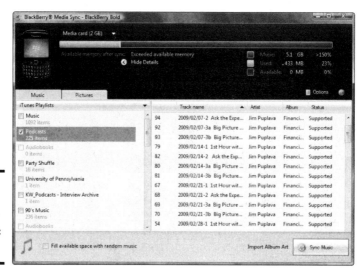

**Figure 1-6:**
The
BlackBerry
Media Sync
screen.

6. **Choose what you want to sync:**

   ✦ *Your music:* From the Music tab (refer to Figure 1-6), you have the option to select your iTunes playlist that you want synchronized with your BlackBerry.

   Note the size of your playlist and make sure that the microSD card on your BlackBerry has enough capacity to store all of your selections. Look at the top of the screen for details on exactly this type of information. You can see on the top of the screen indicators for

   • The total and percent used memory your microSD card has

   • The available space

   • The total size of the selected playlists

   • Warning indicator, if you exceed the capacity

   ✦ *Update pictures:* From the Pictures tab, select your desktop pictures folders that you want synchronized into your BlackBerry. This screen has two inner tabs, BlackBerry Device Pictures and Computer Pictures (shown in Figure 1-7). The direction of this sync is from your computer to your BlackBerry. Click the BlackBerry Device Pictures tab (see Figure 1-8). The Import to Computer button in the bottom of the screen clearly indicates the direction of the sync, which is copying from your BlackBerry to your computer.

7. **Click the Sync Music, Sync Pictures, or Import to Computer button (lower right).**

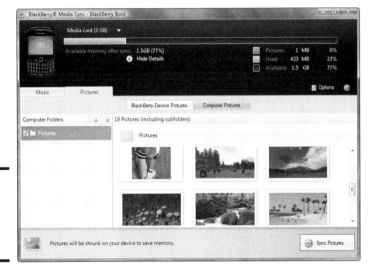

**Figure 1-7:**
Sync your desktop pictures here.

**Figure 1-8:**
Import BlackBerry pictures to your computer here.

# Transferring Media Files Using Windows Explorer

When you're in a hurry, running Desktop Manager and opening Media Manager can be a drag. One quick option is to copy your media files directly to microSD using the familiar Windows Explorer on your PC. Follow these steps:

*1.* **Connect your BlackBerry to your computer using the USB cable that came with your device.**

Make sure that you have the microSD card in your BlackBerry before you do this. When connected, the BlackBerry screen displays a prompt for enabling mass storage mode, as shown in Figure 1-9.

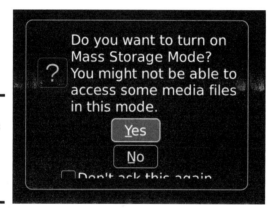

**Figure 1-9:** Choose Yes to enable mass storage mode.

*2.* **Select Yes.**

Another screen appears on the BlackBerry asking for your password.

*3.* **Type your BlackBerry password.**

The device is now ready to behave like an ordinary flash drive. A screen appears on your PC, as shown in Figure 1-10.

*4.* **Choose Open Folder to View Files.**

The familiar Windows Explorer screen opens. Notice that you have several default folders, including documents, music, pictures, and videos, as shown in Figure 1-11.

*5.* **Browse to the file(s) you would like to move onto your BlackBerry, and then drag and drop or copy and paste them to the correct folder on your smartphone.**

**Figure 1-10:**
Choose
Open Folder
to View
Files.

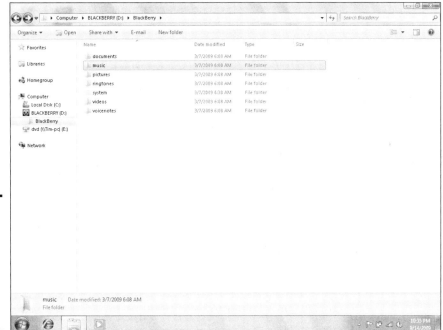

**Figure 1-11:**
Your
BlackBerry
media
files are
organized
into several
folders.

For example, if you would like to move a music file onto your
BlackBerry, find it on your PC and then drag and drop or copy and paste
it into the music folder on the Removable Disk that is your Blackberry,
as shown in Figure 1-12.

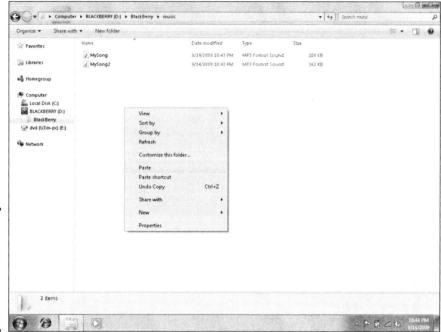

**Figure 1-12:**
Move your
media file to
the correct
folder
on your
smartphone.

You can also use Windows Explorer to move or delete files on your
BlackBerry's microSD card by following these steps:

1. **Connect your BlackBerry to your computer using the USB cable that
   came with your device.**

   The BlackBerry screen displays a prompt for enabling mass storage
   mode. (Refer to Figure 1-9.)

2. **Select Yes.**

   Another screen appears on the BlackBerry asking for your password.

3. **Type your BlackBerry password.**

4. **Choose Open Folder to View Files.**

   The Windows Explorer screen appears. Refer to Figure 1-11.

5. **Browse to the BlackBerry folder that has the file(s) you want to move
   or delete.**

   • *To delete a file:* Highlight the file(s) you would like to delete, and then
     press the Delete key on your keyboard.

   • *To move a file:* Highlight the file(s) you would like to move, and then
     drag and drop or copy and paste to the new folder.

# Chapter 2: Acquiring and Listening to Music

## In This Chapter

✔ **Finding and downloading your favorite music**

✔ **Accessing streaming music**

✔ **Using the Media Player**

✔ **Transferring music between BlackBerry smartphones**

In addition to sending and receiving e-mail, looking up contacts, and keeping track of your appointments, you can use your BlackBerry to listen to music wherever you are — on the bus, at the gym, or simply walking around town. The newer BlackBerry models even support streaming music via Bluetooth, so you don't need to have those cords dangling around.

To focus on getting music onto your BlackBerry, see Book V, Chapter 1. Our focus in this chapter is on using your BlackBerry smartphone to play music.

We also talk about some amazing BlackBerry applications that allow you to listen to music live. You can stream music from the Internet, stream it from your BlackBerry, or stream it from a radio station. You will have so many different types of music to listen to that you will never stop singing and dancing with your BlackBerry.

## Getting Great Music with iTunes

The iTunes program manages a variety of media files. (See Figure 2-1.) Using iTunes, you can download music, movies, TV shows, podcasts, and streaming radio stations. You can easily sort media, create playlists, and play your favorite songs.

iTunes is designed to play music and videos on your desktop and with Apple's iPods. However, you can use BlackBerry's Media Sync tool to synchronize your BlackBerry songs and playlists with iTunes. (See Book V, Chapter 1 for more info on using Media Sync and iTunes.)

Before you can use iTunes songs on your BlackBerry, you need to have iTunes installed on your PC or Mac. You can find the software by steering your Web browser to www.itunes.com. Follow the instructions onscreen to download and install the software on your computer.

## Creating an iTunes account

The iTunes Store is your portal for previewing, purchasing, and downloading all kinds of media files. Through the store, you can browse top downloads, search music billboards from around the world, and search for a specific song, group, or album. With the vast selection of music to choose from, iTunes makes it easy to buy songs for around a dollar a piece. Before you can purchase anything, you must have an iTunes account.

To create an iTunes account, follow these steps:

1. **On your computer, open iTunes.**

2. **From the menu that appears at the top of the screen, choose Store.**

3. **Select Create Account.**

4. **Follow the Create Account instructions to create an account.**

   You're ready to purchase music and other media.

## Finding music to purchase on iTunes

Perusing the iTunes store is as simple as pointing and clicking:

1. **Open iTunes. (See Figure 2-1.)**

2. **On the left side of the screen, click iTunes Store.**

3. **Type the name of a song, musician, or album into the Search bar in the upper-right corner of the screen.**

   You can also browse for music using one of the categories on the main page.

4. **Once you find the song you want to download, click Buy Song.**

5. **The song downloads to your computer's hard drive.**

   You can sync the new song with your iPod if you have one. You can also sync the song with your BlackBerry. For information on syncing your music on your BlackBerry, see Book V, Chapter 1.

**Figure 2-1:**
iTunes
Store.

# Using LimeWire to Find Music

LimeWire is a peer-to-peer file-sharing program. That means that individual users can share files with each other directly using the LimeWire software. In addition to music, you can use LimeWire to share programs, video, and other files. The software is very simple to use, and with a high-speed Internet connection, you can quickly download several songs at a time. What's even better, downloading and sharing files is free.

LimeWire directly downloads songs to a folder on your computer. To get songs onto your BlackBerry, see Book V, Chapter 1. If you also use iTunes, you can store LimeWire songs in an iTunes folder so that you can use Media Sync to transfer them to your BlackBerry.

Many of the files you will find using LimeWire are illegal to download. That's because it's illegal to download copyright-protected content. If you want to stay on the right side of the law, you must turn on copyright filtering. With copyright filtering options set, you can download from independent and unsigned artists. To change your copyright filtering settings, choose Tools➪Options➪Search➪Don't Let Me Download or Upload Files Copyright Owners Request Not to Be Shared.

As with iTunes, you must download and install LimeWire if you want to trade files. Direct your browser to www.limewire.com to find out more about the software.

LimeWire's basic installation is free. To have more features, you need to purchase the Pro version of the software. One advantage of paying for the premium software is that you can have a quicker download.

# Streaming and Caching Music

Streaming music on your BlackBerry smartphone is similar to listening to a radio. The music is not stored on your device, but is constantly received from a remote source. Unlike a radio, which uses radio frequencies, your BlackBerry smartphone receives the music over the data channel. This is the same channel that you use to check your e-mail or browse the Web. As long as you have purchased a data plan for your BlackBerry, you can listen to streaming music.

Streaming music over a non–Wi-Fi connection is data intensive. In plain English, that means that, unless your plan includes unlimited data transfers, you may have heart palpitations when you get your first bill. Check the fine print on your contract before you stream audio or video.

## Listening to streamed music with Pandora

Pandora (www.pandora.com) is a service that provides both a Web site to stream music to your PC and an application to stream music to your BlackBerry smartphone.

You can create your own stations from either your desktop PC or your BlackBerry. You can think of a station as a channel that plays a specific genre of songs. For example, you can set up an easy-listening station that will have mellow music or a comedy station that has only comedians telling jokes. The stations that you create, not the music files, are available on both the Web site and on the Pandora for BlackBerry application. This means you will always have your personal station available no matter which device you use to listen with. From Pandora's site (www.pandora.com) you can also purchase songs. For more information on Pandora and other music players, go to Book VI, Chapter 5.

Streaming media requires a constant data connection that can quickly reduce the life of your BlackBerry device's battery.

## Streaming and caching with Slacker

Slacker, shown in Figure 2-2, is a streaming music application for both the desktop and BlackBerry smartphone. In addition to creating your own personal stations, Slacker has a wide selection of stations created by professional DJs.

What is unique with Slacker for the BlackBerry is that the app allows you to cache your favorite stations. Unlike streaming, where each song is sent continuously to your device, *caching* allows you to save the songs that are in each station (which may include around 500 songs) to your BlackBerry's media card.

The two major advantages of caching over streaming is that you can listen when you have poor or no signal at all, and because caching does not need to keep a constant connection to play songs like streaming does, it can save up to five times of your precious battery life. For more information on Slacker and other music players, go to Book VI, Chapter 5.

**Figure 2-2:** Slacker for the BlackBerry.

# Using BlackBerry Media Player

Earlier in this chapter, we discuss a couple of software options that allow you to purchase and download music. In Book V, Chapter 1, we show you how to transfer music from your PC or Mac to your BlackBerry smartphone. Now it's time for the fun part: listening to your music.

## Accessing Media Player

BlackBerry bundles all its media players under one program called Media Player. In the Media folder, you can access the following stuff:

✦ Music

✦ Videos

✦ Ring tones

✦ Pictures

✦ Voice notes

To access the Media folder and play your music, follow these steps:

1. **Press the Menu key.**

   A list of all the icons appears on the Home screen.

2. **Scroll and click the Media icon.**

   The Media folder opens.

3. **Highlight Music and click to open it.**

   A list of options to choose from appears. (See Figure 2-3.)

   - *All Songs:* This option displays a list of all your songs and allows you to quickly search for a specific song.

   - *Artists:* This option groups songs by artist.

   - *Albums:* This option groups songs by albums.

   - *Genres:* This option groups songs by genre (such as Latin, Techno, R & B, and so on).

   - *Playlists:* This option allows you to create and play personal playlists.

   - *Sample Songs:* By default, BlackBerry provides a few sample songs that you can play.

   - *Shuffle Songs:* This option allows you to play songs in random order.

   - *Now Playing:* This option appears only if a song is currently playing. Click this option to jump to the Media Player main screen.

**Figure 2-3:** Music Player options.

## Playing a song with Media Player

To play a song using BlackBerry Media Player, open it and follow these steps:

1. **Select a song by searching for it by name, or by browsing by artist, album, genre, playlist, or Shuffle, as described in the preceding section.**

   Media Player plays your song. (See Figure 2-4.)

2. **To adjust the volume, press the Up and Down volume buttons on the side of the BlackBerry.**

REMEMBER

If you pause a song and switch applications, Media Player still runs in the background. All open items that run in the background take up precious battery life. If you don't plan on listening to the music, make sure to stop the song playing or close the program by pressing the Menu key and choosing Close.

**Figure 2-4:**
Music
Player
displaying
the track
info screen.

## Navigating between songs using Media Player

Of course, like any media player, you have access to some basic controls. You can rewind to the beginning of a song, stop the song, pause the song, fast-forward through the song, and skip to a specific part in the song. In addition to these basics, when you press the Menu key when a song is playing, you have these additional features and options (as shown in Figure 2-5):

✦ **Repeat (All Songs):** Use this option to turn on or off repeating either an individual song or a playlist of songs.

✦ **Shuffle:** Use this option to have the BlackBerry Media Player play songs in random order.

✦ **Add to Playlist:** Create and add the song playing to a playlist. A *playlist* is a collection of songs that you group together.

✦ **Set as Ring Tone:** You can associate a song with a contact in your address book. When that person calls you, that song will play instead of your default phone call ring.

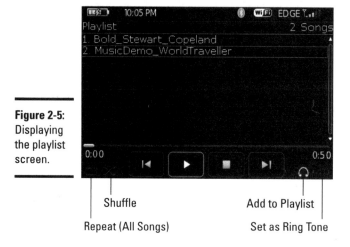

**Figure 2-5:** Displaying the playlist screen.

Shuffle

Repeat (All Songs)

Add to Playlist

Set as Ring Tone

# Transferring Music Files between Two BlackBerry Smartphones

Ever have a friend say, "You should hear this song!" and then pass you his headphones? Well, if you both have BlackBerry devices, there's a better way to do it. You can transfer music files and ring tones between two phones using traditional methods:

✦ **Attach the song to an e-mail message.**

✦ **Send a clip as an attachment to an MMS text.**

When using e-mail or MMS to send a ring tone or music file, the size must be less than 1MB or it will fail to send. To send files larger then 1MB, you need to use another method:

✦ **Instant messaging:** If you have BlackBerry Messenger 5.0 installed, you can transfer music, documents, pictures, and many other files through that program. To download BlackBerry Messenger 5.0 via BlackBerry App World, see Book VI, Chapter 1.

✦ **Bluetooth:** You can also share songs directly between your BlackBerry and a friend's BlackBerry using Bluetooth. In fact, you can send movies, ring tones, pictures, and voice notes using Bluetooth!

Of course, both BlackBerry devices must have Bluetooth, and you must *pair* your Bluetooth device with your smartphone before you can transfer files. Pairing is what happens when you set up your devices so that they can talk to each other using Bluetooth. The pairing process for headsets and other devices is the same as the process for pairing two BlackBerry smartphones. For information on how to pair devices, refer to Book II, Chapter 4.

Transferring files between some other smartphones is possible. Depending on the manufacturer, you may need to have a third-party application installed on your device. Refer to your smartphone's handbook for more information.

## Sending a music file from your BlackBerry to another BlackBerry

If you want to transfer music from your BlackBerry to another Bluetooth-enabled BlackBerry device, follow these steps. (Before a file can be transferred between two BlackBerry smartphones, the two BlackBerry smartphones must be paired together.)

*1.* **On the sending BlackBerry,** *press the Menu key.*

A list of icons appears on the Home screen.

*a. Locate and select the Media icon.*

*b. Select and open the Music menu option.*

*c. Locate and highlight the song you would like to send.*

*d. Press the Menu key and choose Send Using Bluetooth. (See Figure 2-6.)*

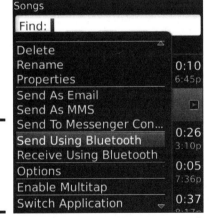

**Figure 2-6:**
Sending a music file over Bluetooth.

*e. Choose the BlackBerry that you wish to transfer the music to. (See Figure 2-7.)*

*2.* **On the receiving BlackBerry,** **prepare for the file transfer by pressing the Menu key and choosing Receive Using Bluetooth from within the Media folder or music application.**

*3.* **On both the sending and receiving devices,** **if Bluetooth is not enabled, you receive a prompt to enable it. Select Yes.**

**Figure 2-7:**
Select your
smartphone.

4. **On the receiving BlackBerry,** after Bluetooth is enabled, a File Transfer Waiting for Connection message appears on the screen.

This message stays on your screen until the sending BlackBerry completes Step 5.

5. **On the sending BlackBerry,** a File Transfer Connecting prompt appears on your BlackBerry Home screen until the receiving BlackBerry accepts the transfer in Step 6.

6. **On the receiving BlackBerry,** an Accept Connection Request From prompt appears when the sending BlackBerry begins the file transfer. Select Yes.

7. **Choose the location you want to save the file. Click Save. (See Figure 2-8.)**

When the progress bar reaches 100%, the file has been transferred and saved to the receiving device.

**Figure 2-8:**
Saving the
file.

# Chapter 3: Taking, Viewing, and Managing Photos

## In This Chapter

✔ Getting ready to say "cheese"

✔ Saving and organizing your pictures

✔ Sharing your photos with other people

✔ Understanding memory limits

**O**h shoot, you are out and about and forgot your camera. Don't worry! Your BlackBerry's there when you need to capture the unbelievable: Grandma doing a handstand, Grandpa doing a cartwheel, or your roommate doing her laundry.

Before you try taking pictures, read this chapter so that you know what to expect and how to get the best shot. We walk you through the easy steps in capturing that funny pose. We also tell you how to store those important pictures, and then show you how to share the joy with your buddies.

## Say Cheese: Taking Pictures

Before you ask someone to pose, examine your BlackBerry (shown in Figure 3-1) first. No matter which model you have, the setup is basically the same.

✦ **Is the camera on?** You see the convenience key on the right side of your BlackBerry? This key can be configured (refer to Book I, Chapter 3) to launch an application of your choice and the default is launching the Camera application. Press it to bring up the Camera application. Alternatively, you can select the Camera icon from the Home screen.

✦ **Is your finger blocking the lens?** The lens is on the back side of your device.

✦ **Do you see an image on the screen?** Pressing the right convenience key again takes the picture. You should hear a funky sound. Neat and easy, isn't it? The picture you just captured is saved in the pictures folder which defaults to `/Media Card/BlackBerry/pictures`.

**Figure 3-1:**
The camera
screen
ready
to take
pictures.

Number of pictures you can capture

Low-light indicator

Flash

Camera key
(right convenience
key)

Hey, you can take a picture of yourself as well. Turn the BlackBerry around (so that you see its back), and you see a mirror right below the lens. Whatever you see in that mirror is what the camera captures.

Itching to take more pictures? Hold those snapping fingers of yours. If you take a few moments first to familiarize yourself with the camera's features, the effort could go a long way.

## Reading the screen indicators

When you open the Camera application, the first thing you see is a screen similar to the one shown earlier in Figure 3-1. The top portion of this screen shows you the image you're about to capture. The bottom part contains icons (starting from the left) that indicate the following:

✦ Number of pictures you can capture

✦ Zoom

✦ Flash

## Choosing the picture quality

The latest BlackBerry smartphones can capture up to 3.2 megapixels (MP) of resolution. Some older BlackBerry models, including the Bold and the earlier Curve and Pearl, only have 2 MP resolution — which is probably enough to suit your purposes. Still, saving at these resolutions requires a lot of space. You can save at a lower quality and save some space on your BlackBerry. You have three resolutions to choose from:

✦ Normal

✦ Fine

✦ SuperFine

The default setting is Normal, which is the lowest quality, but it lets you save the most pictures. The trade-off is that Normal picture quality won't be as smooth or fine compared to the Fine and SuperFine settings. You should choose a setting based on how you plan to use the picture. If you're taking a shot of a breathtaking landscape in which you want to capture every possible detail and print it later, you'll want the SuperFine setting. On the other hand, if you're just taking pictures of your friends' faces so that you can attach them as caller IDs, Normal is appropriate.

Changing picture quality is a snap. Follow these steps:

1. **Open the Camera application.**

2. **Press the Menu key and then select Options.**

3. **Select Picture Quality and then choose the quality that you want.**

4. **Press the Menu key and then select Save.**

   The picture quality you've chosen is now active.

## Zooming and focusing

You need to be steady to get a good focus while taking the shots. Although it's convenient to use one hand while taking pictures, most of the time, you'll get a blurry image if you try that.

When taking pictures, hold your BlackBerry with both hands, one hand holding the device steady and the other pressing the camera button.

Holding the camera with both hands is even more important if you're zooming in. Yes, your camera has digital zoom. The zoom amount varies from model to model, and newer models do not necessarily have higher zoom amounts. For example, the new BlackBerry Pearl 9100 has a 2.5x, whereas the BlackBerry Curve 8520, which is a year older, has 5x digital zoom. Here's what you need to do to focus and zoom:

+ **To focus:** Press the convenience key halfway.

+ **To zoom in:** Slide upward on the trackball or touch screen.

+ **To zoom out:** Slide downward on the trackball or touch screen.

While zooming, the value in the indicator changes from 1x to 2x to 5x and vice versa, depending on the direction you slide.

When zooming on any device but the Storm, your thumb is already on the trackball or trackpad. What a convenient way to take the picture — just press.

Although we went through the trouble of describing the zoom capabilities, we don't recommend using them. Digital zoom (which is what your camera has) gives poor results because it's done through software and degrades the quality of the picture. The higher the zoom factor, the more pixilated the picture becomes.

## Setting the flash

The rightmost indicator on the Camera screen is the flash. The default is Automatic, which shows a lightning bolt with the letter *A*. Automatic means that the camera detects the amount of light you have at the moment you capture the image. Where it's dark, the flash fires; otherwise, it doesn't.

You can turn the flash on or off. When set to off, the lightning bolt is encircled with a diagonal line, just like you see on No Smoking signs. You can toggle the settings on the camera's Options screen, which is accessible by pressing the Menu key.

## Setting the white balance

In photography, filters are used to compensate for the dominant light. For instance, a fluorescent versus an incandescent light could affect how warm the picture appears. Instead of using filters, most digital cameras have a feature to correct or compensate for many types of light settings. This feature is *white balance*. And yes, your BlackBerry has this feature. You can choose from Sunny, Cloudy, Night, Incandescent, Fluorescent, and Automatic. *Automatic* means your camera determines what it thinks are the best settings to apply.

You can change the white balance through the camera's Options screen.

The camera's Options screen is accessible by pressing the Menu key and selecting Options while you're in the Camera application.

Taking, Viewing, and Managing Photos

## Setting the picture size

Aside from picture quality, you can also adjust the actual size of the photo:

+ **Large:** 1600 x 1200

+ **Medium:** 1024 x 768

+ **Small:** 640 x 480

Again, camera settings are accessible through the camera's Options screen by pressing the Menu key and selecting Options from the menu that appears.

## Geotagging

Because your BlackBerry has GPS capability, your location can be easily determined based on longitude and latitude. This information can be added to your media files, including the pictures taken from your BlackBerry. Adding geographic information is referred to as *geotagging*. Now, you don't have to wonder where you took that crazy pose.

Geotagging is disabled by default in your BlackBerry. You can enable it from the camera's Options screen.

If you have longitude and latitude information from one of your photos, you can use one of the free sites on the Web to locate where you were when you took the photo. One such site is www.travelgis.com/geocode.

## Setting the Pictures folder

Pictures you captured from the camera are saved in a pictures folder. The default location of this folder is /Media Card/BlackBerry/pictures. This location is good but in case you want to save it somewhere else, you can with these steps:

*1.* **Select Camera from the Home screen.**

*2.* **Press the Menu key and then select Options.**

Toward the bottom of the Camera Options screen, you should see a Folder button and beside it the path of the current folder where pictures captured by the camera are stored. The Folder button is your gateway to change this folder.

*3.* **Select the Folder button.**

A pop-up screen appears, listing Media Card and Device Memory folders and a list of the thumbnails of pictures stored in the current folder.

You need to change the view to Explore so you can navigate to the folder structure and find the folder you are interested in storing your pictures in.

4. **Press the Menu key and select Explore from the menu that appears.**

This time, the screen only displays a listing of folders starting from Media Card and Device Memory. You should be able to navigate to the subfolders by opening the parent folder listed onscreen.

5. **Highlight a folder, press the Menu key, and select Open.**

Subfolders appear. You may have to repeat this step until you find the folder you want. To go back up the folder hierarchy, simply press the Escape key.

6. **Highlight the folder you want to store your pictures in, press the Menu key, and select the Select Folder option from the menu that appears.**

You're back to the Camera Options screen with the path of the folder you selected displayed beside the Folder button.

7. **Press the Escape key and select Save on the prompt.**

Your setting is saved.

## Working with Pictures

You've taken a bunch of pictures and now you want to see them? Maybe delete the unflattering ones, or perhaps organize them? No problem.

### Viewing

If you take a picture, you want to see it, right? *Viewing* a picture is a common function with your camera. You can see the image you just captured right then and there, as shown in Figure 3-2.

**Figure 3-2:**
The camera screen after taking a picture.

IMG00008-20100419-1825

If you're browsing through your picture folders, you can view a picture by highlighting it and pressing the trackball, the trackpad, or the touch screen.

## Creating a slide show

Want to see your pictures in a slide show? Follow these steps:

1. **While on the Camera screen, press the Menu key and select View Pictures from the menu that appears.**

2. **Press the Menu key.**

3. **Select Slide Show.**

   *Voilà!* Your BlackBerry displays your pictures one at a time at a regular interval. The default interval between each picture is two seconds; if you're not happy with this interval, change it in the Options screen. (Press the Menu key and select Options to get to the Options screen.)

## Trashing

If you don't like the image you captured, you can delete it. Follow these steps:

1. **Highlight the picture you want to trash.**

2. **Press the Menu key and select Delete from the menu that appears. Alternatively, you can press the Del key.**

   A confirmation screen appears.

3. **Select Delete.**

You can also delete an image right after taking the picture; just select the Trash Can icon when viewing the photo. (Refer to Figure 3-2.)

Deleting a picture, or any file for that matter, in your BlackBerry is irreversible. There is no way to recover the deleted file from your smartphone.

## Listing filenames versus thumbnails

When you open a folder packed with pictures, your BlackBerry automatically lists *thumbnails,* which are small previews of your pictures.

A preview is nice, but what if you're trying to search for a picture file and you know the filename? Wouldn't it be nice to see a list of filenames instead of thumbnails? You can view filenames by following these steps:

1. **Go to a picture folder.**

2. **Press the Menu key.**

3. **Select View List.**

That's exactly what you get: a list of all the pictures in the folder. What's neat is that the option also displays the file size. The file size can give you a clue about what settings you used to take the picture. For example, a photo taken at a SuperFine quality produces a much bigger file size compared to one taken at Normal quality.

## Checking picture properties

Curious about the amount of memory your picture is using? Want to know the time you took the photo? You can view a picture's properties as follows:

1. **Highlight the picture from a list.**

While you're on the Camera screen, you can view the list of your pictures by pressing the Menu key and selecting View Pictures.

2. **Press the Menu key.**

3. **Select Properties.**

You see a screen similar to Figure 3-3, which displays the picture's location, size, and last modification.

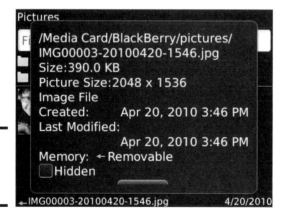

**Figure 3-3:**
Your
picture's
properties.

## Organizing your pictures

Organization is all about time and the best use of it. After all, you want to enjoy looking at your pictures — not looking *for* them. The BlackBerry allows you to rename and move pictures to different folders. Plus, you can create folders, too. With those capabilities, you should be on your way to organization nirvana.

### Renaming a picture file

BlackBerry saves a file when you capture a picture. However, the name of the picture is generic, something like IMG*xxxx,* where *x* is a number.

Make it a habit to rename a photo as soon as you've finished capturing it. It is easier to recognize *Dean blows out birthday candles* than *IMG0029-20081013-0029.* A filename can be composed of any characters or symbols, and there's no limit on its length. It's best though to make the name short but descriptive. Four or five words is about what the screen fits when displaying a list of filenames. When viewing a list, if the filename doesn't fit the screen, the device does not display all words; instead it ends with three dots to denote there are more characters in the name.

Renaming is a snap. Here's how:

**1.** **Display the picture on your screen or highlight it in the list.**

**2.** **Press the Menu key and select Rename.**

A Rename screen appears, as shown in Figure 3-4.

**3.** **Enter the name you want for this picture and then select Save.**

Your picture is renamed.

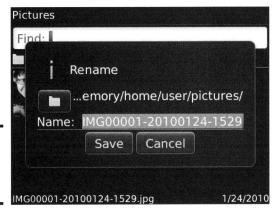

**Figure 3-4:**
Rename
your picture
here.

### Creating a new folder

Being the organized person you are, you must be wondering about the folders we mentioned. Don't fret; it's simple to create one. Here's how:

*1.* **On the Camera screen, press the Menu key and then select View Pictures.**

The screen displays the list of pictures in the current folder and an Up icon for you to navigate up to the folder above this folder.

*2.* **Select the Up icon to navigate to the main folder where you want your new folder to be created.**

You should be *within* the folder where you want your new folder to be created. If not, repeat this step and navigate to that folder.

*3.* **Press the Menu key and then select New Folder.**

*4.* **Enter the name of the folder and then select OK.**

Your folder is created.

### Moving pictures

There are many reasons for moving pictures between folders. The most obvious reason is to organize your pictures. Want to try it? Follow these steps:

*1.* **On the Camera screen, press the Menu key and then select View Pictures.**

The screen displays the list of pictures in the current folder. If the picture you want to move is not in this folder, click the Up icon to navigate up to other folders.

*2.* **Highlight the picture you want to move, press the Menu key, and select Move.**

The screen that follows allows you to navigate to the folder where you want to move this picture.

*3.* **Click the Up icon and navigate to the folder where you want to move this picture.**

*4.* **Press the Menu key and then select Move Here.**

Your picture is moved.

## Sharing your pictures

Where's the joy in taking great pictures if you're the only one seeing them? Your BlackBerry has several options for sharing your bundle of joy:

*1.* **On the Camera screen, press the Menu key and select View Pictures.**

*2.* **Highlight a picture you want to share.**

*3.* **Press the Menu key.**

**4.** **Select from the choices listed here:**

- *Send as Email:* Go directly to the Message screen for composing e-mail, with the currently selected picture as an attachment.

- *Send as MMS:* Similar to Send as Email, this opens a Compose MMS screen with the currently selected picture as an attachment. The only difference is that MMS first displays Contacts, letting you select the person's phone number to receive the MMS before going to the Compose screen. Another difference is that in the MMS it sends a tiny version of the picture.

- *Send to Messenger Contact:* This option is available if you have BlackBerry Messenger installed. This functions similar to Send as MMS, but only displays contacts you have in BlackBerry Messenger. It uses BlackBerry Messenger to send a tiny version of the picture file.

- *Send Using Bluetooth:* This allows you to send the picture to any device capable of communicating through Bluetooth.

You may see other ways to send a picture file if you have other IM clients installed. For example, if you have Google Talk installed, you will see Send as Google Talk.

## Setting a picture as caller ID

Wouldn't it be nice if, when your girlfriend was calling, you also could see her beautiful face? You can do that. If you have a photo of her saved in your BlackBerry, follow these steps to use her picture as caller ID:

**1.** **Select the Media icon from the Home screen and select Pictures.**

**2.** **Navigate to the location of your pictures.**

**3.** **Highlight the photo you want to appear when the person calls.**

If you don't have pictures of your friends, now's the time to start snapping away.

**4.** **Press the Menu key and select Set as Caller ID.**

The photo is displayed onscreen with an inner portrait-size rectangle that displays a clear view, with the outside opaque. The clear view represents the portion of the photo that shows up as caller ID. You can slide the trackball, trackpad, or touch screen to make sure that you capture the face. Cropping it is the last step.

**5.** **Press the Menu key and choose Crop and Save.**

Your contacts list appears.

**6.** **Select the contact you want this picture to appear for.**

A message indicating that a picture is set for that contact appears. You're set.

### Setting a Home screen image

Suppose you have a stunning picture that you want to use as the background image for your BlackBerry. Follow these steps to set the image:

1. **Select the Media icon from the Home screen and select Pictures.**

2. **Navigate to the location of your pictures.**

3. **Highlight the picture you want to use as your Home screen image.**

4. **Press the Menu key and select Set as Home Screen Image.**

    You can always reset the Home screen image by going to the Menu screen and selecting Reset Home Screen Image.

## Setting Camera Memory Options

The camera in your BlackBerry is a piece of hardware and a computer program. As such, the good people at RIM (Research In Motion) incorporated some parameters that you can set so that you can enjoy your camera fully while not affecting other features that share the same memory.

You should know the following two options:

+ **Device Memory Limit:** The amount of device memory your camera can use. The values may vary based on your BlackBerry model, but they generally range from 5–25MB (1MB is 1024K). To get a feel for how many pictures this is, look at the properties of an existing picture in the format you take most often and note its file size. The size of the picture depends on the format you use to take it.

+ **Reserved Pictures Memory:** The amount of memory BlackBerry reserves for the camera to store pictures. Possible values are 0MB, 2MB, 5MB, 8MB, 10MB, and 12MB. This value can't be set higher than the Device Memory Limit value.

You can set Device Memory and Reserved Pictures Memory limits using the steps below:

1. **On your BlackBerry Home screen, select the Media icon.**

2. **Choose Pictures.**

3. **Press the Menu key and choose Options.**

4. **Choose the Device Memory and Reserved Pictures Memory limits you want.**

5. **When you're done adjusting the settings, press the Menu key and choose Save.**

# Chapter 4: Recording, Viewing, and Managing Video

*W*ow, something amazing is happening right now, and you're without your video camera. If only you carried one in your pocket. Wait, you do carry one because you have your BlackBerry! That's right; with your BlackBerry, you can record your unbelievable scene in full motion.

Before you start that BlackBerry camera rolling, read this chapter so that you know what to expect and how to get the best shot. We also walk you through the easy steps in capturing that funny moment, and then show you how to store those videos. Finally, we show you how to share those epic films with your friends.

## Say Action: Capturing Video

The Camera application is a multipurpose one. You can use it to take videos as well as still photos. Here are the quick-and-easy steps to use Video Camera mode:

1. **Press the right convenience key (bottom key on the right side of your BlackBerry) to bring up the Camera application.**

   Right convenience key opens the Camera app by default. See Book I, Chapter 3 on how to customize the convenience keys.

2. **Press the Menu key and select Video Camera. (See Figure 4-1.)**

   The screen appears like a viewfinder on a typical digital video camera, as shown in Figure 4-2.

3. **Press the trackball, trackpad, or touch screen to start recording.**

   Your camera starts recording. You can easily pause recording by pressing the trackball, trackpad, or touch screen, and then repeat this step to resume recording. Once you're done recording, you can proceed to the next step to stop recording.

**Figure 4-1:**
Toggle
to Video
Camera
mode here.

4. **Press the Escape key to stop recording.**

Your recording is now saved in the videos location, which defaults to
`/Media Card/BlackBerry/videos`, and your video camera is now
back to its initial screen, as shown in Figure 4-2, ready for another
recording. The filename of your new video will be written like *VID0029-
20100913-0029*. The middle portion is a numeric representation of the
current date, starting from year, month, and day. So *20100913* means
September 13, 2010. Refer to the section "Renaming a video file" in the
latter part of this chapter to change your video filename to something
descriptive.

**Figure 4-2:**
Your
BlackBerry
becomes a
digital video
camera.

The controls you see on the screen are all context related. When you first
launch the video camera, all you see is the Record button with the large
white dot at the bottom of the screen. (Refer to Figure 4-2.) When you select
the Record button, the video camera starts taking video, and the only available control is a Pause button.

Press the Escape key to show the rest of the controls, as shown in Figure 4-3. The controls are the familiar buttons you see on a typical video recorder/ player. From left to right, they are as follows:

✦ **Record:** Continue recording.

✦ **Stop:** End the current recording.

✦ **Play/Pause:** Play the current video you just recorded or pause the playback.

✦ **Rename:** Rename the video file.

✦ **Trash:** Get rid of the video file of the current recording.

✦ **Send:** Share your current video recording. You have the option to send it as e-mail, as MMS, through BlackBerry Messenger, or through Bluetooth. If you have IM clients installed, such as Google Talk or Yahoo! Messenger, the IM client will be listed as one of the options for sending the video file.

**Figure 4-3:** The video camera controls.

Record  Stop  Play/Pause  Rename  Trash  Send

# Customizing the Video Camera

Even with its size, your BlackBerry has a few settings you can tweak to change the behavior of the video camera. Like every other BlackBerry application, to see what you can customize, don't look anywhere else but the application's Options screen — in this case, the Video Camera Options screen.

Follow these steps to get to the Video Camera Options screen:

1. **Open the Camera application.**

2. **Press the Menu key and choose Video Camera. (Refer to Figure 4-1.)**

3. **Press the Menu key and select Options.**

   The Options screen appears, as shown in Figure 4-4.

4. **Adjust your settings:**

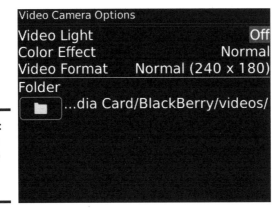

**Figure 4-4:**
Customize
your video
camera
here.

The available options are quite easy to digest, but in case you need a little help, here's what you can tweak:

- *Video Light:* In case it's a little dim, you can turn the video camera's lights on.

  Dropped something in a dark alley? This video light is a good alternative to a flashlight when you need one.

  Video Light is the flash that you use when taking still pictures. It stays lit when you set the setting to On and open the Video Camera. The default is Off.

  Using the light drains your battery.

- *Color Effect:* The default is Normal, which is "what you see is what you get." If you are in the mood for effects, you can choose from either Black & White or Sepia.

- *Video Format:* This option is the screen resolution size. The default here is Normal, but if you are planning to send your video to friends through MMS, you can choose the MMS mode, which has the smaller size and is optimal for MMS. Normal is set at the maximum size of your BlackBerry screen. You should see one of the values below, depending on your model:

  **480 x 360:** BlackBerry 8900, BlackBerry Bold (9700), BlackBerry Tour (9600)

**640 x 480:** BlackBerry Pearl (9100)

**480 x 320:** BlackBerry Bold (9000)

**320 x 240:** BlackBerry Curve (8500)

**360 x 480:** BlackBerry Storm 2

- *Folder:* Change the default location where your BlackBerry saves the video file. The default location is `/Media Card/BlackBerry/videos`.

5. **Press the Escape key and select Save on the prompt that follows.**

   This last step saves your settings.

## Watching Your Recorded Videos

What good is recording videos if you can't see them, right? *Viewing* your videos is a breeze with these steps:

1. **On your BlackBerry Home screen, select the Media icon, and then choose Videos.**

2. **Select the video you want to watch.**

   The video will load and begin playing in full screen.

3. **Press the trackball, trackpad, or touch screen to bring up the video player options.**

   You have the standard options: Skip Backward, Play/Pause, Stop or Skip Forward (as shown in Figure 4-5).

Figure 4-5:
Your video
camera's
playback
options.

Progress bar          Play/Pause          Skip Forward

        Skip Backward          Stop

You can fast-forward or rewind if you pause the video, and then select the progress bar and roll the trackball, slide the trackpad, or slide your finger on the touch screen, left and right. Left will rewind and right will fast-forward.

4. **When you're done watching the video, press the Menu key and choose Close to return to the list of videos on your device.**

## Working with Videos

So you've recorded a bunch of videos and know how to watch them, but videos take a lot of media card space and maybe you want to delete the bad ones to free up needed space, or perhaps organize them all? No problem. We show you how to do it all in this section.

### Deleting videos

If you don't like the video you recorded, you can delete it. Follow these steps:

1. **From the BlackBerry Home screen, select the Media icon and choose the Videos option.**

2. **Highlight the video that you want to get rid of.**

3. **Press the Menu key and select Delete from the menu that appears, as shown in Figure 4-6.**

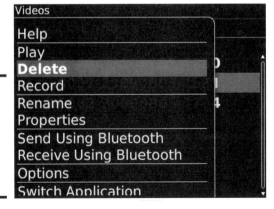

**Figure 4-6:** Choose Delete to get rid of that unwanted video.

You can also press the Del key.

A confirmation screen appears, as shown in Figure 4-7.

4. **Select Delete.**

**Figure 4-7:**
Choose
Delete and
that video is
history.

You can also delete a video right after recording it; just select the Trash icon when viewing the video. (Refer to Figure 4-3.)

Unlike trash in a Mac or Recycle Bin in Windows, files you delete in your BlackBerry can't be restored. The deleted file is gone for good.

## Checking video properties

Curious about the amount of memory your videos are using? Want to know the time you recorded the video? You can view a video's properties as follows:

*1.* **On the BlackBerry Home screen, select the Media icon, and then choose the Video option.**

*2.* **Highlight the video in the list.**

   While you're on the video Camera screen, you can view the list of your videos by pressing the Menu key and selecting View Videos.

*3.* **Press the Menu key.**

*4.* **Select Properties.**

   A screen displaying the video's location, size, and last modification appears, as shown in Figure 4-8.

## Organizing your videos

Organization is all about finding the best use of your time. Let's face it; you want to enjoy looking at your videos, not looking *for* them. The BlackBerry allows you to rename and move videos to different folders. Plus, you can create folders, too. With these capabilities, you should be on your way to organization nirvana.

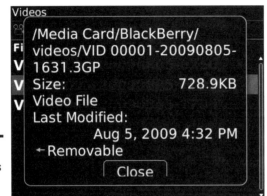

**Figure 4-8:**
Your video's
properties.

### Renaming a video file

BlackBerry saves a file when you record a video. However, the name of the video is generic, something like VID *xxxx,* where *x* is a number.

Get in the habit of renaming a video as soon as you've finished recording it. It is easier to recognize *Bob trips and falls* than *VID0029-20101013-0029.* You can use any characters and symbols in your name, and there's no limit to the name's length. But a short descriptive name works to your advantage. About four to five words fits well when you're viewing the names listed in the video screen.

Renaming is a snap. Here's how:

*1.* **On the BlackBerry Home screen, select the Media icon and choose Videos.**

*2.* **Highlight the video you want to rename.**

*3.* **Press the Menu key and select Rename.**

   A Rename screen appears, as shown in Figure 4-9.

*4.* **Enter the name you want for this video and then select Save.**

   Your video is renamed.

### Creating a new folder

Being the organized person you are, you must be wondering about the folders we mentioned. Don't worry, it's easy to create a folder using these steps:

*1.* **On the BlackBerry Home screen, choose the Media icon.**

*2.* **Press the Menu key and choose Explore.**

**Figure 4-9:**
Rename
your video
here.

3. **Select Media Card or Device Memory, depending on where you want to store your videos. Then navigate to the main folder where you want your new folder to be created.**

   You should be *within* the folder where you want to create your new folder. If not, repeat this step and use your trackball, trackpad, or touch screen to navigate to that folder.

4. **Press the Menu key and then select New Folder, as shown in Figure 4-10.**

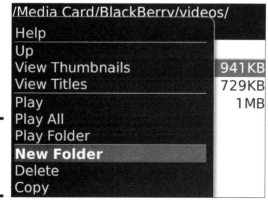

**Figure 4-10:**
Choose to
create a
new folder.

5. **Enter the name of the folder, and then select OK.**

   Your folder is created, as shown in Figure 4-11.

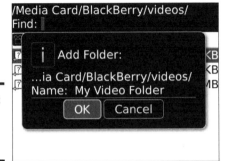
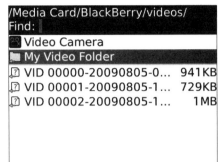

**Figure 4-11:**
Enter your
new folder
name.

## Moving videos

There are many reasons for moving videos between folders. The most obvious reason is to organize them. Want to try it? Follow these steps:

*1.* **On the BlackBerry Home screen, choose the Media icon.**

*2.* **Press the Menu key and choose Explore.**

*3.* **Navigate to the video you want to move, press the Menu key, and select Move, as shown in Figure 4-12.**

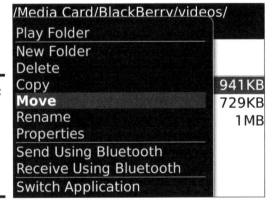

**Figure 4-12:**
Choose
Move to
move your
video to
another
folder.

The screen that follows allows you to navigate to the folder where you want to move this video.

*4.* **Choose Up and select the folder where you want to move the video.**

*5.* **Press the Menu key, and then select Move Here, as shown in Figure 4-13.**

Your video is moved.

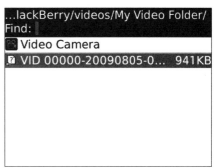

**Figure 4-13:**
A
successful
move!

## Sharing your videos

What's the point of recording great videos if you're the only one watching them? Your BlackBerry has several options for sharing your works of art. You can attach a video to an e-mail or MMS, send it through Messenger, or transmit it via a Bluetooth connection. Here's how:

1. **On the BlackBerry Home screen, select the Media icon and choose the Videos options.**

2. **Highlight a video you want to share.**

3. **Press the Menu key.**

4. **Decide what you want to do with your video:**

   - *Send as Email:* Go directly to the Message screen to compose an e-mail, with the currently selected video as an attachment.

   - *Send as MMS:* Similar to Send as Email, you open a Compose MMS screen with the currently selected picture as an attachment. The only difference is that MMS first displays Contacts, letting you select the person's phone number to receive the MMS before going to the Compose screen.

   - *Send to Messenger Contact:* This option is available if you have BlackBerry Messenger installed. This functions similar to Send as MMS, but displays only contacts you have in BlackBerry Messenger. It uses BlackBerry Messenger to send a tiny version of the picture video.

   - *Send Using Bluetooth:* This allows you to send the video to any device capable of communicating through Bluetooth.

You may see other ways to send a video file if you have other IM clients installed. For example, if you have Google Talk installed, you see Send as Google Talk.

## Watching Streaming Video

The latest generation of BlackBerry smartphone can stream video from Web sites, such as YouTube and Google Video, right through BlackBerry Browser and media player. How cool is that? Your BlackBerry should have support for video streaming if the smartphone has OS version 4.3 or later.

You can find the OS version by selecting Options from the Home screen and then selecting About. It's typically written in the third line and reads like v5.0.0.664. The first two numbers after the letter *v* is what's important; in this case, 5.0 and should support video streaming.

Streaming video uses a lot of data. Make sure you have an unlimited data plan, or are aware of the data charges you might rack up, before you stream video on your BlackBerry.

Follow these steps to watch streaming video on your BlackBerry:

1. **Open BlackBerry Browser and go to a streaming video site.**

   One example is `http://m.youtube.com`.

2. **Select a video you want to watch.**

   The BlackBerry media player opens and begins *buffering* — downloading — the video. Because the speed of your connection varies, buffering is a way to ensure that you don't get interrupted for every split second that your connection slows down. With buffering, a certain amount of video will be downloaded ahead of time. The video starts playing as soon as enough of it is downloaded.

3. **When you're done watching the video, press the Menu key and choose Close.**

   You return to the Web site you were viewing.

You also have a couple other alternatives to streaming video through your BlackBerry smartphone:

✦ A very cool piece of hardware called a Slingbox lets you stream cable TV, DVDs, and other types of video media that you have hooked up at your home through your BlackBerry.

✦ Several third-party vendors have created applications that provide a more customized experience for viewing popular streaming video sites like YouTube on your BlackBerry. If you search BlackBerry App World (see Book VI, Chapter 1) for keywords like *YouTube* or *Google Video*, you should be able to find these applications.

# Book VI

# BlackBerry Applications

The 5th Wave     By Rich Tennant

"Hold on, Barbara. I'm sure App World has a BlackBerry application for just such a situation."

# Contents at a Glance

# Chapter 1: Finding Applications

## In This Chapter

✔ **Introducing BlackBerry App World**

✔ **Finding BlackBerry applications**

✔ **Using and managing BlackBerry applications**

Applications (also known as *apps*) and games allow you to get work done or have fun, depending on your mood. You can choose from hundreds of games and applications that are available though BlackBerry App World or you can visit third-party mobile application Web sites.

While some apps are free, others may cost you a fee. Either way, if you haven't explored your app options lately, you might be surprised by all the cool tools and fun stuff out there. For example, with the introduction of GPS, there are a whole host of apps you can download that can help you find where you are and where you want to go. In addition, multimedia apps are a lot of fun when you want to watch a YouTube video, listen to music, or even watch TV — on your phone. If you're into activities, you can download an app to track your workouts and diet regimen. Heck, there's even an app that acts as a personal trainer. The possibilities are endless.

In this chapter, we walk you through how to use BlackBerry App World. You find out how to download and install the App World application, search for apps, and manage the apps that you acquire. You also find out about third-party BlackBerry app stores. Finally, we give you some tips and tricks for using your new BlackBerry apps.

## Figuring Out What You Want to Do

Before you go to the trouble of downloading BlackBerry App World, you might want to know what kinds of apps are available.

Um . . . every kind. If you want to browse apps before downloading App World, visit `www.blackberry.com/appworld` on the browser of your home or work computer. Here's a list of some of the more popular categories:

✦ **Business:** You can get everything from a UPS app to a call time tracker, mileage tracker, remote printing app, and a business card reader.

✦ **Education:** From e-flashcards to task lists, WikiMobile and dictionaries, you can get educational apps to help you stay prepared for the next big test any time, anywhere.

✦ **Entertainment:** This category of apps is a lot of fun and contains apps that are also useful, such as a flashlight app, which we could have used a few weeks ago when we couldn't find our house keys.

✦ **News:** Apps in this category give you all the news that's fit to view on your minicomputer that doubles as a phone.

✦ **Sports and Recreation:** If you're into fantasy sports, are a fanatic for a particular team, or just like to keep on top of all your sports, you'll like the apps in this category.

✦ **Themes:** Sick of the default themes on your BlackBerry? Themes apps allow you to customize everything about your BlackBerry display so that it reflects your true personality.

✦ **Utilities:** You can find tools that will read your e-mail and text messages out loud (especially useful if you're driving and want to stay safe and hands-free). Another app protects your personal data if your phone is ever lost or stolen. An incredibly popular utility is called Tether. This app allows you to access the Internet on your computer by using your BlackBerry's Wi-Fi connection. Genius, isn't it?

There's a ton more to look for, from shopping to health and fitness. We promise you that if you spend five minutes perusing your app options you will find at least one app about which you'll say, "Yeah, I could use that." This fact is true even if you're absolutely sure you don't need any apps at all. To see more info on finding an app once you have App World installed, see the upcoming "Finding an app with BlackBerry App World" section.

# Finding and Installing Apps

With hundreds of apps out there, you won't have a hard time finding an application that will help you get work done, complete a task, or have fun. You can install apps quickly without having to connect your BlackBerry to your computer. This simple installation method is called an over-the-air (OTA) installation.

## Visiting BlackBerry App World

Research In Motion has made it very simple for BlackBerry owners to find every kind of application using BlackBerry App World. BlackBerry App World is an application that allows you to browse through several app categories, quickly find the top downloaded apps, search for an app by name, or even choose from one of the featured apps. You may already have BlackBerry App World installed on your BlackBerry. To find out, press the

Menu button and browse the icons on your Home screen. If you don't have it installed, the following section tells you what you need to know.

If you're on a limited data plan or hate using BlackBerry Browser, you can access BlackBerry App World on your computer (at `www.blackberry.com/appworld`) and browse categories, check ratings of apps, and e-mail a link to your BlackBerry. As long as you also have App World installed on your BlackBerry, you simply click the link in the e-mail to install the app. Genius.

## Installing BlackBerry App World

**Book VI**
**Chapter 1**

If BlackBerry App World was not installed on your BlackBerry smartphone when you first received the device, you need to download it. BlackBerry App World can be downloaded only over the air (OTA). You can use the same method to download and access any application that is available through BlackBerry Browser.

**Finding
Applications**

Here's how to download BlackBerry App World:

*1.* **From your BlackBerry Browser, go to `www.blackberry.com/appworld`.**

Or use your desktop computer to steer your browser to `www.blackberry.com/appworld` and choose Download BlackBerry App World. An e-mail is sent to the e-mail address associated with your BlackBerry smartphone. You can then open the e-mail on your BlackBerry and click the link to download the app.

*2.* **Scroll down and click the Download Now button.**

*3.* **Select a download language and click Next.**

*4.* **Click Download, as shown in Figure 1-1.**

**Figure 1-1:**
Downloading
BlackBerry
App World.

When the application is downloaded, you're prompted to reboot the BlackBerry. You can choose to reboot now or reboot later.

5. **If you plan to use BlackBerry App World right away, choose Now.**

   If you choose Later, you must reboot your BlackBerry first before you can use App World. When you do, pick up at Step 6.

6. **Press the menu button to open the BlackBerry Home screen.**

7. **Open the Downloads folder.**

8. **Locate the BlackBerry App World icon on your Home screen and click to open it. (See Figure 1-2.)**

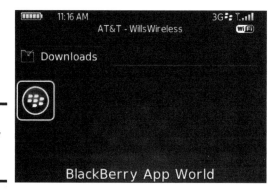

**Figure 1-2:**
BlackBerry
App World
icon.

The first time you use BlackBerry App World, scroll down to accept the legal terms.

The main page for BlackBerry App World appears onscreen, which shows Featured Items, as shown in Figure 1-3.

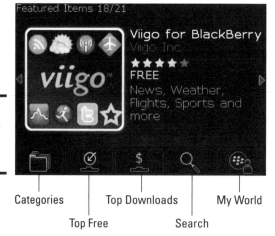

**Figure 1-3:**
BlackBerry
App World
main page.

## Moving around your apps

On your BlackBerry smartphone, the two default folders are Applications and Downloads. All applications that you install go into the Downloads folder. The Applications folder contains the native BlackBerry applications.

You can move icons to the Home screen or other folders by following these steps:

1. **Scroll to highlight the icon.**

2. **Press the Menu button and select one of these options:**

   *Move:* Change the position in the current folder.

   *Move to Folder:* Place the icon in a different folder.

## *Finding an app with BlackBerry App World*

BlackBerry App World offers you these five different ways to find an app:

✦ **Featured Items:** Browse featured apps that are being highlighted by RIM. You can scroll to the left or right to view them and select them to download.

✦ **Categories:** Browse app categories and subcategories to find apps based on a task you need to perform. Each app is assigned a category.

✦ **Top Free:** Click this option to find the top 25 downloaded applications that are free.

✦ **Top Paid:** If you can't find a free application to do what you need, try looking at the top 25 paid applications.

✦ **Search:** This option allows you to search for an application by a keyword. BlackBerry App World searches the name and description of the application as well as the company.

You can find free apps, free trials, and paid applications in BlackBerry App World. If you search for the keyword *free*, you can find not only free apps, but also any app that happens to have the word "free" somewhere in the name or description. For example, a paid application called Free Wi-Fi Café Spots will appear in the search results.

You can find the app you want by either clicking it from one of the featured applications, browsing the categories, choosing one of the top 25 downloaded, or simply searching for it by name.

## Reviewing an app before you buy it

It's important to know what you're getting *before* you download. That's not just because we don't want you wasting your money. There are a few other reasons. For one thing, buggy apps suck up a lot of storage space on your BlackBerry. Apps that don't do what they're supposed to do can also be time wasters. There are also stories in the news every day about malicious apps. Bottom line? It pays to do your homework. Here are some tools offered by BlackBerry App World to help you get the info you need:

✦ **Overview:** If you scroll down on the Details page, you can find a brief overview of what the application is and its key benefits. This information helps you know if it's the type of application that you want to install.

✦ **Reviews:** Before downloading an app, read what other people have to say about it. Each application is given a star rating, with five stars being the highest. Don't let ratings fool you, though. Opinions are just that — an application that you find useful, someone else may not.

✦ **Screenshots:** Everyone likes to see what they are buying before they actually commit to it. Screen captures allow you to check out what the application looks like before you install it. You can usually tell at a glance whether an application has an easy-to-use interface that you'll like.

 Once you've downloaded, installed, and used an app, you can share your insights with other potential users. Go back to BlackBerry App World, find the app, and click the Recommend button. After you finish your recommendation, others will see it in the Reviews section.

## Downloading and installing an app

To install an app of your choice, follow these steps. (We're using PeekaWho in our example.)

*1.* **Search for the app you want.**

*2.* **Click the app's name to go to its Details page.**

Here, you may have a few options. See Figure 1-4.

*3.* **Click one of the following buttons:**

• *Download:* This button appears if the app is free. Click the button if you like what you see. After you download the application, this button goes away and a Run button appears.

• *Download Trial:* For applications that are not free but provide a free trial, this button appears. After you download the trial version, this button disappears.

 Trial versions could have several different limitations on them. See Book VII, Chapter 3 for more information.

• *Purchase:* This button appears for apps you must purchase. To purchase an application, you must have a PayPal account.

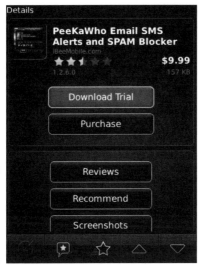

**Figure 1-4:**
The
application's
Details
page.

By the time this book is released, App World may allow other types of payments.

Once you choose one of these options, your download begins, as shown in Figure 1-5.

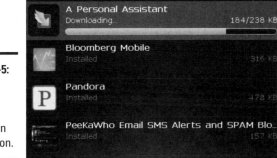

**Figure 1-5:**
My App
World
down-
loading an
application.

4. **After you download and install the app, you may see a button labeled Run. Click this button to launch the application.**

   After downloading an application, you have a few ways to run the application. You do not always need to click Run in BlackBerry App World.

If you don't click Run, you have a few other options for accessing your app:

✦ Go to your Downloads folder and click the application icon you want to run.

✦ Open BlackBerry App World and go to My World. Then highlight the application you want to run. You can do this either by pressing the Menu key and choosing Run or by clicking the app and clicking the onscreen Run button.

## Setting up shop so you can purchase apps

With the hundreds of free applications that you can download, sometimes the one that you want requires payment. BlackBerry App World makes it easy to buy applications.

If you don't currently have a PayPal account, you can create one using a desktop computer. You can't create an account from the BlackBerry smartphone. To create a PayPal account, visit www.paypal.com. PayPal requires you to link a bank account or credit card to your PayPal account. Make sure that you use the same e-mail address for your PayPal account that is associated with your BlackBerry smartphone.

## Purchasing an app from BlackBerry App World

To purchase an application though BlackBerry App World, follow these steps:

*1.* **Locate the app you want and click the Purchase button.**

See "Downloading and installing an app," earlier in this chapter.

By default, the e-mail address that you use with your BlackBerry appears.

*2.* **If necessary, enter your e-mail address to log in to PayPal.**

By default, BlackBerry displays your primary e-mail address in the PayPal login screen. This may not be the e-mail address that you used to set up your PayPal account. This e-mail address must be the same address that you set up your PayPal account with. Check this field carefully before proceeding.

*3.* **Enter the password that you use for PayPal.**

*4.* **Click the Log In button.**

A summary page of your purchase appears. (See Figure 1-6.)

*5.* **Click the Buy Now button to complete your purchase.**

**Figure 1-6:**
The PayPal
login
screen and
Payment
screen.

# Uninstalling an App

Now that you're app savvy, you may decide to download more apps, and the more apps you download, the more likely you are to download an app that's a dud, isn't what you want, or that you never use.

Fear not! You can dump a bad app(le) pretty darn quickly and easily. If you don't use an app very often or you know it doesn't do what you want to do, you should uninstall it sooner rather than later. That's because some apps are space hogs. Removing them can actually increase battery life. In addition, removing applications definitely saves space for more applications, pictures, videos, and data.

You have different uninstalling choices, depending on how you downloaded the app.

To uninstall an application that you downloaded via App World, follow these steps:

1. **Go to the BlackBerry Home screen.**

2. **Open the Downloads folder.**

3. **Click the BlackBerry App World icon.**

4. **Click the My World button.**

   A list of installed apps appears.

5. **Highlight the app you want to uninstall.**

6. **Press the Menu key. From the menu that appears, select Uninstall. (See Figure 1-7.)**

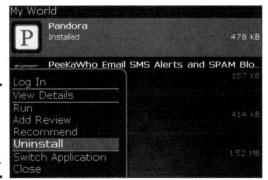

**Figure 1-7:**
My World
uninstalls
your app
just like that.

If you didn't use BlackBerry App World to download an application, it won't display in My World. For example, if you downloaded an app using BlackBerry Browser or installed it using Desktop Manager, you need to follow different steps to uninstall it.

The following steps can be used to uninstall *any* app, whether you downloaded it from BlackBerry App World or not:

*1.* **Go to the BlackBerry Home screen.**

*2.* **Locate and select the Options icon (it looks like a wrench).**

*3.* **Select Advanced Options.**

*4.* **Select Applications.**

*5.* **Scroll to and highlight the application you want to uninstall.**

*6.* **Press the Menu key and choose Delete. (See Figure 1-8.)**

*7.* **Click the Delete button.**

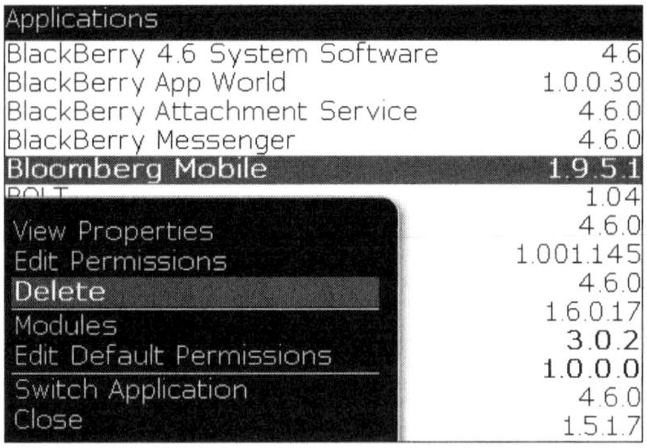

**Figure 1-8:**
Uninstalling
an
application
without
BlackBerry
App World.

TIP

If the BlackBerry Application World app gives you trouble, you can uninstall it and reinstall it using the steps described here.

## More Ways to Find What You're Looking For

What if you can't find what you're looking for in BlackBerry App World? Don't worry — you're not out of luck. You can still try to find what you're looking for in many other Web sites. Some Web sites even specialize in providing BlackBerry apps for users just like you.

What's awesome about these Web sites is that they are sometimes much easier and faster to use than BlackBerry App World. You can purchase apps with a credit card instead of using PayPal.

Here's a list of third-party sites that offer BlackBerry apps:

✦ **Handango:** www.handango.com Offers a wide range of different applications for your BlackBerry smartphone, including games and tools. (See Figure 1-9.)

✦ **ShopCrackBerry:** www.shopcrackberry.com This is a great place you can get applications, but unlike other stores, this one also lets you purchase BlackBerry accessories. (See Figure 1-9.)

✦ **BPlay:** www.bplay.com Contains a great selection of games, themes, ring tones, and applications for the BlackBerry smartphone. BPlay often has special offers, including $1 deals.

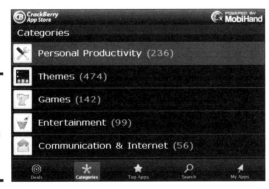

**Figure 1-9:** Third-party app stores give you the best of all worlds.

Mobile carriers have their own app stores. You can access your carrier's app store by clicking an icon on your BlackBerry or by visiting the Web site of your carrier. Because these applications are tied directly to your mobile carrier, purchasing applications is even easier. Your carrier adds the charge for the app to your monthly bill. Here are two examples of carrier app stores:

+ **AT&T:** AT&T's Media Mall is your Web-based one-stop shop for app pur-chases. You can access it through BlackBerry Browser. (See Figure 1-10.)

+ **Verizon:** Verizon's app store for the Storm is a pre-loaded application that Verizon customers can use to get applications. (See Figure 1-10.)

+ **T-Mobile:** T-Mobile's web2go, which can be found as a bookmark in the Web browser or an icon on the Home screen, is a place where T-Mobile users can get apps.

+ **Sprint:** Sprint Software Store is a place where Sprint users can down-load games, business tools, ring tones, and other applications for a BlackBerry smartphone.

Check your local carrier for information on how to access its app store.

**Figure 1-10:**
AT&T Media
Mall and
Verizon
Application
Center.

Free apps are not what service providers are known for. Expect to pay for any apps you find from your service provider.

## Knowing more about your BlackBerry applications

It's not a competition to see who can install the most applications. However, if you're an app junkie, you should know a few things before you start down-loading as many applications as you can.

First, you should understand what those prompts are that you receive when you first install an application. Next, you should know how to find information about the application, like its description, size, and what types of access it has to your BlackBerry smartphone data. Battery life is also a major concern when using any mobile device, so we show you how to save some of that, too.

When the application is downloaded, you may be prompted with a few screens (see Figure 1-11.):

✦ **Trusted Application:** If you're trying to install an app that isn't a Research In Motion trusted application, you're prompted to agree that you allow the application to possibly access sensitive functions on your BlackBerry such as phone, GPS, and the Internet.

✦ **Permissions to Access *x*:** Depending on the application, it may require access to other parts of your BlackBerry. For example, if you install an application that displays phone numbers that you can call, it needs access to your phone. If you deny this permission, the application won't allow you to make a call when you try to.

✦ **Http Connections:** Some applications need to communicate with their Web sites and/or external servers in order for the application to work.

If you do not allow an application to access some parts of your BlackBerry, the application may not work. However, do not allow applications that you do not trust. Unfortunately, there is no way really to know which applications you should trust. We recommend reading reviews of an application so that you can find out what other people are saying about the app and make a judgment call on installing and trusting the application. Applications can send personal information to someone. This information can be then used in mischievous ways.

If your BlackBerry is connected to an Enterprise Server, your administrator may have set application permissions using software configurations that suppress your prompts. For more info on deploying applications in Enterprise, see Book VIII, Chapter 3.

Some applications may prompt you to open the Permissions screen. (See Figure 1-12.) This screen is where you can see exactly what the application is trying to access and the type of access it needs. Press the Menu key and choose Save in order for the application to work properly.

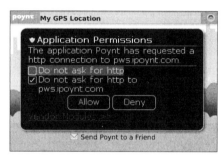

**Figure 1-11:**
Trusted
Application
prompt
(top left),
Permission
prompt
(top right),
External http
Connection
prompt
(bottom).

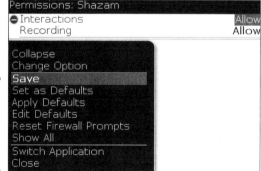

**Figure 1-12:**
Permissions
screen
for an
application.

In the event that you denied one of the prompts for an application and you realize that it isn't working properly as a result, you can manually change application permissions. You don't need to remove and reinstall the app; you just need to follow these steps:

*1.* **Go to the BlackBerry Home screen.**

*2.* **Select Options (the wrench icon).**

*3.* **Select Advanced Options.**

*4.* **Select Applications.**

5. **Select the application you want to change the permissions for.**

   Start typing the application name to reduce the number of items shown.

   The applications that are in bold have had their application permissions changed. This allows you to quickly find the applications that you have allowed to have access to other parts of your BlackBerry smartphone.

6. **Press the Menu key for a list of options:**

   You can see a few of the options that you have to choose from:

   - *View Properties:* This contains information about what the application is and what it does. It also includes the name of the company that created it and the version of the application that is installed.

   - *Edit Permissions:* Here is where you can see and edit permissions that an application has. You can change all permissions in a group or selectively choose which item you want to change.

   - *Delete:* From this menu, you can remove an application from your BlackBerry smartphone. This also deletes any information that was stored in the application and that is not on your removable memory.

   - *Modules:* These files make up that specific BlackBerry application. By highlighting one of the modules and choosing View Properties, you can see more information, including the size and date it was created. Note that this is a list of all modules on the BlackBerry smartphone, not just of that application.

7. **Choose Edit Permissions.**

8. **Press the Menu key and choose Apply Defaults.**

   Choosing this option changes all the permission settings back to their original state.

You should not change any other permissions, other than the ones an application requests. Changing these permissions may allow an application that you do not want to have access to other parts of your BlackBerry smartphone that you don't want it to access.

## Tips and tricks

Now that you are a BlackBerry application master, there are only a few more things that you should know about managing BlackBerry applications. These tips help you save battery life and space on your device.

- ✦ **Save battery life by closing applications.** A BlackBerry smartphone can run multiple applications at the same time. They run in the background as you use another application. Each application that is running in the background takes up a little bit more of your battery. If you play Brick Breaker and hit the Esc key to check your messages, the application is

**Book VI
Chapter 1**

**Finding
Applications**

still running in the background. To close an application, press the Menu key and choose Close (or Exit, depending on what the developer named that function). In some applications, including Slacker, instead of Close you may see something like Shut down Slacker. (See Figure 1-13.)

**Figure 1-13:**
Shut down
Slacker
Radio.

✦ **Remove or limit hidden applications.** Some applications are still secretly running in the background, even if you press the Menu key and select Close or Exit. If you choose Switch Application, they don't even show up on the ribbon. Unfortunately, there is no way to tell which of these applications are silently running.

Applications that need to notify you based on e-mails, such as the Facebook or MySpace apps, need to be running all the time so they can alert you when you have new notifications. This process, even though it's very small, still takes away from your battery life. The more of these applications you have running, the quicker your battery will run out.

The only way to stop this from happening is to uninstall the applications you do not use. This also helps with freeing up space on your BlackBerry smartphone.

✦ **Find out how large your apps are.** If you download a lot of applications, you may start running out of room. You may want to know how much room you'll free up if you delete an application. To find out the size of an application, follow these steps:

1. *Go to the BlackBerry Home screen.*

2. *Select Options (the wrench icon).*

3. *Select Advanced Options.*

4. *Select Applications.*

5. *Press the Menu key and choose Modules.*

6. *Type the application name that you want to look for.*

7. *Highlight the first module.*

   Note: An application may have more than one module.

8. *Press the Menu key and choose View Properties.*

9. *Find Size.*

   Note: The size is in bytes. 1024 bytes = 1MB. (See Figure 1-14.)

Some applications have more than one module; therefore, you will need to find out the size of each one and add them up to know the total size of the application. If an application has more than one module, repeat Steps 7-9 for each module and add up the sizes.

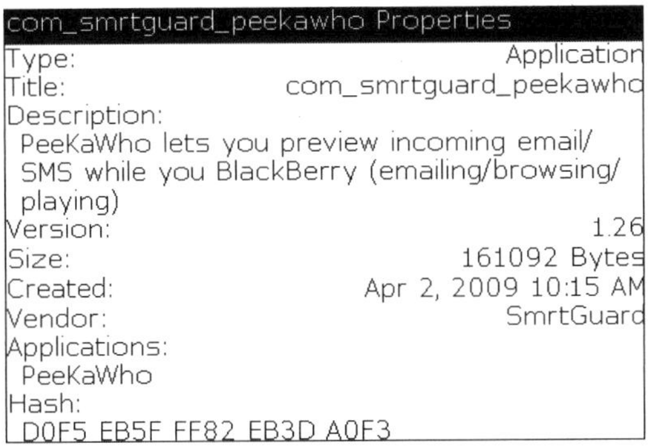

**Figure 1-14:**
The
BlackBerry
Application
Properties
screen.

```
com_smrtguard_peekawho Properties
Type:                              Application
Title:             com_smrtguard_peekawho
Description:
 PeeKaWho lets you preview incoming email/
 SMS while you BlackBerry (emailing/browsing/
 playing)
Version:                                 1.26
Size:                         161092 Bytes
Created:              Apr 2, 2009 10:15 AM
Vendor:                          SmrtGuard
Applications:
 PeeKaWho
Hash:
 D0F5 EB5F FF82 EB3D A0F3
```

# Chapter 2: Networking Like a Social Butterfly

## In This Chapter

✔ Using social networking apps on your BlackBerry

✔ Finding out about handy features of some apps

✔ Communicating through social networking apps

Social networking tools have become a popular way for people to keep in touch, find jobs, promote businesses, get a date, or just make new friends. Along with the various social networking venues come various communications tools and practices. These include, for example:

✦ **Information sharing:** By updating your status and profile and posting pictures and information about things that you're doing, friends and family know what is going on in your life. Such sharing keeps friends and family better informed without the need to send individual messages to each person.

✦ **Personal messaging:** E-mail is so yesterday. When communicating through the social networking sites, you still send messages like e-mail, but you learn more information about a person to make your message more personal. You can start a conversation by checking out your buddy's pictures to see where he has been lately or comment on his status update.

✦ **Professional networking:** Business is all about who you know. That's why social networking sites are becoming a key way to find jobs and keep those business relationships growing. You can also see who your friends know, and that may provide you with a competitive advantage for securing the job you want.

✦ **Location tracking:** Some social networking tools use the BlackBerry smartphone's GPS to provide even more functionally. These applications track where you are and share that information with your friends. You can set up alerts that notify you when someone special is close.

All the different networking sites have their own communication tools, and you may find that managing the e-mails, status updates, and other types of messaging becomes overwhelming. In this chapter, we present some of the social networking applications that you can use with your BlackBerry smartphone so that you no longer have to sit behind a computer to keep on top of the notifications, updates, alerts, and e-mail messages from the social networks that you are a part of.

We assume in this chapter that you already have a MySpace, Facebook, LinkedIn, and Twitter account. If you haven't already plunged head-first into the world of social networking but you'd like to, hurry off to a desktop computer and visit the social networking sites of your choice to set up your profile. We don't recommend using your BlackBerry to set up a profile. (The BlackBerry is good for a lot of things, but setting up a lengthy profile that represents your online identity for the whole world to see is not one of them.) When you're all set up, come back to this chapter and find out how you can sync your social networking identity with your BlackBerry.

## Hooking Up Your BlackBerry with Social Networking Apps

You have two main options for using mobile versions of social networking sites on your BlackBerry smartphone. One is to use BlackBerry Browser and surf the Web like you would visit any other Web site. The second is to download the mobile app to your BlackBerry smartphone. Downloading the app provides additional benefits, such as allowing you to directly upload pictures or quickly send a message to someone in your network. If you want to download an app associated with a social networking site, you have these two choices:

+ **Use BlackBerry App World,** an application that helps you find, acquire, and manage other applications for your BlackBerry smartphone. (Book VI, Chapter 1 tells you how to get and use BlackBerry App World.)

+ **Use BlackBerry Browser** to download directly from a link on the social networking site.

We recommend that you use BlackBerry App World for downloading social networking apps whenever possible. BlackBerry App World decides whether you have enough room on your BlackBerry smartphone — before downloading — so you don't run out of space on your device. Another advantage of using BlackBerry App World is that you receive upgrade notifications. When an upgrade is available, an icon on the main screen appears, as well as a message that looks like an e-mail that links you to download the new version.

## Deciding Which Social Networking Apps You Want to Retrieve

You can easily retrieve the social networking apps you want in one of two ways (see the preceding section), but choosing which one(s) you want to add to your BlackBerry smartphone may be more of a challenge. Because you have limited space for apps on your BlackBerry, you should add apps that you think you'll make the best use of.

A social networking app is always listening for new messages. The app runs in the background without you knowing it, so your battery will be used more, resulting in the need to charge it more often. If you notice your battery is constantly running low after installing an app that runs in the background, you should consider removing the app. If your battery life returns to normal, then you know that was the issue. Instructions on how to remove an application can be found in Book VI, Chapter 1.

In this section, we give you an overview of various social networking applications, including MySpace, Facebook, LinkedIn, ÜberTwitter, and OpenBeak. We tell you about each app's key features and functions, as well as give you a review from our perspective.

## MySpace

www.myspace.com

MySpace is a social networking Web site that gives each user his or her own Web page (or space) within the MySpace network. You can create a personal profile using MySpace's templates, upload photos and videos, play games, and send and receive messages to other members of the MySpace community. In addition to personal pages, MySpace also hosts pages for celebrities and musicians so that members can keep track of their favorite artists.

You can do just about anything MySpace-related using the MySpace app for the BlackBerry smartphone, shown in Figure 2-1. The application integrates MySpace's notifications within the BlackBerry so that you can send and receive messages on the go. Other features of MySpace include:

✦ **A full messaging interface:** Using the MySpace address book, you always have your friends' contact information when you need to send them a message. You can also post messages to your friends' pages for everyone else to see, just as you can in the full Web site version.

✦ **Real-time status and mood updates:** Members can choose a picture of a smiley face or other cartoon that represents how they are feeling as well as a quick blurb about something they want to share. These are called *mood indicators* and *status updates*.

✦ **Connection management:** You can use your BlackBerry to search for a friend, add a friend to your MySpace community, or respond to friend requests.

✦ **Picture uploads:** From your BlackBerry smartphone, you can simply take a picture and upload it directly to your MySpace page.

✦ **Notifications of events:** You never have to worry about missing a message, status update, or picture comment. You can set up a custom notification so that when anyone posts something to your page, you will be notified.

Having MySpace on your BlackBerry makes it easy to stay in touch with friends, stay up-to-speed on what they're doing, and stay in tune with how they feel. The unique music portal also lets you follow your favorite bands, and includes the ability to view upcoming tour information and concert dates. The animated mood icons are unique features that allow you to quickly express yourself.

**Figure 2-1:**
MySpace
status
update
(left) and
MySpace
managing
photos
(right).

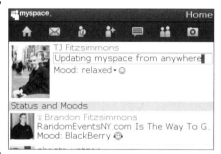

MySpace also has a BlackBerry-friendly Web site. This site makes it easier for you to do things that are more difficult to do using the MySpace app. If you want to upload pictures, for example, you can visit the mobile site by going to `http://m.myspace.com` instead of trying to use the app.

## Facebook

`www.facebook.com`

Facebook is a social networking Web site that helps you find and connect with others. You can add friends and colleagues and send them messages, share pictures and video, and exchange wall posts, status updates, and comments.

Key features of Facebook include

✦ **Regular status reporting:** Update your status, view the updates of others, and comment on your friends' status updates.

✦ **Photo sharing:** Upload photos and post them to your Facebook page immediately, using your BlackBerry.

✦ **Online communication:** Download your Facebook address book to your BlackBerry so that you can send messages and reply to any of your Facebook contacts, 24/7, wherever you are.

✦ **Time and contact management:** Because the Facebook app is integrated with the native BlackBerry Calendar and Address Book, you can sync your appointments and contacts. This functionality allows you to see your Facebook Events and friends' birthdays in your BlackBerry Calendar.

Facebook for BlackBerry is an essential tool for any BlackBerry user with a Facebook account. It delivers the main parts of the Facebook Web site to your phone so that you can use them quickly and easily wherever you are. Another great BlackBerry integration feature is the ability to sync your Facebook friends' profile pictures to your BlackBerry address book. If you receive a call from a Facebook friend, his or her profile picture will automatically appear on your BlackBerry screen.

The Facebook for BlackBerry app doesn't support Facebook Chat. You also can't use the app to search for friends.

## LinkedIn
www.linkedin.com

It's not a secret that the working world has a lot to do with relationships and who you know. That is why it is important to ensure that you always maintain some type of communication with people that you meet. LinkedIn is the best tool for this. It is not a social networking site like Facebook or MySpace in which you post pictures and status updates on everyday life. LinkedIn is specifically focused around your professional career.

LinkedIn is like a personal Rolodex that is never outdated. Any time a LinkedIn contact updates his or her e-mail addresses, phone numbers, job history, or other information, this updated information is right there at your fingertips. With pictures, resumes, and status updates, you can find out more about your contacts and leverage that information while you conduct your business or search for other opportunities. LinkedIn doesn't have a BlackBerry app, but you can still use your BlackBerry to stay up to speed on LinkedIn.

From your BlackBerry you can quickly find a contact that you know or have just met. Depending on what the contact has posted on LinkedIn, you may be able to find information that helps you strike up an interesting conversation. By updating your LinkedIn status from your BlackBerry, you can tell people where you are in your business travels. This way, your business partners nearby can reach out to set up a personal meeting. Other key features of LinkedIn help you

✦ **Quickly find updated information such as e-mails in your LinkedIn address book.** From your BlackBerry, type the first or last name of the contact in the LinkedIn search bar. You can also search by the person's place of employment.

✦ **Filter and see all the latest updates by your contacts.** These updates include info about projects they're working on, conferences and events they're attending, or general information on what's happening in their lives and career.

✦ **Invite and add people you've recently met to your LinkedIn address book.** From the Web site, click Invite and type the person's e-mail address to send the request.

✦ **Search through the LinkedIn membership by keywords, name, company, or title.** Suppose you're on your way to a job interview. If you want to find out if you know anyone who already works for the company, you can search for that company's name to find a list of employees at the company who have LinkedIn accounts.

If you're on Twitter, be sure to link your Twitter account to LinkedIn. Then whenever you post a tweet to Twitter — using the Twitter Web site or any other application that updates your Twitter page — LinkedIn is also updated.

## Twitter

www.twitter.com

Twitter is a Web site that allows users to post, or *tweet,* anything that they wish — as long as they can do so in 140 characters or less. Tweets can also include links to pictures, videos, or Web sites.

Twitter connects you with friends and family, as well as musicians, politicians, artists, actors, and other celebrities. Companies also use Twitter to promote their business and to offer special discounts.

You can post updates to your Twitter account without downloading an application; however, what you can do is very limited. Here are your options:

✦ **Visit Twitter's mobile site, http://mobile.twitter.com.** From here, you can view your Twitter page and quickly post updates.

✦ **Set up Twitter updates using SMS/text messaging.** Twitter allows you to send Twitter updates via text message. To find out how to set up your phone to work with Twitter, visit www.twitter.com and search for SMS on the Help page.

✦ **Download a third-party app.** See the following section for more information.

## Twitter-related apps

The following information tells you how to do more with Twitter. These apps are for Twitter addicts — people with multiple Twitter accounts and people who use Twitter as a professional marketing tool.

### ÜberTwitter

www.ubertwitter.com

*ÜberTwitter* is an app that allows you to tweet from anywhere. In addition to the usual (text, pictures, and links), you can use GPS to update your location. Imagine you're in a little café and you had a wonderful warm apple pie. You can update your tweet page with "I just had the most delightful apple pie." If you're also using ÜberTwitter, your exact coordinates are attached so others that subscribe to your Twitter feed can visit that place to enjoy the same dessert. Some people might call that TMI (too much information), but if you're a restaurant critic with the Twitter handle @PieSnob, your followers may consider your GPS coordinates a necessary revelation.

**Book VI
Chapter 2**

Networking Like a
Social Butterfly

ÜberTwitter integrates with Twitter, but it can also optionally update your Google Talk status with your last tweet, making your tweets reach an audience wider than your Twitter subscribers. Other key features of ÜberTwitter include:

✦ **Accessing multiple accounts.** You can have multiple Twitter accounts for your personal or business use. ÜberTwitter allows you to manage all your accounts to send and view tweets from each account.

✦ **Making your location known.** By default, your GPS location is attached to every one of your tweets. This can be a crucial component to your business's marketing strategy. If you don't want your location recorded, you can always turn off this setting or have it prompt you every time you post.

✦ **Using integrated picture and video uploading and viewing options.** If you see something that you want others to see, all you need to do is take a picture and quickly upload and share it on Twitter. Give your followers the feeling that they're right there with you!

If you're a Twitter junkie and tweet everywhere you go, then ÜberTwitter (shown in Figure 2-2) is a must-have application for your BlackBerry. ÜberTwitter is more powerful than any desktop client because of its picture/video uploads and its ability to send your GSP coordinates (to indicate where you are tweeting from).

**Figure 2-2:**
ÜberTwitter
home page
(left) and a
Tweet page
(right).

## OpenBeak

www.orangatame.com

Use OpenBeak to post text, links, pictures, and video anywhere you are. You can also read what your friends are doing, comment on their posts, or send them a direct message. OpenBeak (shown in Figure 2-3) offers the following features:

✦ **View timelines.** Some people post multiple messages per week, day, or even per hour. OpenBeak allows you to view a timeline of all of the recent postings so you won't miss anything from someone you're following.

✦ **See new messages.** No matter where you are, when you get direct messages, you will be notified so that you can quickly respond.

✦ **Receive customized notifications.** If you have a favorite song or ring tone, you can use it as your customized OpenBeak notification.

To conserve battery life, if you receive many tweets, keep the sound for each notification short. For example, a quick beep is better than playing a 30-second song clip. This word of advice applies to all notifications, from tweets to e-mail notifications and incoming text message notifications. You can find out how to change your settings in Book I, Chapter 3.

**Figure 2-3:**
OpenBeak
tweeting.

# Engaging Your Social Networks from Your BlackBerry

Everyone — or just about everyone — is a social networking junkie. Families (including grandma and grandpa) keep in touch with status updates; friends share pictures of last night's party; music junkies follow their favorite bands; and professionals search for career opportunities using social networking tools. If keeping in touch is important to you, you need to know the basics of using your BlackBerry to access your social networks when you can't do so from home.

## Following your favorite band on MySpace

If you have a favorite musician, comedian, or filmmaker, you can use your BlackBerry smartphone to keep up-to-date on what this artist is up to. By adding the person/group as a MySpace friend, you can look up upcoming shows, find pictures, and write and read comments about the artist. Adding friends, people, and groups is easy and can be done from both your desktop and MySpace mobile application.

**Book VI**
**Chapter 2**

Networking Like a
Social Butterfly

Follow these steps to add your favorite group and get more information about them:

1. **Go to the Downloads folder on your BlackBerry smartphone and click to open the MySpace app.**

2. **Press the Menu key and select Search.**

3. **Enter the name of your favorite band or artist, such as U2 or Aerosmith.**

The search results may display more than one choice. To identify the correct group, you need to open the group to see the following items: The banner on top changes from MySpace to MySpace Music, and Upcoming Shows is now an option to choose.

4. **Scroll to and highlight the group you wish to add and click the track-ball to open the group's page.**

5. **Scroll down to the Add to Friends option and select it to add the group.**

6. **Click Add to Friends.**

7. **Type in the letters you see in the CAPTCHA entry box, as shown in Figure 2-4.**

*CAPTCHA* stands for Completely Automated Public Turing Test to Tell Computers and Humans Apart. It's a test that requires you to type the numbers and letters that you see onscreen.

You can find more details about the artist by clicking the artist's name in your friends list.

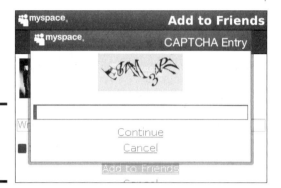

**Figure 2-4:**
MySpace
CAPTCHA
screen.

## Sharing your pictures with your Facebook friends

No longer do you need to wait to come home from a vacation or hanging out with friends to post pictures to Facebook. If you have a BlackBerry with a camera, you can use your BlackBerry to post pictures instantly and tag your friends as things happen.

1. **Go to the Downloads folder on your BlackBerry smartphone and click to open the Facebook for BlackBerry app.**

2. **Scroll to the top, press the trackball or trackpad, and select Camera, which is the third icon from the left.**

   The camera application opens.

3. **Take a picture by pressing the trackball or trackpad.**

4. **The Facebook application reappears, and you have these options:**

   - *Caption:* Enter a brief description of the picture.

   - *Album:* You can place the picture in an existing album or create a new one.

   - *Tag This Photo:* Tagging a photo allows you to choose people from your Facebook friends and highlight them so a link to their Facebook profile is created for others to see.

5. **Select Upload to send your picture to your site.**

Make sure that the pictures you post are appropriate. You cannot delete pictures that you post from your BlackBerry from your BlackBerry. To delete pictures you will need to log into Facebook from your computer.

## Inviting your new client to connect on LinkedIn

It is all about timing. After you meet someone, you can't wait too long to add the person to your LinkedIn network. Your BlackBerry allows you to choose the right time to manage your connections.

Here are the steps to add someone to your network:

1. **Click the BlackBerry Browser icon on the Home screen.**

2. **Press the Menu button and choose Go To or simply type** http:// m.linkedin.com **into the address bar.**

3. **Sign in with your e-mail and password.**

4. **Choose Invite.**

5. **Complete the form with the person's first name, last name, and e-mail address. Add a subject header and a message.**

6. **Click Send Message.**

Your message should be personal. Include info about where you met or what you talked about. This personal touch will not only help the recipient remember you, but it will help you to continue building the relationship.

## Posting a picture on Twitter with ÜberTwitter

With only 140 characters to convey a message on Twitter, it's not a surprise that pictures can help you say a lot. If you tweet from Twitter.com, there is no way to attach a picture. That's why ÜberTwitter's picture-posting ability comes in handy. It extends all of Twitter's main features and adds a bunch of other cool stuff.

Follow these steps to post a picture with a tweet in ÜberTwitter:

1. **Go to the Downloads folder on your BlackBerry smartphone and click the ÜberTwitter icon to open it.**

2. **Press the Menu key and select New Tweet.**

3. **Using 120 characters or less, enter your tweet.**

Craft your message carefully because your tweet must be 20 characters shorter than normal. A video or picture takes up 20 characters of tweet real estate. You only have 120 characters of space to type your message.

4. **Press the Menu key and select one of the following options:**

   • *Add Picture:* You are magically taken to the pictures that are already stored on the device.

   • *Take Picture:* Your camera magically starts up so that you can take a picture to store it on the device.

5. **Select the picture you want to post.**

6. **Press the Menu key and choose Send Tweet.**

## *Reading the tweets of others with OpenBeak*

When you have time to kill or just want to read what other people are talking about, using the Public Timeline feature on OpenBeak is great. It displays the messages of the last few people who tweeted. These tweets give you a slice of life on Twitter — conversations between two users, a picture of world events, or links to interesting sites. To see these posts, simply do the following:

1. **Go to the Downloads folder on your BlackBerry smartphone and click the OpenBeak icon to open the app.**

2. **Press the Menu key and select Everyone near You.**

3. **Scroll down to read the tweets.**

To respond to someone's tweet, highlight the message, press the Menu key, and choose Reply. Type your response and click Send Tweet.

# Chapter 3: All Work and No Play Makes for a Boring BlackBerry

## In This Chapter

✔ **Finding games for your BlackBerry**

✔ **Choosing games worth paying for**

✔ **Getting free games**

**C**hecking your e-mail, viewing your calendar, evaluating your tasks, using the memo pad — what fun is that? Take a break from checking your e-mail and play a game — one of the hundreds of games available on your BlackBerry smartphone. In this chapter, we introduce you to a few of the top games available through BlackBerry App World and elsewhere on the Web. Choose from all different categories, such as:

✦ **Arcade and action games:** Play classic games from PacMan to the latest games like Guitar Hero.

✦ **Cards and casino games:** Don't have time to fly to Vegas for a chance to have fun and play games? Don't worry.

✦ **Puzzles:** From Sudoku to Tetris, this category has all your puzzle needs covered.

✦ **Sports games:** No matter what sport you're into — golf, baseball, bowling, football — there's a game for you.

✦ **Strategy games:** Games that require you to figure out the right move — chess, SimCity, and even Tic-Tac-Toe — are all available to download and play.

Some games are free and some you need to pay for. Some of the paid games have a free trial period so you can play before deciding to buy. Purchasing them is easy, and most of the time, it can be done right from your BlackBerry.

In Book VI, Chapter 1 we discuss the different sites and application stores that you can use to search and install the applications or games that you want. We also provide step-by-step instructions on how to install the apps and troubleshoot some of the issues you may have. In this chapter, we cut down on your searching and provide the info on the best games.

We recommend that you use BlackBerry App World for downloading games and other applications whenever possible. That's not because we don't love BPlay, CrackBerry, Handango, or any of the other sources for games that we talk about in this chapter — we do. However, BlackBerry App World automatically determines whether you have enough room on your BlackBerry smartphone — before downloading the game — so that you don't run out of space on your device. Other app providers simply can't do that.

Another advantage of using BlackBerry App World is the upgrade notifications you receive in your BlackBerry messages. When an upgrade is available, an icon on the main screen appears, as well as a message that looks like an e-mail that will link you to download the new version.

 Not shutting down a game correctly can reduce your valuable battery life. To shut down any application, including games, use the BlackBerry Menu button and select Close or Exit, depending on the application.

# Getting Started as a BlackBerry Gamer

You can download games in several ways to your BlackBerry. All the methods and detailed steps for using them are outlined in Book VI, Chapter 1.

## BlackBerry App World

For most of the games featured in this chapter, the best way to get them is through BlackBerry App World.

BlackBerry App World is an app that helps you find, acquire, and manage other games and applications for your BlackBerry smartphone. For details on how to install and use different sites and App World, read Book VI, Chapter 1. With the BlackBerry App World app installed and open, simply follow these steps:

1. **Click the Search icon.**

   It's the fourth icon on the bottom of the screen.

2. **Type the name of an application or a part of the application name.**

   For example, if you type **Battleship**, a list of games with Battleship in the name appears.

3. **Select the application you want, read more about it, and click the Download or Purchase link.**

 You don't need to know the exact name of a game you're searching for. In fact, if you want to just see what's out there, you can use the search feature to discover games. For example, if you type **ship** in the BlackBerry App World's search box, you see a long list of games, from Battleship to GI Joe, and more.

## Third-party sites

BlackBerry App World has a limited number of games. You may also be able to get better pricing for the same game from another site. We suggest that you check out all sites before making your purchase. Here is a list of other app and game providers you might want to check out:

✦ **CrackBerry:** `www.crackberry.com` The one-stop shop to get information on the latest trends in BlackBerry, help with issues, reviews on devices, and a store to purchase accessories, applications, and devices.

✦ **Handango:** `www.handango.com` Dedicated to downloads specifically for smartphones. With their detailed categories and subcategories it makes finding what you need easy.

✦ **Download.com:** `www.download.com` Everything you need to download applications for your BlackBerry, Windows, or Mac computer.

When downloading applications from places other than App World, you will not have the ability to see how much memory you have available. It's best to check the amount of memory before installing applications to ensure that an application will install. To find out how much memory you have, do the following:

1. **From the BlackBerry Home screen, select the Media folder.**

   It's the one with the blue triangle Play button icon.

2. **Click the BlackBerry button for the menu and select Memory Use.**

3. **If, under Device Memory the total amount of free space is less than 2MB, don't install additional applications.**

## Finding games using BPlay.com

BPlay.com is a Web site that allows you to download games, themes, and more for your BlackBerry. BPlay offers a well-organized Web site to help you quickly find what you're looking for. It also displays the top 15 games so that you can easily see what's hot right now. These games aren't free, but you can often get free trials of games.

If you choose to use BPlay.com to download games, follow these general steps:

1. **From your desktop computer or BlackBerry Browser, go to `www.magmic.com`.**

2. **Find the item you want to download.**

   You can choose from the Search Products option at the top of the screen, view Top Games (this option appears on the left side of the screen), or browse the game selection by selecting the Games tab in the middle of the screen.

Book VI
Chapter 3

All Work and No Play Makes for a Boring BlackBerry

3. **Select the game you wish to purchase. Depending on the game, you may see two options:**

   - *Buy It Now:* Purchase the game by using your credit card.

   - *Bill to my Phone:* By entering your e-mail, mobile number, and BlackBerry PIN, the charge will be added as a one-time charge to the bill you receive from your mobile service provider.

   You will receive an e-mail from BPlay with a link to download the game, as well as any codes you will need to register the game.

4. **Open the e-mail on your BlackBerry and follow the directions to install the game.**

# Finding Our Favorite Games That Cost a Little Something

Without further ado, the following sections list our favorite games that you can purchase from BPlay and BlackBerry App World.

## Guitar Hero

www.magmic.com

Guitar Hero is a popular game on so many different entertainment consoles, and it's also available for the BlackBerry! In order to be successful at this game, you have to hit the BlackBerry keys at the same time as the rhythm of the music to simulate playing the guitar. There are different levels to challenge your skills. If you can't stop playing Guitar Hero at home, then you're not going to be able to put down your BlackBerry.

The game comes with 15 songs to start with, but you can always download more songs if you get bored for free from the application. Here are some additional features to the game:

✦ **New songs:** Weekly, there are three new songs that you can download. This means that you'll always have something new to challenge your guitar skills.

✦ **Customizable venues:** A *venue* is the backdrop or stage that simulates the concert experience. Changing your venue doesn't change the skill level, but it is fun to play from a gigantic outdoor stadium to an intimate club every now and again.

✦ **Customizable characters and guitars:** Choose from a few different characters and types of guitars. You can choose one of four authentic Gibson guitars. Game play is the same with each guitar, but it's always interesting to feel like you are playing a new instrument.

✦ **Additional features:** The goal is to get the best score, not just because it makes you feel good but because high scores unlock different types of guitars, and other perks. You can earn achievements and break records from previous plays. See Figure 3-1.

**Figure 3-1:**
Playing a game with Guitar Hero.

## Who Wants to Be a Millionaire

www.magmic.com

The television show *Who Wants to Be a Millionaire* is a quiz show that's been broadcast all over the world for years. If you love the thrill of trivia, then you're sure to enjoy the BlackBerry version of the game, shown in Figure 3-2. Just like the TV game show, the mobile version challenges you with questions to win a million dollars.

If you master this game, you won't really win a million dollars, but it is still a lot of fun.

You are challenged to answer 15 questions that keep getting harder after each round. If you don't know an answer, don't worry — just like the TV show you have four lifelines: Ask the Audience, Phone a Friend, 50/50, and Ask the Expert. If you answer questions before time is up, the extra time is banked for the million-dollar question. That means that if you're smart enough or lucky enough to get to the end, you will have extra time to answer.

This game offers many of the same aspects as if you were playing the game for real. Such items are

✦ **A gazillion questions:** Okay, maybe not a gazillion (maybe only hundreds), but you can play a whole lotta rounds of this game before you hit a rerun question. The questions fall into several categories, so you get a diverse range of trivia to challenge your knowledge.

+ **Realistic sounds and animation:** It feels like you're really sitting in the *Millionaire* hot seat. With sounds and animations from the TV show, your heart will beat, your palms will sweat, and you'll have a great time playing.

**Figure 3-2:**
Who Wants
to Be a
Millionaire
questions
(left) allow
you to
poll the
audience
(right).

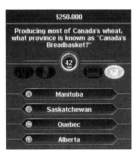 

## Street Fighter II Champion Edition

www.magmic.com

Wow. This is where gaming technology has taken us — you can play Street Fighter (shown in Figure 3-3) on your BlackBerry! If you ever played Street Fighter in the arcade, you will love love *love* this game. The graphics and background sounds are exactly as in the original arcade game. If only the BlackBerry keyboard were big enough to fit two hands so that you could play against a friend.

You can choose one of the 12 characters of the other original arcade version of the game, as well the four bosses. All the special moves that you remember will work for each player. It's a bit difficult to get the move out every time on the keyboard, but it's still a ton of fun to play.

Here are some of our favorite features:

+ **Arcade graphics and background sounds:** Each background, character, and move looks just like the original game. Unfortunately, you won't hear all the sounds you're used to, but all the background sounds are just like you remember.

+ **Auto save:** Don't worry about not having hours to play to beat the game. The auto save feature lets you save who you beat so you won't need to start from the beginning to try and win the game.

+ **Special characters moves:** There's no need to worry about learning new combinations for special moves. All the original moves to throw a fireball or upper cut or blow flames out your mouth will still work.

**Figure 3-3:**
The Street Fighter game has all the same characters you love (left) and playing is just like old times (right).

## Texas Hold'em King

www.magmic.com

The Texas Hold'em poker game for the BlackBerry (shown in Figure 3-4), is just as much fun for the serious poker player as it is for those of you who just dabble. However, unlike real poker, you do not win real money, but try to win as much fake money as possible.

The game is just like the actual poker game, with a total of seven cards (two per person) and five community cards. The goal is to make the best five-card poker hand out of all seven cards. It is a game of knowing the odds, knowing your opponents, and knowing how to bluff by betting on a bad hand.

Texas Hold'em King is a great-looking game that allows you to play by yourself or join a networked game where you can play poker with people from around the world. There are over 100,000 poker members in the Texas Hold'em community to play tournament games with.

**Figure 3-4:**
Playing Texas Hold'em.

There are many Texas Hold'em poker games for the BlackBerry, but Texas Hold'em King for the BlackBerry is one of the most feature-filled Texas Hold'em games around. It includes great sounds and graphics and has several key features:

# Getting the inside edge with Street Fighter II

In Street Fighter II, each character has the same set of six moves: three punches and three kicks. In addition, each character has its own special moves. You can win the game without learning the special moves, but it will be much harder. It is also really cool when you can beat your opponent by throwing a fire ball or electrocuting him.

The keys used to move around the screen are: A and D to move forward and back, W and S to jump and kneel, and Q and E to jump forward and back. To fight, use any of these keys to punch and kick your way to the top:

✔ **Punches:** Broken into Quick punch (U), Medium (I) and Power( O). Quick punches

don't do a whole lot unless you use several in succession. Medium punches pack a bigger wallop, but aren't as fast. The mega punch takes longer to use, but it will knock the power out of your foe.

✔ **Kicks:** Broken into Quick kick (J), Medium (K), and Power (L). If you need speed and agility, throw several quick kicks to throw off your opponent, and use the slower, more powerful punches when you have thrown your foe off guard.

For a list of special moves, you can visit: `www.wegame.com/guides/street-fighter-2-moves`.

✦ **Online multi-player action:** Take advantage of the multi-player mode, which challenges you to outsmart your opponents as you win your way to the top of the leader board. As you play, don't forget to use the chat feature to try and bluff your opponents.

✦ **Online Web statistics:** Check the online leader board to see how you stack up to the rest of the poker-playing community. Use this place for your bragging rights or post a message to specific players.

✦ **Unlock special features:** As you win money points in the game you also have the chance to unlock special tables and event venues that you can play at. The special tables and events give you a new look for your table and cards.

Games with online play take additional battery life to play, so you may notice that your battery will run out quicker than if you play against the computer. Be sure to charge your smartphone more frequently when you've been playing this game.

If you don't already have one, you may want to contact your mobile carrier to add an unlimited data package to your service plan. Internet gaming requires a lot of data to pass back and forth; therefore, if you are not on an unlimited plan, your monthly carrier bill may be high.

If you don't know the rules of Texas Hold'em, feel free to look online for sites that explain it in detail. A good site to check out is `http://poker.about.com/od/poker101/ht/holdem101.htm`.

## Sudoku

`www.magmic.com`

When you need a brain teaser, try Sudoku. Sudoku is a game that is easy to play, but challenging to complete. (See Figure 3-5.) You no longer have to carry around a pencil and erase the numbers you want to change.

The goal of the game is to fill in all the blanks with the numbers 1–9 without repeating the same digit in a row or a column. The grid is 9 x 9, with nine 3 x 3 boxes within in. To start, you get a few numbers on the board; however, the games get harder as fewer numbers are given to begin with.

Whether you are new to Sudoku or a Sudoku aficionado, you will love this BlackBerry version because it comes with several features:

✦ **More puzzles than you can shake a stick at:** There are 200 puzzles with four different difficulty levels, from easy to genius. The difficulty is measured by how many numbers are provided at the beginning of the game.

✦ **Downloadable puzzles:** If the 200 boards that come with the game aren't enough for you, you can download new games daily.

✦ **Customizable board themes:** If you're tired of the same Sudoku board, you can change its appearance. This may not help you finish the board any quicker, but it's a nice feature.

✦ **Rate yourself:** If you think you're good at Sudoku, you can find out for sure. After you submit a completion time for a game, go online and check out how you compare against others. Keep on trying until you become a Sudoku master.

**Figure 3-5:**
Sudoku
is a great
challenge
for those
who like
puzzles.

# Getting Free Games That Are Worth Playing

The thing about games you pay nothing for is that sometimes they're worth what you pay for them. Don't worry, though. You don't have to spend a bundle to buy games. We've picked some of the best free games around so that you don't have to go to the trouble of downloading a dud.

## Ka-Glom

www.magmic.com

Ka-Glom (shown in Figure 3-6) is a game that requires both a little strategy and a little luck. It's like Tetris meets the Blob, but with explosions. Maneuver the blocks and the Ka-Glom jelly around. When you get four like Ka-Gloms in a row, you start an explosive reaction of all like colored blocks surrounding the exploding Ka-Gloms. As the blocks fall, new Ka-Gloms are created, starting new explosions. You gain points and bonus points for blowing up more and more blocks.

This is one of those classic games that you just can't put down. Here are some of its best features:

✦ **Several playing modes:** You can play in one of three modes: Normal, Survival, and Time Attack.

✦ **Access to hidden modes:** As you explode more and more blocks and increase your score, you unlock different ways to play the game.

If you try to download this game from BPlay.com, you have to pay for it. Download it free from BlackBerry App World.

**Figure 3-6:** Ka-Glom is a fun free game.

## PacMan

www.bennychow.com

There is nothing better than playing a classic game on the latest technology. By just using your trackball or trackpad, you can navigate PacMan around the maze, eating pellets and dodging Blinky, Pinky, Inky, and Clyde.

PacMan on the BlackBerry is definitely a game that you can play for hours, challenging yourself to get to the next board and get the high score. This free game doesn't take up a lot of space in your BlackBerry memory and provides hours of fun.

Some features you'll enjoy are

✦ **Full-screen resolution:** No matter the type of device you're using, you get the full-screen experience.

✦ **Arcade sounds:** Just like the original game, it has all the sounds you remember.

✦ **High score list:** Unlike the original game, this version remembers your high score, so you can keep trying to beat it.

## Bubble Defense (free demo)

www.secondgeargames.com

Bubble Defense (shown in Figure 3-7) is a strategy game that's easy and fun to play. Place different defense towers around the street to shoot down different types of enemy bubbles. The goal is to stop the bubbles from making it all the way through the road. For each bubble that makes it, you lose a little bit of life.

Challenge yourself by placing the towers in different locations and mix and match different types of towers to provide the best defense. The demo version is a fully functional game, but to get more levels you need to purchase it.

Some of the features of this game include:

✦ **More boards than you'll know what to do with:** There are 200 enemy waves that you need to destroy to win this game.

✦ **Four towers:** Choose from four different types of towers that you can change to develop different ways to kill your enemy. Mix and match your towers. Don't forget you can use the points in the game to sell and buy different towers. Depending on the placement and location, some towers will work better, but it all depends on you.

✦ **Five levels of difficulty:** Finding the game a bit too easy? Try one of the other skill levels. You can choose from easy to insane. Insane is not something that you'll really be able to win, but it's a lot of fun trying.

## Spreading out your warships

When you're playing Golden Thumb's Warship, don't lay ships next to each other side by side, but scatter them as best as you can. If one ship is hit, then the next shot will be in one of the surrounding squares. If the surrounding square has a different ship, then two of your ships will be hit.

**Figure 3-7:**
Bubble
Defense
is a great
strategy
game.

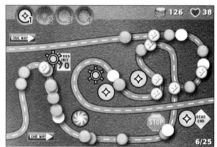

## *Golden Thumb's Warship*

www.goldenthumb.ca

Golden Thumb's Warship (shown in Figure 3-8) is one of the easiest games to play for the BlackBerry — and yes, it's free. Just like the board game Battleship, which you likely played sitting across from your little brother or best friend back in the day, you play against a real opponent. (In this case, your opponent is on the other side of the Internet.)

Each player takes turns by placing ships horizontally or vertically on the board. Then the shooting begins. Some of the key features of this game are:

✦ **Online game play:** Score points by playing against opponents just like you.

✦ **Player chat:** Use the chat feature to talk with your opponent.

✦ **Arcade sounds:** Hear the splashing of the water when you miss your target, an explosion when you hit a ship, and a sinking sound when you sink one.

You can play Warship only when you have access to the Internet. The game randomly chooses someone who is also playing the game. You can't play against the computer as in many other one-on-one games. Because you need

Internet coverage to play this game, you may want to make sure you have unlimited Internet access on your service plan. Otherwise, you might have a bit of a surprise the next time you look at your bill.

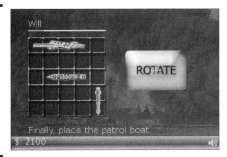 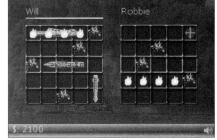

**Figure 3-8:** Play Golden Thumb's Warship against a live opponent across the Internet.

## Blackjack Spin³

www.spin3.com

Don't have time to fly to Vegas to play blackjack? Well, now you can play blackjack (also known as twenty-one) at any time — for free. Okay, maybe you won't win real money, but it's still fun to play.

Feel like you're really in the casino with the arcade style screen (see Figure 3-9) and dealer's voice when you win or bust. The other features of this game include:

✦ **Large, easy-to-read cards:** Unlike some other games, the cards and text are large enough that you can read them with no issues.

✦ **Arcade sounds:** Get audio feedback when you win hands. It makes you feel like you're really in Las Vegas.

**Figure 3-9:** Blackjack Spin³ is just like being in Vegas.

# Chapter 4: Managing Time and Resources

## In This Chapter

✔ Finding apps for managing files and e-mail

✔ Using the BlackBerry to keep your finances in order

The BlackBerry smartphone traditionally has been a tool that increases productivity by allowing quick access to all the items you need to organize your life — and we're talking about more than your address book and calendar. Time to put the games down, turn off the music, and get back to business.

Some productivity apps are fully integrated with native BlackBerry apps, seamlessly improving performance and increasing options. Others are apps that you select an icon to use.

If you need to do something, believe us, there is an app to help you do it. In this chapter, we tell you about a few of our favorite time-saving productivity apps.

## Finding the Apps

You have two main options for getting productivity apps on your BlackBerry smartphone. One way is to use BlackBerry Browser and surf the Web like you would visit any other Web site. The second is to download the app to your BlackBerry smartphone. If you want to download apps, you have these two choices:

✦ **Use BlackBerry App World,** an application that helps you find, acquire, and manage other applications for your BlackBerry smartphone. (Book VI, Chapter 1 tells you how to get and use BlackBerry App World.)

✦ **Use BlackBerry Browser** to download directly from a link on a Web site.

We recommend that you use BlackBerry App World for downloading apps whenever possible. BlackBerry App World decides whether you have enough room on your BlackBerry smartphone — before downloading — so you don't run out of space on your device. Another advantage of using BlackBerry App World is that you receive upgrade notifications. When an upgrade is available, an icon on the main screen appears, as well as a message that looks like an e-mail that links you to download the new version.

# Getting Productive

*Productivity tools* are any tools that save you time, allow you to multitask, and handle the other aspects of your otherwise busy life. You may think that having a calendar on your smartphone improves your productivity — and it does. But you're just scratching the surface. In this section, we open your eyes to a whole host of tools that will change the way you work.

## FileScout

http://emacberry.com

Aside from being a very handy file explorer for your device, the many functions built in to FileScout (shown in Figure 4-1) make this a must-have app for both power users and novices alike.

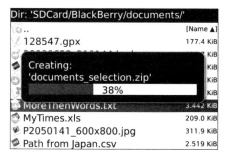

**Figure 4-1:** Viewing a picture's properties and creating a zip file.

FileScout performs common file operations, such as copying, moving, renaming, and deleting. In addition, it locally compresses (or *zips*) and decompresses (or *unzips*) e-mail file attachments. FileScout edits and resizes images, creates playlists, edits files and their attributes, searches for files and directories, and more. All the functions are done within the application.

Less experienced users as well as "abusers" will appreciate FileScout's simple method for deleting sample media from your BlackBerry without the fear of breaking your device.

Here are some of the other things you can do with this app:

✦ **Manage files.** Browse both your internal and external memory to rename, delete, and move files.

✦ **Zip and unzip files.** If you have more than one attachment that you would like to send to a contact, you can group them together and let FileScout zip them for you.

✦ **Edit images.** Resize and rotate pictures right on your BlackBerry.

✦ **Manage media.** Create a playlist that spans multiple folders. No longer will you have to keep all your music organized in one folder.

With FileScout, you can zip multiple files together on your BlackBerry and send them as a single e-mail attachment. This action reduces the total file size and allows you and the recipient the convenience of one attachment instead of several. Follow these steps to zip files with FileScout:

1. **Open the FileScout app on your BlackBerry.**

   Check your Downloads folder for the app.

2. **Browse to the files you want to send.**

   The files might be on a media card or somewhere else in your BlackBerry system, depending on where you put them.

3. **Highlight a file that you want to add to the zip file and press the spacebar.**

   The icon changes to a green circle with a white arrow, as shown in Figure 4-2.

4. **Repeat Step 3 for all the files you want to include in the zip file.**

5. **Press the Menu key and select Zip<selection>.**

   A new zip file is created.

6. **Create an e-mail message as you normally would.**

7. **Press the Menu key and select Attach File.**

8. **Browse to your saved zip file and attach it.**

9. **Send the message.**

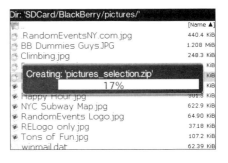

**Figure 4-2:** Selecting five pictures with FileScout.

## Using BlackBerry App World to find productivity apps

You can download apps a variety of ways, but the best way is to use the BlackBerry App World app. This app informs you if you have enough space on your BlackBerry to handle a new app (before you install it), notifies you if an updated version of the app is available, and offers plenty of free and inexpensive apps. If you haven't already installed this app, stroll back to Book VI, Chapter 1. Once you have BlackBerry App World installed, it's happy hunting time.

You don't need to know the exact name of an app. You can use the search feature to retrieve a list of close matches based on what you want to do. For example, if you type **time** in the BlackBerry App World search box, you can find time-management apps, stop watches, and world clocks.

## Forward Reply and Edit

www.bberryApp.com

Forward Reply and Edit is one of those applications that's integrated into your BlackBerry so that all you need to do is press the Menu key in an e-mail to use it, instead of having to open the application first.

Imagine this scenario: You want to forward an e-mail to someone, but before you send it, you want to change, remove, or add something to the message body. If you're using the native BlackBerry e-mail program, you're in trouble. Enter Forward Reply and Edit, a free BlackBerry app that integrates with your BlackBerry messages and allows you to do exactly what its name suggests: Open an e-mail message and choose to either forward the message with edits or reply to the message with edits. Forward Reply and Edit allows you to fix spelling mistakes, add content, delete content, add comments, and more.

The only hitch is that the e-mail won't forward an attachment if it's attached in the original message.

Here are two great ways you can use the Forward Reply and Edit features:

✦ **Forwarding:** Imagine that someone sent you an e-mail with directions to a restaurant, but the name of the restaurant is wrong. You can now forward that message to another friend and make the correction without having to retype the message all over again.

✦ **Replying:** Say your boss sends you an e-mail that has a bulleted list of action items. You now can reply with a status update on each item on the list without having to retype the whole list.

With Forward Reply and Edit, you can edit the body of an e-mail using the native BlackBerry e-mail application. First, install the app, and then follow these steps:

*1.* **Locate the app.**

If you're opening the app for the first time, an Application Permission prompt appears.

*2.* **Select Yes and follow the registration instructions.**

*3.* **Open the message you want to forward or reply to.**

*4.* **Press the Menu key and select Forward & Edit, as shown in Figure 4-3.**

Or you could select Reply & Edit.

*5.* **Make your changes and edits to the e-mail, as shown in Figure 4-3.**

*6.* **Send the message as if it were any other message.**

**Figure 4-3:**
Forward
& Edit
(left) and a
forwarded,
edited
e-mail
(right).

 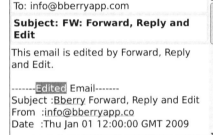

# SmrtAlerts

smrtguard.com

If you've ever used Gmail or Microsoft Outlook on your desktop computer, you probably get a little pop-up message whenever you receive a new e-mail. Well, SmrtAlerts (shown in Figure 4-4) does the same thing for your BlackBerry!

This app is great because when you receive a message on your BlackBerry, you don't need to stop what you're doing to figure out who it's from, whether it's important, and so on. The SmrtAlerts pop-up screen allows you to quickly see the sender, subject, and some of the body of the message, so you can tell at a glance whether it's a nice message from mom or a thumbs-up on the new account you've been slaving over for months. SmrtAlerts doesn't just work for e-mail messages, either. It works for SMS and MMS messages, as well. This program is always running, so there is no need to click an icon to make it work. On top of all that, SmrtAlerts comes with a spam blocker.

SmrtAlerts is a simple but must-have application. CIO magazine named it one of the top ten, must-have BlackBerry applications, and we agree with CIO.

Take a look at a few of the features that make SmrtAlerts such a great application:

✦ **Get the most important information.** When you receive an e-mail, a pop-up displays the sender's name, the e-mail subject, and the first few lines of the body text. If a picture is associated with the sender, it appears in the pop-up box. From the pop-up, you can open the message, delete it, or go to your inbox.

✦ **Configure pop-ups.** You can customize the pop-ups to your liking. For example, you can disable pop-ups when you're in a specific application. Imagine you have a mobile book reader on your BlackBerry. As you're reading a book, you don't want to be disturbed by pop-ups about incoming messages, so you tell SmrtAlerts not to show you pop-ups when you're using this app.

✦ **Block spam.** SmrtAlerts lets you create a whitelist (desirable e-mails) and a blacklist (undesirable e-mails) to filter e-mails and e-mail pop-ups. With the blacklist, you have the option to autodelete junk. See Figure 4-4. For those of you who get hundreds of junk messages a day, this is a very useful application.

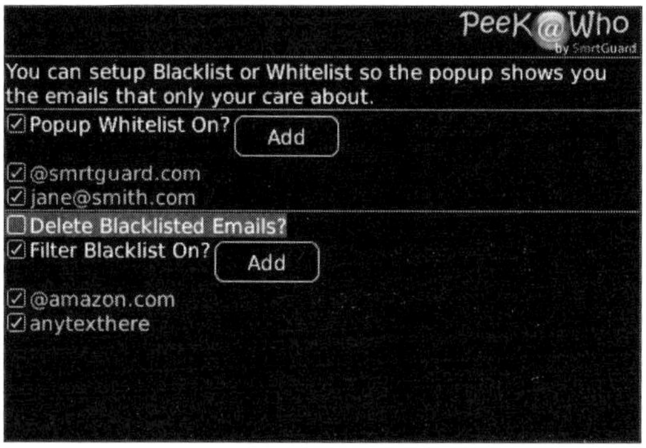

**Figure 4-4:**
Set up both whitelists and blacklists with SmrtAlerts.

> Peek @ Who
> by SmrtGuard
>
> You can setup Blacklist or Whitelist so the popup shows you the emails that only your care about.
> ☑ Popup Whitelist On? [ Add ]
>
> ☑ @smrtguard.com
> ☑ jane@smith.com
> ☐ Delete Blacklisted Emails?
> ☑ Filter Blacklist On? [ Add ]
>
> ☑ @amazon.com
> ☑ anytexthere

# gwabbit for BlackBerry

www.gwabbit.com

If you need an application to keep your address book updated and add new contacts as you exchange e-mails, then look no further. gwabbit (shown in Figure 4-5) is the tool that you need. This application is integrated with your BlackBerry, so once the application is running, it automatically reads each e-mail you open and determines if it should create a new contact or update an existing contact. gwabbit, LLC offers two versions. The free version alerts each contact that you added him or her to your Address Book when you use it. The premium version is significantly less annoying — it doesn't send e-mails to each contact you add.

gwabbit is an award-winning app that has both a desktop and BlackBerry version. When you receive an e-mail, gwabbit checks the e-mail address and then scans for the sender's signature. If you don't have the contact in your address book, gwabbit displays a prompt and asks you if you want to add the contact using all the information the app can recognize from the signature. If the contact already exists, then gwabbit matches it with what is already in your address book and prompts you to update it if the information has changed.

Book VI
Chapter 4

Managing Time and
Resources

This is definitely a "must have" if you like to keep your contacts up to date. It's a very simple tool to use, and it makes it easy to keep your information up to date. Here are some of the perks of gwabbit:

✦ **Uses sophisticated semantic technology**. gwabbit uses technology to identify signature blocks in e-mails. The technology interprets the signature so that it can be added directly into your address book. The engine that processes this information is always updated to match wits with the different ways people create signatures.

gwabbit only uses text to create or update a contact. It can't create or update contacts if the sender's signature is an embedded picture.

✦ **Updates contacts as their information changes.** Your colleagues may switch jobs, telephone numbers, titles, and more. Other colleagues may stay with the same firm, but the company might change its name. gwabbit alerts you of the change and then automatically updates your address book with the latest information.

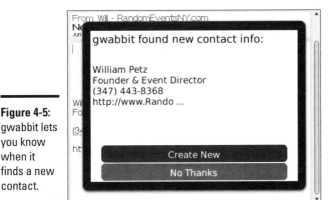

**Figure 4-5:** gwabbit lets you know when it finds a new contact.

## LaterDude Pro

www.mobileutil.com

Imagine that a colleague sends you an e-mail and says to contact him again in one week. Well, you can quickly create a calendar entry (by selecting a menu option) with the e-mail information. Now, in addition to never missing meetings, you don't miss follow-up contacts either. LaterDude Pro (shown in Figure 4-6) is a great way to keep on top of everything.

LaterDude is like a personal assistant for your BlackBerry. It reminds you to call, SMS, or e-mail someone by creating a new calendar entry or task for you. Fully integrated with all aspects of the native BlackBerry apps, LaterDude offers you a menu option to add a calendar event when you're in the Call Log, checking your e-mail, sending or receiving text messages, or updating your address book. With a few simple clicks, you can set up an appointment or reminder.

Using LaterDude provides these advantages to you:

+ **Increases your efficiency.** Because you don't need to manually enter reminders, you definitely save time.

+ **Makes certain that you will not forget anything anymore.** Using LaterDude ensures that you never miss an appointment or a task. Even small tasks, such as picking up milk from the grocery store (I got it, honey!), are easy to enter into LaterDude.

**Figure 4-6:** LaterDude Pro reminds you about all your errands, appoint-ments, and follow-ups.

## Managing Your Money

Maybe you think you've got your financial productivity needs met with the calculator that came packaged with your BlackBerry. Uh, no. In this section, we take you beyond the basics with some really nifty tools that can help you stay thrifty.

# Ascendo Money

www.ascendo-inc.com

If you're on a budget (and who isn't?), you need to track expenses. This is particularly important if you're on a travel budget or per diem for your job. With Ascendo Money (shown in Figure 4-7) on your BlackBerry, it's easy to keep up to date with your finances. The quick-entry shortcut allows you to quickly enter your transactions. You can synchronize and manage budget information from your PC, as well. Overall, the features and ease of use make this one of the best money management apps around.

**Figure 4-7:**
Ascendo
Money
makes it
easy to
organize
transactions
and
visualize
spending
patterns.

| Ascendo Money | | | |
|---|---|---|---|
| Search: | | | |
| Checking ($) | | | |
| **Payee** | **Date** | **Amount** | **Balance** |
| Healthy Living | 02/15 | -124.00 | 957.65 |
| Best Western | 02/10 | -125.00 | 1,081.65 |
| UGC Cinemas | 02/10 | -56.00 | 1,206.65 |
| Gas Station | 02/10 | -35.00 | 1,262.65 |
| Electric Comp .. | 02/09 | -220.00 | 1,297.65 |
| IKEA | 02/08 | -525.00 | 1,517.65 |
| Nordstroms | 02/08 | -265.00 | 2,042.65 |
| **Ending Balance:** | | | **957.65** |

Ascendo Money is like having a personal banker and accountant all rolled up in one. You can create checking, credit card, and savings accounts, as well as other financial items to manage your income and expenses. Ascendo Money analyzes the data, provides spending reports, and suggests budgets — and even forecasts your balances.

Ascendo Money allows you to set customizable reminders for recurring transactions, and you can use these other great features:

✦ **Transaction categories:** You can set transaction categories with budget limits to better manage your spending. For example, if you want to allocate $100 per month toward entertainment, you can quickly see how much money you have left in your budget.

✦ **Charting:** Ascendo offers bar charts (see Figure 4-8), pie charts, and reports to analyze your finances and help you visualize where you're spending the most money per month or per year.

✦ **Shortcuts:** Quickly enter transactions with time-saving shortcuts.

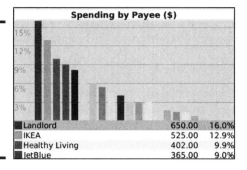

**Figure 4-8:**
You can easily see where your money is going with Ascendo's charts.

| Spending by Payee ($) | | |
| --- | --- | --- |
| Landlord | 650.00 | 16.0% |
| IKEA | 525.00 | 12.9% |
| Healthy Living | 402.00 | 9.9% |
| JetBlue | 365.00 | 9.0% |

## Personal Assistant

http://pageonce.com

If you like to keep close track of your accounts, you will love Personal Assistant by Pageonce. Personal Assistant simplifies keeping track of accounts and helps you stay on top of everything without logging into multiple sites.

Personal Assistant allows you to view all of your online accounts in one place. No more remembering multiple account name and passwords. With just one secure username and password, you can view detailed information about your finances, utilities, travel, e-mail, shopping, and social networking accounts on a single screen.

The key is to have a secure username and password. This is not a time to opt for your name and the password 1234! Your password should be impossible to guess, with a random combination of letters, symbols, and numbers.

Here are some key features of the app:

✦ **Generates info on multiple accounts.** Keep track with alerts and a display of your banking accounts, frequent flyer points, mobile bill and minutes, NetFlix queue, eBay bids, and more. With the Pageonce app, you can find out if your flight is running on time, check the number of minutes remaining on you cellphone plan, and even access your checking account balance.

✦ **Sends alerts when important changes occur.** Know which of your
accounts requires your attention at a glance (see Figure 4-9) or get
e-mail notifications when something out of the ordinary happens on
your accounts. For example, if an account that generally has very little
activity suddenly receives a $1,000 charge, you will receive a notification
so that you can determine whether the transaction is fraudulent.

**Figure 4-9:**
Get
summaries
and
overviews
of your
account
activity with
Personal
Assistant.

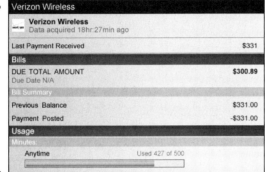

| Verizon Wireless | |
|---|---|
| **Verizon Wireless** Data acquired 18hr:27min ago | |
| Last Payment Received | $331 |
| **Bills** | |
| DUE TOTAL AMOUNT Due Date N/A | $300.89 |
| Bill Summary | |
| Previous Balance | $331.00 |
| Payment Posted | -$331.00 |
| **Usage** | |
| Minutes | |
| Anytime | Used 427 of 500 |

**Book VI
Chapter 4**

**Managing Time and
Resources**

# Chapter 5: Using Apps to Stay in Touch and in the Know

## In This Chapter

✔ Finding apps to keep you up to date with current events

✔ Using instant messenger apps

✔ Examining streaming audio and video options

Tired of always being the last one to find out about breaking news? Want to send instant messages from anywhere? Wish your BlackBerry could keep you entertained during your commute? If you answered yes to one, or all, of these questions, this is the chapter for you! We introduce some mobile news applications, mobile IM applications, and streaming audio and video applications.

## Getting the Application That Will Keep You Informed

Book VI, Chapter 1 introduces the different sites that you can use to browse to find the applications you are looking for. This chapter provides information on several types of applications that you can find in BlackBerry App World or other mobile sites under categories such as Work & School, Business, and Music & Audio. Just as with any other application, you have two main options to get these applications on your BlackBerry smartphone:

✦ **Use BlackBerry App World,** an application that helps you find, acquire, and manage other applications for your BlackBerry smartphone. (Book VI, Chapter 1 tells you how to get and use BlackBerry App World.)

✦ **Browse a mobile Web site from your BlackBerry or computer.** Book VI, Chapter 1 provides a list of sites that you can view from your BlackBerry or computer that you can use to download these types of applications.

We recommend that you use BlackBerry App World for downloading social-networking apps whenever possible. BlackBerry App World decides whether you have enough room on your BlackBerry smartphone — before downloading — so you don't run out of space on your device. Another advantage of using BlackBerry App World is that you receive upgrade notifications. When an upgrade is available, an icon on the main screen appears, as well as a message that looks like an e-mail that links you to download the new version.

# Getting News and Information

In our world of 24/7 news cycles, receiving updated news and information has become a necessity. You may need to stay updated for your job or you may be a newshound. Either way, in this section we introduce some great news and information apps.

## Viigo for BlackBerry

www.viigo.com

Viigo is one of the first free applications you should install on your BlackBerry if you want information fast in an amazing interface. You can keep up on the latest news and trends, follow your favorite sports teams, watch YouTube videos, and more. It's a one-stop shop for many of your information needs.

Viigo for BlackBerry is an award-winning app that delivers news, weather, social networks, blogs, podcasts, flight info, sports scores, financial news, entertainment, and more.

Viigo, shown in Figure 5-1, packs many features into a great-looking tool. Some of the features include:

✦ **Customizable modules:** Modules incorporate customizable channels and features to make your Viigo experience unique. Modules provide local and national news, local weather, your choice of podcasts, and real-time sports scores for your favorite teams.

✦ **Social networking options:** Viigo makes it easy for you to share news and information with your social networks. See something you like? Quickly post it to Facebook, Twitter, Del.icio.us, or more.

**Figure 5-1:**
Viigo allows you to customize the news and information you want based on your interests and location.

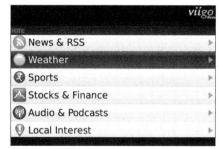
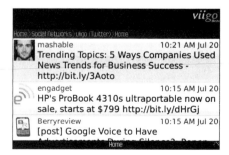

## Staying hands-free

A growing number of states are setting strict punishments for driving while talking, texting, or otherwise fiddling with cellphones. Distracted driving is no joke, and studies show that people have become increasingly addicted to playing with their gadgets while on the road — despite the fact that playing with your smartphone can have the same effect on your ability to concentrate that drinking can have.

All of the apps we discuss in this chapter (and in Book VI in general) are fun. You may even call them must-have apps. Just remember to use them sensibly and only when you're not supposed to be doing something else very important — like driving your car.

Viigo can be downloaded only from BlackBerry app world or by going to `http://viigo.com` from your BlackBerry or desktop.

### Setting up Viigo alerts

To set up Viigo alerts, you need to create a rule and then create a BlackBerry sound profile. The profile can alert you by playing a tone or vibrating. You can set this tone to a different tone than the one you hear when you receive a phone call or e-mail message. These are the steps to set up both the Viigo alert and a BlackBerry sound profile:

1. **From your BlackBerry Home screen, select Downloads.**

2. **Select Viigo.**

   If you can't find the application, visit Book VI, Chapter 1 for instructions on how to find and download it.

3. **Navigate to the category where you want to set up an alert.**

   For example, choose News & RSS.

   When you create Viigo alerts, you must set alerts on each individual item. For example, if you want to be alerted every time the word "free" appears in any of the 10 feeds that you subscribe to, you must set up an alert on each feed with the keyword *free*.

4. **Scroll to the item you want to set up the alert on. Press the Menu key and select Set Channel Alert.**

5. **(Optional) Scroll to Enter keywords.**

6. **Enter a keyword that you want Viigo to monitor.**

   When new feeds appear with your keyword, Viigo triggers an alert. See Figure 5-2.

**Figure 5-2:**
Adding a
Viigo alert.

### Setting a profile for Viigo alerts

The second part of setting up a Viigo alert is to set up the BlackBerry profile. The BlackBerry profile is the place where you assign a tone, vibration, or a combination of both to be triggered when your Viigo alert is triggered. To set up a profile, follow these steps:

1. **From your BlackBerry Home screen, select Profiles.**

   To find out more about setting up profiles, visit Book I, Chapter 3.

2. **Scroll down and select Advanced.**

3. **Scroll and select the Active profile.**

4. **Scroll and select Viigo Alerts.**

5. **Select the ring tone or vibrations your BlackBerry smartphone will use when you receive a new Viigo alert.**

6. **Press the Menu key and select Save.**

## AP Mobile

www.apnews.com

Rather than browsing a news aggregator or RSS feed from a site such as CNN, Yahoo! News, or Google News, why not go straight to the source? AP Mobile, shown in Figure 5-3, is your one-stop shop for all of your news.

Designed specifically for your BlackBerry, AP Mobile is a free app with local, regional, international, and even some wacky news items. You will never wonder what's happening around the world — or around your neighborhood — again.

AP Mobile is the news you choose. With AP Mobile, you can be the first to know about today's important events wherever in the world you are. Stay up to date with breaking stories, browse dynamic photo galleries, and access thousands of trusted local news sources when you want it, where you want it. Several key elements of this application make it one of the ultimate tools for your BlackBerry. For example, the sports section includes everything you need to follow your favorite teams. Also, with the breaking news notification, you won't miss anything.

Here are some other perks:

✦ **News and photo galleries:** With over 40 categories, AP Mobile provides a vast number of articles. Bonus: Outstanding photos!

✦ **Customization:** Filter out the news channels you don't want to see.

✦ **Offline viewing:** Stories are available for offline viewing, so that if you're on an airplane or subway, you can read previously published news.

✦ **Social networking capabilities:** Share stories with your friends through e-mail, Twitter, Facebook, or another social networking app.

**Book VI
Chapter 5**

Using Apps to Stay
in Touch and in
the Know

**Figure 5-3:**
AP Mobile
gives you
headlines
and photo
galleries.

## Forbes.com

www.forbes.com/mobile

We do our best to keep up to date on what is going on in business. Forbes.com Mobile Reader for BlackBerry, shown in Figure 5-4, is an exceptional product that provides a stunning interface and easy navigation so that you can get the latest business and financial news.

Reaching over 20 million visitors each month, Forbes.com is the first choice for business news. Forbes publishes more than 4,000 articles every day. You can get breaking news, stock market updates, feature articles, videos, and custom programs. Here are some of the app's key features:

✦ **Customize tabs:** There are nine channels of mobile content, including Business, Technology, Markets, and Personal Finance. You can customize these channels by adding and removing categories in each channel.

✦ **Get information whenever you need it:** The app offers real-time stock quotes and weather. In addition, you can download headlines and full-text articles so that you can read Forbes anytime and anywhere.

✦ **Share content across social networks:** Share articles with your friends by e-mail, Facebook, Del.icio.us, or other networks.

**Figure 5-4:** Get real-time stock quotes and customize news channels with Forbes. com.

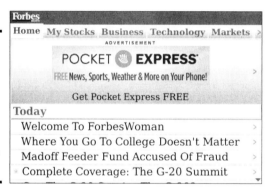

Forbes Mobile Reader is a great application that lets you know what is going on in the news. However, you may want to filter out some of the news that you don't need to know about. With Forbes Mobile Reader, you can customize your news so that you read only what you want to. Follow these steps after downloading the Forbes Mobile Reader for BlackBerry app:

*1.* **From your BlackBerry Home screen, select Downloads.**

*2.* **Select Forbes.com Mobile Reader.**

If you can't find the application, visit Book VI, Chapter 1 for instructions on how to find and download it.

*3.* **Navigate to the category you want to filter out news from.**

For example, choose Business.

*4.* **Press the Menu key and select Configure tab. (See Figure 5-5.)**

A list of subcategories appears.

**Figure 5-5:**
Configuring
the
Business
tab in
Forbes
Mobile
reader.

*5.* **Scroll to the topics about which you wish not to receive news articles. Deselect each item by pressing the Menu key and choosing Select.**

You will no longer get any news feeds pertaining to any items that you have unselected.

When selecting or deselecting an item while configuring your tabs, the tab name always says Select. It doesn't say Deselect. In order for you to know if the item is selected, look to the left of the subcategory word. If there is a box with a check mark, it is selected.

*6.* **Once you are done configuring the tab, press the Back button and select Save.**

# Getting Instant Gratification with IM

Instant messengers (IM) allow you to have real-time conversations with friends and contacts without having to send a text message or make a phone call. Unlike e-mail, IM allows you to see whether the other person is online so that you know if you can expect an instant response. IM also comes in handy if you have limited texting on your service plan. BlackBerry comes with an app, called BlackBerry Messenger, which allows you to message with only other BlackBerry users. If you want to send IMs with anyone — whether they're using a smartphone or a home computer, check out the apps in this section.

## IM+ All-in-One Mobile Messenger

www.shapeservices.com

With the world moving from phone calls and e-mail to instant messages, IM+ All-in-One Mobile Messenger (shown in Figure 5-6) does an amazing job of incorporating all your favorite desktop IM programs into one easy, well-organized, and reasonably priced mobile app. IM+ includes a unique feature called *push mode* that allows you to stay connected while also saving precious battery life. It's like magic. Magic that you can download from BlackBerry App World, that is.

IM+ All-in-One Mobile Messenger lets you chat in real time within any of the following IM programs from your BlackBerry smartphone, anywhere in the world, in any language, and with any mobile operator:

| | |
|---|---|
| AIM iChat | MSN/Windows Live Messenger |
| Yahoo! Messenger | ICQ |
| Jabber | Google Talk |
| MySpace IM | Facebook |
| Skype | Twitter |

The app also allows you to send low-cost text messages worldwide.

Some carriers offer unlimited texting plans; however, you may want to verify that your plan also includes unlimited international texting. If your plan doesn't include international texting, then you will benefit from IM+'s texting feature.

The app also gives you the kind of control you would expect from a desktop IM program. You can add, delete, and rename contacts, as well as group them by the types of IM services, online status, or alphabetically. You can perform quick contact searches. When you receive a new IM, a pop-up message appears at the top of your BlackBerry screen. Simply open your e-mail application and select the IM, just as you would open an e-mail message.

If you use IM+ in *push mode* (that's when you don't have IM+ running), IMs are automatically sent as e-mails and *pushed* to your BlackBerry inbox. That way, you can use your native BlackBerry e-mail client to IM any time. Some other features are:

+ **Send files and voice notes:** When you want to exchange more than just text messages back and forth, you can send a quick voice note. You can also do the usual — like send a picture or some other file.

+ **Stay Online mode:** Your IM identity always appears to be online, even when you lose coverage. That way, you never miss an instant message, even if you can't respond instantly.

Book VI
Chapter 5

Using Apps to Stay
in Touch and in
the Know

**Figure 5-6:**
IM and chat
with all your
friends any
time.

## Nimbuzz

www.nimbuzz.com

Nimbuzz is a great BlackBerry and desktop app that brings all your social networking and IM applications together into one. As shown in Figure 5-7, Nimbuzz is a mobile social messenger that combines instant messaging, geographical coordinates, and voice over IP (VoIP) services. The free application lets you connect and interact with your buddies across popular communities, including Skype, Windows Live Messenger, Yahoo! Messenger, ICQ, Google Talk (Orkut), AIM, and social networks including Facebook and MySpace.

Regardless of what IM client they use, all your contacts are combined in a single contacts list. Then you can call, chat, share files, and share locations with all your friends using a single app.

Here are some of the app's features:

✦ **All-in-one mobile instant messaging and VoIP calling:** Use one application on your BlackBerry to chat with all your contacts across a vast number of social networks and IM clients. The desktop version of the app allows you to speak to people with VoIP just as you would on the telephone. You need to have a microphone installed on your desktop computer, of course.

✦ **File and location sharing:** Exchange pictures, music files, and other documents. You can also quickly send your location to your buddy.

✦ **Available for mobile devices, PCs, and Macs:** Unlike other BlackBerry applications, Nimbuzz allows you to chat and manage your users from both the desktop computer (both PC and Mac) and your BlackBerry.

## AIM for BlackBerry smartphones

www.aim.com

AIM (AOL Instant Messenger) is a free online service that lets you communicate and stay connected with your buddies in real time; see Figure 5-8. America Online (AOL) has its own proprietary app for the BlackBerry so that you can have the same AIM experience on your BlackBerry that you're used to having on your desktop computer.

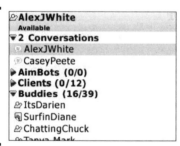

**Figure 5-8:**
AIM for
BlackBerry
lets you
find your
buddies and
chat with
them.

AIM for BlackBerry smartphones gives you on-the-go IM options with all the desktop features you already love. Best of all, AIM is fully integrated with all of your native BlackBerry applications, including the BlackBerry unified Inbox and Address Book. That means your AIM buddies and your contacts are neatly synched. Here's what you get from AIM for BlackBerry:

✦ **Desktop Buddy sync:** AIM for BlackBerry supports all your buddies and groups from the desktop. Changes made from either the BlackBerry or the desktop client immediately take effect everywhere, seamlessly.

✦ **Share photos:** Include pictures in your chat sessions with ease.

✦ **See Buddy info:** View your friends' Buddy icons and Buddy info from your BlackBerry. Sometimes it's hard to remember what someone's screen name is. Because AIM for the BlackBerry includes Buddy icons and info, you can always remember who is who on your buddy list.

## Palringo

www.palringo.com

Ever chat with someone and say, "I wish you could hear this!" or "I wish you could see this!"? Well with Palringo, shown in Figure 5-9, it's possible. You can take a picture or record a voice memo and send it to one of your IM buddies. Palringo is free, and it works with all the most popular IM and social networking sites, including MSN, AOL, Google Talk, Yahoo! Messenger, and Facebook Chat.

You can turn your BlackBerry into a walkie-talkie and share photos using picture messaging to show any of your contacts what you're looking at.

Palringo also enables you to share your location. In addition, you can chat one-on-one or in a group. Here are some other features:

✦ **Update your location:** See where your friends are and let everyone know where you are.

✦ **Make group announcements:** Create a group so you can send messages to everyone at the same time. IM everyone at once to make sure all your friends get the same message. This feature saves tons of time.

✦ **Use chat rooms:** Find a group of people that you have something in common with. Then enter the chat room to see what is going on and have a conversation.

**Figure 5-9:**
Palringo allows you to get the word out and chat with people who share your interests.

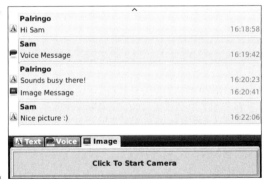

## Broadcasting Straight to Your BlackBerry

As we mention in Book V, Chapter 2, you can use your BlackBerry to listen to downloaded music and streaming media. In this section, we discuss a few great streaming media options.

## Slacker Radio

www.slacker.com

Put away your MP3 player. All you need is Slacker installed on your BlackBerry. You can listen to high-quality songs, even when you don't have mobile coverage. If you use Slacker plus, which provides lyrics and the capability of skipping as many songs as you want, it's like having your own private DJ and karaoke machine. With over 2.4 million songs to choose from, what more could you ask for?

Slacker is what radio was meant to be: personal, portable, and free, whether you're listening from your home computer or using your BlackBerry.

### Setting up Slacker to cache music

There are many applications that provide streaming audio. The two disadvantages of streaming audio are:

✦ It requires a constant Internet connection.

✦ It drains your battery.

Slacker can stream audio like the rest, but it also downloads and stores hundreds of songs at a time on your BlackBerry into a file that only Slacker Radio can play. This capability allows you to listen to your music without an Internet connection and saves precious battery life. Slacker calls this feature Cache Playing Station.

One downside of Cache Playing Station is that it takes up a ton of space on your BlackBerry. You should install additional memory to your BlackBerry before using this feature. Check your user manual for the type of media card you can install on your BlackBerry or go to Book I, Chapter 5 to help determine which media card is supported in your Blackberry.

Once you have this feature set up, Slacker downloads hundreds of songs over a matter of minutes (depending on your connection) at night when your device is charging, so you always have a new collection of songs to listen to.

To set up a Cache Playing Station, follow these steps:

*1.* **From your BlackBerry Home screen, select Downloads.**

*2.* **Select Slacker Radio.**

If you can't find this application, visit Book VI, Chapter 1 for instructions on how to find and download it.

3. **Select a station to cache.**

   For example, you could choose Top Stations. To find a Top Station, on the main screen of Slacker, scroll to and select Top Stations.

4. **Scroll and highlight the station you want to cache. Press the Menu key and select Cache Highlighted Station. (See Figure 5-10.)**

   A prompt tells you how many stations you can cache due to space on your BlackBerry.

5. **Select Yes.**

**Figure 5-10:**
Selecting
a station
to cache
in Slacker
Radio.

### Setting up Slacker to synchronize your cache while you sleep

Slacker automatically synchronizes your cache stations when it is connected to power. This means any new stations you have added or changed. When Slacker synchronizes you will have the music that you intended. To set this up follow these steps:

1. **Open the Slacker app.**

2. **From any screen on the Slacker interface, press the Menu key and select Settings.**

3. **Scroll down to Auto Refresh Cached Stations options and select it to switch the option to On.**

4. **Click the Back button to return to the Main screen.**

## Pandora Radio

www.pandora.com

Pandora is a personalized Internet radio service. Tell Pandora the name of a favorite artist or song, and the application delivers a radio-style listening experience tailored to your musical taste.

With Pandora for BlackBerry, you can create up to 100 distinct stations — one to match every mood and setting. The application is also linked to your account on the Web. Any stations you create on the Web are immediately reflected on your BlackBerry and vice-versa. Never run out of music on the go again. One downside: After listening to Pandora for a specified number of hours a month, you have to either pony up some money to listen more or you're shut out of the system for the rest of the month.

## iheartradio

www.iheartradio.com

Just like the name says, we heart iheartradio, shown in Figure 5-11. You will enjoy the flexibility to listen to different morning shows, and you can still listen to your local stations when traveling. When you feel like a little change, you can choose the randomizer and it will just pick something for you to listen to.

The iheartradio app is Clear Channel Radio's free mobile application. The app offers access to more than 350 radio stations, including the best in local radio, genre-defying channels, talk radio, and traffic reports available on demand.

**Figure 5-11:** Check out the iheartradio channels and listen up.

## Shazam

www.shazam.com

How many times are you hanging out with friends and someone says "Who sings this song?" or "What is this song called?" Now you can be the hero. At the press of a button, Shazam listens to the song and provides the answer. If it's a song you really like, you can purchase it right away from Shazam, shown in Figure 5-12. Abracadabra? Shazam!

Shazam is the world's most popular mobile music discovery service. Launch Shazam on your BlackBerry and hold the device up to a song that you hear playing — wherever you are — on the radio, TV, in a store, or at a bar.

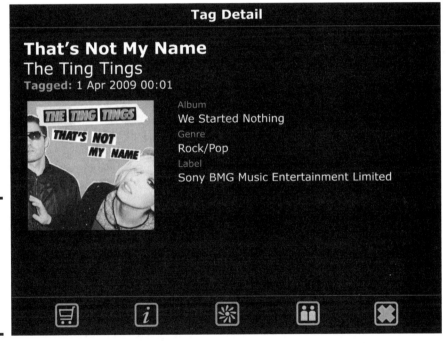

**Figure 5-12:** Shazam identifies songs and lets you purchase them right away.

## SlingPlayer Mobile

http://mobile.slingmedia.com

Wow, cable TV on your BlackBerry! We're huge fans of Sling Media, the makers of SlingPlayer Mobile, shown in Figure 5-13, and you will be, too, if you love the idea of being able to fully control and watch cable TV, DVDs, and pre-recorded shows anywhere from your BlackBerry. And if you don't have cable, don't worry. You can watch your regular TV service, too.

All you need to purchase is the Slingbox hardware that connects to the Internet and source you want to watch. Prices range from $179 to $299. You can purchase it online or in many retail stores. You pay no additional monthly fees (over and above your monthly cable bill and your monthly wireless bill, that is). The app costs less then $30.

Unlike the other apps in this chapter, you can't download it from BlackBerry App World. Instead, you must go to Sling Media's Web site to download it.

Here's what you get:

✦ **More value:** Get more value out of the TV services you already pay for by accessing them more often and in more places. Watch while you're on vacation, at a friend's house, or just on your way to work — that is, if you take a subway or carpool.

✦ **Record and watch shows:** Use your phone to program new recordings on your DVR. You can even just navigate your program guide, surf channels, and fast-forward through a show, just like you do at home.

✦ **Don't pay monthly fees:** Watch TV from your any computer or your mobile device and never have to worry about a monthly fee. All that is required is the purchase of the hardware and software, and you are ready to watch TV with no other charges.

**Figure 5-13:** You can watch streaming cable TV with SlingPlayer Mobile.

Imagine you're driving home and your favorite sports team is playing. The game is being broadcast on cable TV and is not available on the car radio. If you can't watch, you can listen in with SlingPlayer Mobile. An additional benefit of using SlingPlayer Mobile in Listen Only mode is that it saves battery life.

To listen instead of watch a channel, follow these steps:

*1.* **From your BlackBerry Home screen, select Downloads.**

*2.* **Select SlingPlayer Mobile.**

   If you can't find the application, visit Chapter 1 of this minibook for instructions on how to find and download it.

*3.* **From SlingPlayer's main screen, select the Connect button to start your streaming video.**

*4.* **As you're watching the video, use the Menu button and select Listen.**

*5.* **To switch back to seeing the show, use the Menu button and select Watch. (See Figure 5-14.)**

**Figure 5-14:**
Listening
to your
television
with
SlingPlayer
Mobile.

# Chapter 6: Knowing What's Around You and Where You Are

## In This Chapter

✓ Getting from point A to point B

✓ Checking the weather

✓ Translating words and phrases

**W**hether you're looking for a restaurant across town or planning a cross-country road trip, tools for the BlackBerry can help you find places to go and directions and information about what to expect when you arrive. You can find everything from ways to plan your trips to a local place for lunch.

BlackBerry smartphones use GPS (global positioning system) technology to give you real-time information about what's around, how far you are from where you want to be, and how to get there. The type of BlackBerry smartphone you have determines what type of GPS is used. If your BlackBerry has a GPS receiver, your device uses satellites to find where you are. (If you are unsure if your BlackBerry is GPS enabled, you can check out the device chart in Book I, Chapter 5 to find out). If your device doesn't have a GPS receiver, it can still use cell triangulation technology to determine where you are. With cell triangulation, your BlackBerry uses three cellular towers to help determine your location. The GPS receiver, however, provides quicker and more accurate results.

Because BlackBerry smartphones are connected to the Internet 24/7, GPS maps are updated when new streets are built and some even provide real-time traffic info. Depending on the application, turn-by-turn directions via voice or text make sure you never have to look down at a piece of paper again.

## Locating the Apps That Will Locate You

In Book VI, Chapter 1, we discuss the different sites and application stores that you can use to search for and install the applications or games that you want. We also provide step-by-step instructions on how to install the apps and troubleshoot some of the issues you may have.

GPS applications can be found in most application stores and in BlackBerry App World under Maps & Navigation. However, there are many other applications not in that category that take advantage of the features of GPS. One such application is ÜberTwitter, which sends your location when you post information to your Twitter site. This way, people who read your posts can find out exactly where you are. For more information on ÜberTwitter, visit Book VI, Chapter 2.

We recommend that you use BlackBerry App World for downloading apps whenever possible. BlackBerry App World decides whether you have enough room on your BlackBerry smartphone — before downloading — so you don't run out of space on your device. Another advantage of using BlackBerry App World is that you receive upgrade notifications. When an upgrade is available, an icon on the main screen will appear, as well as a message that looks like an e-mail that will link you to download the new version.

## Getting Maps, GPS, and Local Info

There are many apps that can help you make the most of GPS to find out where you are, as well as how to get to your destination. Then you can find cool stuff to do.

### WHERE on the BlackBerry

www.where.com

Everyone knows the adage, "Wherever you go, WHERE you are." WHERE, shown in Figure 6-1, is a free app available that puts local information in the palm of your hand. You can get weather reports, local news, restaurant reviews, the address of the station with the cheapest gas, traffic updates, and the ability to connect with other users. Of course, you also get maps and directions.

Here are some of the app's key features:

+ **Comprehensive search results:** The app aggregates local search results from multiple sources, giving you comprehensive information: weather, news, restaurants, gas prices, movies, and anything else you might be looking for.

+ **Integration with key native BlackBerry apps:** The app integrates results with your Address Book so that you can save and easily access search results. In addition, the search results are synched with Google Maps so that directions are always up-to-date.

+ **Interactivity:** The app allows you to connect with and get input from other users, so you're getting the inside scoop from locals who know.

✦ **Razor-sharp GPS functions:** The app uses GPS and your wireless network to figure out where you are and give you directions to your destination. If you've ever received directions to no-man's land, then you'll really appreciate this functionality because it won't steer you in the wrong direction.

Say you're visiting Times Square, New York City, for the first time and decide you want a slice of pizza. WHERE makes it easy to feed your craving using these steps:

1. **From your BlackBerry Home screen, open the Downloads folder and locate the WHERE app icon.**

2. **Open the WHERE app.**

   WHERE automatically tries to figure out your location using GPS.

   If you don't have GPS, or GPS is unavailable, you can select Enter My Location Manually to enter it.

3. **Once your location is set, type** pizza **in the search bar at the top of the WHERE screen and press the track ball or trackpad.**

   WHERE searches for nearby pizza restaurants and lists them, starting with the restaurant closest to your current location.

4. **Select a restaurant to see more details.**

   From here you can call the restaurant, visit its Web site, see it on a map, read reviews, and even add a review of your own.

WHERE does a really good job of incorporating many different applications into one. It's one of the only applications that provides a way to find the closest Starbucks, Zipcar, and the cheapest gas station. We also like the Citysearch integration. Overall, WHERE does a good job providing a wide range of items.

**Figure 6-1:**
WHERE
gives you
everything
you need,
from the
cheapest
gas station
near you
to movie
reviews.

## Poynt

www.poynt.com

Poynt, shown in Figure 6-2, is a convenient, free, and timesaving service that helps you find local businesses, restaurants, and movie theaters. Simply enter your search term to get phone numbers and addresses of local businesses. If you don't know what a movie is about, Poynt also allows you to read movie reviews and watch the video trailers.

In addition to the basics — the ability to pinpoint businesses of all kinds, the integration of the app with your BlackBerry's native mapping apps, and the integration of the app with your Address Book — Poynt has the following key features:

✦ **Make purchases/reservations.** Poynt allows you to purchase movie tickets and book reservations without ever dialing the phone.

✦ **Find people.** Poynt helps you to find people nearby, using the convenient people search tool, which also includes reverse lookup by phone number and address.

✦ **Add events.** The app allows you to quickly add movie listings and restaurants to BlackBerry Calendar.

Poynt is the perfect complement to your favorite mapping application to find what you want. If you travel outside the U.S., it also works in Canada.

Use the reverse call feature when you receive a missed call from a number you don't recognize — so you never wonder who called you again.

**Figure 6-2:**
Poynt points you in the right direction.

## foursquare and the future of mobile marketing

We're going to be honest — foursquare isn't for everyone. It's an app for people on the bleeding edge, people who want to take social networking to the next level. Part game, part social networking, in our opinion, foursquare is part of a growing niche in mobile marketing. Here's how it works. Download the app from the foursquare Web site (`http://foursquare.com`) and install it on your BlackBerry. You need to set up a username and password, and you should read the privacy information.

Every time you "check in" to a location, foursquare tells your friends where you are. The app uses the same GPS and cell tower triangulation features of other mapping apps to figure out where you are. You can earn funny badges and awards if you're diligent about checking in wherever you go. Eventually, you could become the "mayor" of your favorite destination. While the designation of mayor doesn't give you any *actual* perks, it does earn you a little respect among fellow foursquare users.

Although it may seem creepy that an app "knows" where you are at all times, it's similar to Google Maps — the app can't track you, you can hide your location, and you can choose who you share information with.

You may wonder what value this app has, and we can't really say that it has any value other than the fact that it's fun, especially if several of your friends also use foursquare and you compete with each other to see who can win the most badges. As for the future, we think more and more apps are going to have a geo component to them. Look for businesses to use geo options to market to customers in real time. As an example, say you check in at your favorite restaurant and the next thing you know, you get an alert to come on in and get a dollar off your burger.

## Google Maps for mobile

`www.google.com/mobile/maps`

Google Maps for mobile, shown in Figure 6-3, is a free app that makes it easy to access maps when you're on the go. Google Maps for mobile makes it easy to figure out your location with or without GPS, get driving directions, transit directions, and phone numbers and addresses for local businesses. Google Maps uses cellphone towers to triangulate your location if GPS is unavailable, so you don't need to have GPS for this feature to work.

Here's what you get with Google Maps for mobile:

✦ **Google Latitude:** See your friends' locations and status messages. Start Maps and then select Join Latitude. You have complete control over your privacy settings, too.

✦ **MyLocation:** See your location on a map. Google can find you even if your GPS isn't working or activated.

After you figure out where you are, you can do a Google search to find what you're looking for. Hungry? Type what you want to eat into the search box and see what restaurants are nearby.

✦ **Driving, transit, and walking directions:** Get routes and schedules to travel via car, subway, bus, or on foot. Thanks to MyLocation, you don't have to enter your starting point.

✦ **Address Book integration:** Show a contact's address on Google Maps or get directions to a contact's address from your current location, all from the options menu in your Address Book.

✦ **Street View:** View street-level imagery of businesses and turns in directions.

**Figure 6-3:**
Google
Maps tells
you where
you are and
shows you
how to get
anywhere
you want
to go.

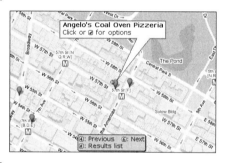
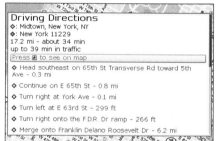

Say you've just landed at JFK International Airport and the first place you want to visit is Times Square. Sure, you can hop in a taxi, but what if the driver doesn't know how to get there? Google Maps can guide you in a few easy steps.

*1.* **Open the Downloads folder and locate the Google Maps icon.**

*2.* **Open the Google Maps app.**

Google Maps figures out your location using GPS — or using cellphone tower triangulation if GPS is not available.

*3.* **Press the Menu key and select End Point from the menu.**

*4.* **Select Enter an Address.**

You can type in addresses, locations, landmarks, and more to help you find what you are looking for.

*5.* **Type** Times Square, New York **and press the trackball or trackpad.**

6. **Choose Get Directions.**

   Google Maps gives you a nice list of directions from where you're standing straight to Times Square.

   If you want to get really fancy, choose Show on Map, and then press 8 for Street View turn-by-turn directions.

Google Maps is undoubtedly the best, free, mapping application available. This is one of the first applications that any BlackBerry smartphone owner should install.

## TeleNav GPS Navigator

www.telenav.com

TeleNav GPS Navigator, shown in Figure 6-4, uses GPS to give you turn-by-turn directions to anywhere you want to go. The app is free for 30 days and then you pay a subscription each month. Because TeleNav GPS Navigator is connected to the Internet, you don't have to worry about maps or business listings going out of date as they would in your car GPS. You also get real-time access to information like daily gas prices, traffic alerts, business reviews, and weather reports. Now if only you could get it to drive the car, you'd be all set.

Here are some of the features you get with TeleNav:

✦ **Sophisticated interface:** With full-color, 3-D, moving maps and voice and onscreen directions, you get such a sophisticated GPS system that you'll wonder why anyone pays for a dashboard-mounted GPS system. TeleNav also has speech recognition, which means that you can tell TeleNav where you want to go without typing a thing — and it will get you there.

✦ **Superior use of GPS:** Miss a turn? No problem. TeleNav automatically reroutes you so that you end up where you intend to go — and not in downtown Sketchville.

✦ **Bonus features:** Updated maps and business listings with reviews, as well as real-time traffic alerts, daily gas prices, weather, and commuter alerts.

✦ **Online trip planning:** Planning a trip? TeleNav makes getting there easy. Just sign in at www.telenav.com, enter your destination address, and select Send to Phone to have it show up in the My Favorites section of TeleNav on your BlackBerry.

**TIP**

Why buy a standalone GPS when you can carry around a mobile version with even more functionality? We adore this product because of the turn-by-turn voice directions. TeleNav also alerts you if you're about to hit traffic.

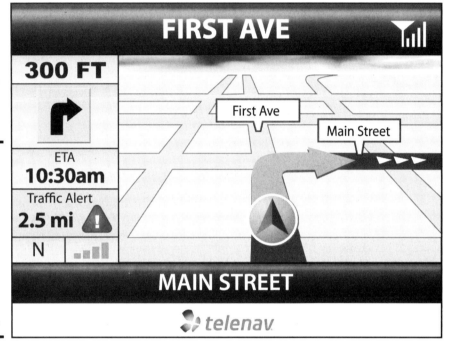

**Figure 6-4:**
TeleNav GPS Navigator navigates you to your destination with traffic alerts and helps you find businesses.

# To Boldly Go Where No BlackBerry Has Gone Before (Travel)

If your travel destinations are a little more far-flung than the next town over, BlackBerry still has you covered. You can find out weather information and get other travel-related support on the go. In this section, we give you all the info you need to use apps that make travel easy. Trying to figure out what to pack? You're on your own.

## The Weather Channel for BlackBerry

www.weather.com/mobile

The Weather Channel is the leading weather provider on all digital platforms, and The Weather Channel for BlackBerry provides accurate, local weather information and enhanced weather features — for free! It is easy to use and accurate — it can pinpoint your location for detailed conditions (which is really convenient if you aren't near a window and can't just see for yourself).

Get current conditions and expert forecasts. Going beyond just the weather, this app gives you personalized weather layers and neighborhood landmarks that you will find useful whether you're a busy parent, have a daily commute, travel a lot for business, or are an outdoor enthusiast. (And who doesn't fit into at least one of these categories?)

You can do a lot with this app, including:

✦ **Customize maps.** Add weather layers, including radar, clouds, snowfall, and other data.

✦ **Get info at a glance.** The Weather Channel app opens to an At a Glance dashboard, which shows the current weather details for your area. It also displays a red pop-up menu for any severe weather alerts in your area.

✦ **Get info for all your locations.** Travel a lot? Have friends and family all over the place? You can save up to 20 locations with personalized labels so you know the weather virtually everywhere — locally, nationally, and internationally.

✦ **Get real-time traffic info.** Never wonder again. Plan activities with metro traffic maps and incident details.

✦ **Integrate the app with your Calendar and Address Book:** Use the Get Weather option in your Address Book and Calendar menus to see what the weather is like where a contact lives or where an event is taking place.

**Book VI Chapter 6**

**Knowing What's Around You and Where You Are**

The Weather Channel for BlackBerry also has advanced mapping features that you should really check out. With so many detailed layers, you'll never wonder about the weather again.

## WeatherBug Elite

http://weather.weatherbug.com/aws-corporate/default.asp

WeatherBug Elite, shown in Figure 6-5, is the source for truly live, local weather information. WeatherBug Elite gives you current weather conditions, seven-day and hourly forecasts, *NWS* (National Weather Service) alerts, detailed radar and maps, daily national outlook broadcast video, weather camera images, and more.

WeatherBug Elite is the first — and currently only — weather app on BlackBerry App World to leverage BlackBerry push technology, allowing you to receive severe weather alerts via BlackBerry e-mail as soon as that information becomes available.

In addition, WeatherBug Elite integrates deep within the BlackBerry OS, providing you with access and customization of their weather data. WeatherBug Elite is the premium, ad-free version of WeatherBug for mobile smartphones.

Here's what you get:

✦ **Alerts via e-mail:** You can get real-time weather alerts with BlackBerry push technology.

✦ **Deep integration:** WeatherBug embeds the weather forecast in the events saved to your BlackBerry Calendar. It also includes the current weather for all your BlackBerry Contacts, as well as weather information embedded in BlackBerry maps.

✦ **Cool graphics:** You'll be happy with the detailed alert maps with vector polygons and full NWS alert messages.

✦ **Customization:** You can get precise weather conditions with GPS, as well as conditions in saved locations and an unlimited number of static locations.

There are several versions of the WeatherBug app, all of which you can find at BlackBerry App World. We recommend WeatherBug Elite, which is around $5, as it doesn't include the annoying ads that come with the free edition.

**Figure 6-5:**
WeatherBug
Elite gives
you weather
conditions
and other
weather-
related
information.

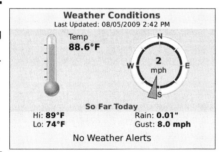

## WorldMate

www.worldmate.com

WorldMate, shown in Figure 6-6, is the leading BlackBerry app for business travelers. It condenses all the information you need — from travel plans to business meetings — into a single interactive itinerary. WorldMate monitors your trip in real time and sends you the information you need when you need it.

Having WorldMate is kind of like having a clairvoyant personal assistant you never need to pay. (The app is free.) For example, if your flight is delayed, WorldMate lets you know and shows you all the alternative flights. Once you land, WorldMate provides the local time and weather. Need a hotel near you? WorldMate suggests hotels that you can book straight from your BlackBerry.

Here are the features we love the most:

+ **Automatic itinerary creation:** Enter the key info and WorldMate takes care of the rest.

+ **Flight alerts and schedules:** After you have an itinerary, you don't have to check on your flight's status. WorldMate alerts you automatically.

+ **Hotel booking:** This is the clairvoyant part. We're not sure how WorldMate does it, but it locates hotels and helps you book a room once you arrive at your destination.

+ **Travel tools:** Get other information, such as weather, currency converters, world clocks, and more.

WorldMate saves time when your flight has a gate change or is delayed. If you're a traveler, you need this app.

**Figure 6-6:** WorldMate has psychic powers and keeps you sane when you're traveling.

## BBTran (also known as FancyTran)

www.bbtran.com

BBTran, shown in Figure 6-7, is a free app that translates words and phrases in more than 40 languages. You can use the app to look up words in your own language that you don't understand, as well. BBTran acts as a unified interface to several online services such as Google Translate, Systran, FreeTranslation.com, Reference.com, and Bing Translator.

BBTran has two editions: Regular and Pro. The Pro edition allows you to use 500+ translations a day, examples of how a word can be used, and message integration. The cost for the Pro edition is $5.99.

BBTran makes translation on the BlackBerry extremely easy because of its integration with other BlackBerry applications. Here are some of the app's key features:

- ✦ **Works with multiple translation services.** You get the most accurate translation because BBTran aggregates translation information from several translation engines. Getting alternative translations helps you understand subtleties in the language and overcomes the issues you sometimes face when you use a computerized translation.

- ✦ **Offers Dictionary mode.** Dictionary mode allows you to look up individual words instead of whole phrases.

- ✦ **Integrates with BlackBerry e-mail/SMS/PIN messages (Pro edition only).** The Pro edition of BBTran adds a Translate Selection with BBTran option to your e-mail, SMS, and PIN menus. This way, you can translate content within a message instead of copying and pasting the message into the BBTran interface.

- ✦ **Lets you choose the look of the interface.** BBTran allows you to customize the font size, font, title bar, and background colors of its interface. This provides your own unique visual experience when using the application.

BBTran is a great way to break (or at least weaken) the communication barrier if you're traveling and don't speak the native language. It's also a heck of a lot more portable than a dictionary.

This app goes by two names. If you're downloading it from BlackBerry App World instead of the manufacturer's site, look for it under the name FancyTran.

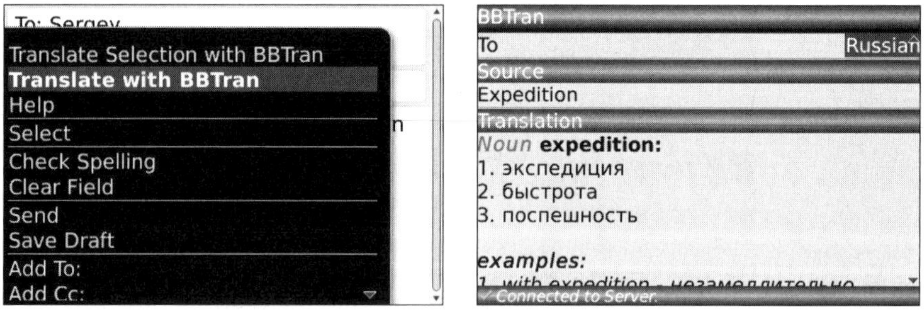

**Figure 6-7:**
BBTran gets you through tough translations.

# Book VII

# Roadmap to Application Development

The 5th Wave          By Rich Tennant

"What I'm doing should clear your sinuses, take away your headache, and charge your BlackBerry."

# Contents at a Glance

# Chapter 1: Getting a Start in BlackBerry Application Development

## In This Chapter

✓ Knowing what you need to know before you develop an app

✓ Choosing the right development platform

✓ Getting to know the tools for each development platform

✓ Creating BlackBerry smartphone themes

So you're ready to develop an application for the BlackBerry? Great!

Three platforms are available for developing BlackBerry smartphone applications. Each platform has something different to offer, so in this chapter, we help you choose which platform is the best fit for your application and skill level.

Building apps isn't something you should consider unless you feel fairly comfortable with coding. Even the drag-and-drop tools work better if you understand what's happening under the hood. Also, a background in user interface design, an understanding of basic Web development concepts, and some fundamental knowledge of graphics editing programs can't hurt.

If you're not sure what kind of application you want to develop, you can start by building a BlackBerry smartphone *theme*. A theme controls the look and feel of the smartphone interface. By changing the icons, colors, fonts, and arrangement of all the items on the Home screen, you can create your own personalized BlackBerry smartphone experience.

If you're creating applications for use on a BlackBerry Enterprise Server, or BES, you're likely to collaborate with BES administrators in deploying apps. To find out the ins and outs of the BlackBerry Server environment, see Book VIII.

## Finding an Application Development Platform

Finding the right development platform is an absolute necessity. Thankfully, it is also pretty easy if you know where to start. BlackBerry applications can be divided into these three main categories, which we discuss in more detail later in this chapter:

✦ **Web-based applications and Web sites:** If you want to create a mobile version of an existing Web site, or you want to create a Web site from scratch that is tailored to a mobile audience, you fit into this category. Developers who are interested in mobile Web development usually build their products for all mobile devices, including iPhones, the Android platform, and various BlackBerry models. However, it's pretty common to start with one type of device and then make tweaks for other types of devices.

✦ **Java-enabled applications:** BlackBerry's native operating system is Java Virtual Machine (JVM). All applications, including both native and third-party applications, are either written in Java or translated into Java to run inside this JVM. You may not be surprised to discover that serious BlackBerry application developers gravitate toward writing their applications in Java because Java is the only language that allows developers a wide range of application programming interfaces available for the BlackBerry.

If you have a great idea for an app and you think it's going to be a complex one, we recommend you use Java to create it. We're not going to lie — of all the options, Java is the most demanding. On the other hand, thousands of developers make their living writing BlackBerry applications using Java. The path is definitely well laid out, with tons of help available. One such book is *Java All-in-One For Dummies,* by Doug Lowe (Wiley Publishing).

✦ **BlackBerry theme development:** Technically, a BlackBerry theme is an app. However, most serious developers don't design themes. If you're a graphically oriented individual or an interface designer with a knack for creating a pleasing and functional user interface, designing themes might be up your alley. Creating a BlackBerry theme is quick and easy. The tools are easy to follow and use. Heck, you don't even need to write code, and you'll see the fruits of your labor right away, literally in minutes. Good themes sell well and are priced relatively the same compared to Java apps that take months to write. And unlike complex Java apps, post-development cost in terms of supporting users is minimal.

To effectively design themes, you should have at least a rudimentary background in graphic design — which means that you should understand using graphic tools like Adobe Photoshop.

Regardless of the development platform you choose, you also need to consider the target end users. Your target users greatly influence your development, testing, and deployment process. You can slice and dice different types of users, but the two main categories to consider are

✦ **Enterprise users:** Enterprise users are users whose BlackBerry smartphones are provided by their employers. Their devices are connected to a BlackBerry Enterprise Server. Typically, these users have a device that is used strictly for business purposes. The apps needed by enterprise users are apps that help them do their daily jobs. For example, you might create an app that can be used by bankers to help them access enterprise information through their BlackBerry devices. The app can be extremely useful when bankers are out in the road meeting clients and need access to certain financial or proprietary information right away. You can also *push* content to your users using message, browser, or cache push technology (more on these topics later in the chapter). You can limit your app development to a few target BlackBerry models used across the enterprise.

✦ **Consumer or personal users:** Consumers use their devices for personal use. They want apps for games or entertainment, social networking, business, and productivity purposes. If you're building apps for the consumer market, you have a wide range of models and operating system versions to target. In order for your app to be visible in the marketplace, you need to spend some time and energy marketing your app and deploying it to all available app stores.

With the three development platforms to choose from and the type of end users, picking the right one for your needs depends on these three factors:

✦ **Your skill level with Java programming:** Some platforms require you to understand Java, while others don't. If you don't know Java and you don't want to learn it, some platform decisions will be made for you.

✦ **Your experience with Web development:** Some platforms don't require you to know anything about Web development, CSS, HTML, or anything else. With WYSIWYG application development, you don't need to know what's going on underneath the hood, but you can still drive the proverbial car.

✦ **The goals and requirements of your development project:** If you want to build a BlackBerry theme, you probably don't need to worry about using a platform that's heavily focused on Web development. Likewise, if you're working for an enterprise, using the tools for enterprise application development makes a ton of sense. If you have no idea what you want to do, figure out your goals and go from there.

In the following sections, we outline your platform options and help you sort them out.

# Introducing BlackBerry Browser Development

The BlackBerry Browser development platform is ideal if you want to create Web applications for the BlackBerry. BlackBerry *Web applications* are simply Web pages that run on BlackBerry Browser. The only difference for you is how to lay out your pages so that they are easy to navigate on a small screen. When deciding on a layout for the Web pages, take into consideration user interactions with a touch screen device.

When we talk about BlackBerry Web apps, we mean Web pages that are designed to display on the BlackBerry smartphones.

BlackBerry Web applications can be used to do so many things, but the most important thing is for them to enable users to obtain information while on the go. For example, a Web page that gives you the status of an airline flight is infinitely more helpful if you can access it from your smartphone when you're on your way to the airport. Of course, established Web sites like Facebook have Web applications that give users the same type of functionality. However, they have totally different Web pages when viewed from a smartphone like BlackBerry than they do when you view them on a computer. Later in this chapter, we describe the BlackBerry device's unique ways of pushing Web pages to the smartphone.

Most of the tools that you use for developing a Web page for a desktop computer can be used for developing Web pages for the BlackBerry. Some of these tools are simple text-editing tools like Notepad, or sophisticated Web development tools like Dreamweaver, Microsoft Visual Studio, or Eclipse. If your application requires JavaScript to run, then you need tools that have JavaScript debugging capabilities. Of course, a countless number of Web development software is available, so whichever one you're comfortable using is okay.

If you like BlackBerry Browser development and you're interested in taking your Web content to the next level, RIM also offers something called Plazmic Content Developer's Kit, which extends your Web development options by allowing you to include *rich media* (such as animation) and *dynamic content* (information that changes its display automatically based on device type and resolution) for the BlackBerry Browser. Here are some items to consider when thinking about BlackBerry Browser development:

+ You don't need experience with Java programming.

+ Web page development knowledge is a must. You may be able to create simple HTML pages without a background in Web development, but to create good page layout, you need to step up your knowledge, including understanding how to use Cascading Style Sheets (CSS). Picking up a little JavaScript is a must if you want to make your site interactive.

✦ JavaScript is also necessary if you want to integrate your Web development projects with other native BlackBerry applications, such as BlackBerry Contacts or BlackBerry e-mail.

# Getting to Know the BlackBerry Widget

BlackBerry Widget is a BlackBerry application that is written using Web technologies, translated into Java, and packaged as a Java application. The Web technologies are HTML, JavaScript, and Cascading Style Sheets (CSS).

If you're used to using Web technologies, this platform will be a cinch to adapt; you can quickly build BlackBerry applications with it. The only difference between writing browser apps and using the BlackBerry Widget is that the Widget can use JavaScript to integrate with native BlackBerry applications such as BlackBerry Contacts or Messages. We talk more about the advantages and disadvantages of using BlackBerry Widget later in this chapter but to illustrate how simple the process is to create a BlackBerry app using BlackBerry Widget Software Development Kit (SDK), here are some general steps:

*1.* **Write an HTML page.**

Make sure the HTML adheres to the BlackBerry Widget guidelines that are provided in the BlackBerry Widget SDK. The page can contain CSS for page layout. It can contain JavaScript for interactivity and for integration to the BlackBerry platform. All Web files must be located in a single folder.

On your Web page, you'll be writing JavaScript scripts and interfacing with BlackBerry JavaScript application programming interface (API). This BlackBerry JavaScript API is provided in the BlackBerry Widget SDK. This API allows you to call native BlackBerry functions, such as invoking the Messages app for sending e-mail.

*2.* **Create a configuration file.**

This is an XML (eXtensible Markup Language) file that describes the application. The configuration file also describes what BlackBerry JavaScript API the application is using. It should be located in the same folder as your Web files.

*3.* **Convert your HTML page to the BlackBerry application file format (.cod and .jad).**

You simply run a program called BlackBerry Widget Packager that comes with the BlackBerry Widget SDK. You tell the program the folder location of your Web files (HTML, JavaScript, graphics, CSS, and configuration XML). This program creates both .cod and .jad files, which are the actual BlackBerry app installation files. The .cod extension is the compiled code document, which is the actual program that runs on your BlackBerry. The .jad file stands for Java Application Descriptor, which contains information about your program.

**Book VII Chapter 1**

**Getting a Start in BlackBerry Application Development**

4. **Test the Web page in a BlackBerry smartphone simulator.**

   On the BlackBerry smartphone simulator, you simply load your `.cod` file. The simulator is the most convenient way of testing an app because installing the app to a device takes some time.

5. **Test the Web app in the simulator and tweak it until you're happy with it.**

6. **Install the Web app on a real device.**

7. **Test and deploy to a BlackBerry smartphone.**

   You can use the BlackBerry Desktop Manager to load your app into a device.

 With integration to native BlackBerry applications, a Web developer who doesn't know how to write Java code can now write BlackBerry applications that behave like Java applications. For example, you can write an app that links social networking friends with the BlackBerry Contacts using simple JavaScripts and HTML pages. For enterprise users, you may want to extend your enterprise contact management system (CMS) to the BlackBerry. You can build a widget that lets your sales team update client information from a BlackBerry contact, which cuts down on data entry time.

# Introducing Java Application Development

All applications running on the BlackBerry smartphone are Java apps. To fully take advantage of the capabilities of the device, including integrating your app with location information, the accelerometer, multimedia, the camera, messages, contacts, MemoPad, other native apps, and system functions, you need to write your app using Java. If you have a great idea for a game that takes advantage of the accelerometer, you absolutely must write the app using Java.

 A smartphone is a computer in a small package. Being small has its limitations — it has a small memory and relatively low processing power compared to a desktop computer. The Java technology used in a smartphone is also designed for small devices. It's called Java Micro Edition, or commonly referred to as JME. JME is a subset of the Java language. When you're developing BlackBerry apps, you're using JME. You need to be aware of these things when using JME:

✦ **Mobile Information Device Profile or MIDP:** This is a standard in the mobile industry for device makers to follow. It defines the type of information available on the device, like the address book and notes. There are sets of JME API that specifically correspond to the MIDP standard. The idea here is that if you write your code using this API, you can run your app on any devices (such as Nokia smartphones) that support such a standard. BlackBerry supports MIDP 2.0.

✦ **Connected Limited Device Configuration (CLDC):** This standard defines minimum device capabilities. Its focus is specifically geared toward network connectivity.

MIDP and CLDC go hand in hand. In essence, MIDP allows you to write applications and services that can be downloaded to network-connected devices. When used with CLDC technology, MIDP acts as a Java runtime environment that runs on the BlackBerry.

The mobile standards body that maintains MIDP, like any other standards body, only reacts to changes in the industry. With the rapid pace of innovations in smartphones, we find that MIDP has a considerable lag time to react to what's really happening in the industry. Clearly MIDP lacks the API feature sets that you need to build great BlackBerry apps. Device makers have their own and more extended sets of APIs, and the BlackBerry is no different. Java developers can leverage RIM's BlackBerry Java API, which includes a User Interface Library, also called a UI Library. The benefit of the UI Library is that it allows you to access various development environments such as Eclipse and NetBeans. Eclipse is one of the most widely used multilanguage software development environments. It allows you to use third-party plug-ins that make it much easier for you to develop applications for the BlackBerry. You also have access to NetBeans, another platform that is used for developing applications in multiple languages.

Java application development is the most commonly used platform for building applications for the BlackBerry because if gives you the most freedom to create your own application — exactly the way you want it to look and work. However, if you choose this method you should consider a few items:

✦ You must have Java programming knowledge.

✦ You may need to develop multiple versions of your application so that the application will work across different BlackBerry models, other smartphone types, and operating system versions. (You don't need to worry as much about creating multiple versions when you're designing a Web app because BlackBerry Browser is relatively stable across various models and OS versions.)

## Comparing Platforms

In this section, we outline some high-level features of each development platform in Table 1-1. Your choice really comes down to some basic factors.

| Table 1-1 | Choosing the Right Development Platform | | |
| --- | --- | --- | --- |
| *Project Description* | *Use BlackBerry Browser?* | *Use BlackBerry Widget?* | *Use Java Development?* |
| My application does not need to interact with native BlackBerry applications, or I want to run my application through a simple Web page. | This is an ideal option. | This option isn't necessary unless you plan on integrating to native BlackBerry apps in the future. | This is a possible option, and yet it requires more effort. |
| My application will have minor interaction with native BlackBerry applications. | You can't use this option. | This is an ideal option. | This is a possible option, and yet it requires more effort. |
| My project requires some heavy-duty Java programming knowledge, or my app requires a lot of BlackBerry native application integration. | You can't use this option. | You shouldn't use this option. | This is an ideal option. |

# Getting Inspired by Application Development Options

In this section, we break down each of the platforms with the different ways you can use them. We also alert you to some information you should know.

## BlackBerry Browser development

Nowadays, more people are using their smartphones to browse the Web and send messages than they are to talk with. There's money to be made as mobile browsing continues to increase in popularity. In fact, high-speed mobile broadband, larger device screens, and devices that render Web sites as if you were viewing them from a desktop computer make the mobile browsing experience much more enjoyable than it was back in the day.

Many companies are developing mobile versions of their Web sites so that smartphone users can view and use sites on a smaller device. Of course, it's important for visitors to access information quickly and efficiently. That means taking special considerations into account for mobile device users.

One way that companies are making mobile browsing easier is to use a standard URL format for mobile versions of a Web site. It's pretty easy — a mobile site may be placed on a subdomain and contain the letter "m" or "mobile" in the URL. For example, the mobile version of the Facebook Web site is `http://m.facebook.com`.

Even if you don't choose to create a mobile subdomain of your Web site, many smartphone companies (including RIM) are also improving the experience. For example, BlackBerry has enhanced the Web browsing experience significantly. BlackBerry handheld software allows users to view a full Web page by default, so that even if users aren't looking at a site with "m" in the URL, they can scroll around the page and zoom in on areas for a closer look, shown in Figure 1-1.

**Figure 1-1:**
This nonmobile-formatted site renders on a BlackBerry exactly as it does on the desktop.

You can use BlackBerry Browser development to create a Web page that redirects a user to a mobile site based not only on the browser that is being used, but also on the model of BlackBerry smartphone being used. For example, a BlackBerry Bold user might be directed to a different page than a BlackBerry Storm user — all in the name of user optimization!

You can use BlackBerry Browser to develop an application in four ways. (However, Message Push and Cache Push can be used only if the BlackBerry is connected to a BlackBerry Enterprise Server.)

✦ **Web page:** Users access a Web site by clicking a link or going to their BlackBerry Web browser and typing in the URL.

✦ **Message Push (for BES):** A user receives what looks like an e-mail. It's actually a link to a selected URL.

✦ **Channel push:** An icon appears on the BlackBerry. When the user clicks it, it opens BlackBerry Browser and goes to a specific URL. The user can change the icon for read or unread content.

✦ **Cache push:** An icon appears on the BlackBerry. Information is stored in a cache, which allows the information to be viewed even if the device is not receiving a signal.

### Developing Web page content

If you know how to create a Web page, you already know how to develop for a BlackBerry. The best part is that you won't need special tools or formatting, as long as you develop using one of the following languages that BlackBerry Browser can handle. You also have the option to redirect users to a Web page that was formatted for mobile use. These options give the freedom to build sites that are optimized for a smaller screen and wireless speeds (which are usually *just a little bit* slower than your typical desktop broadband speeds).

BlackBerry Browser development supports the following:

✦ Standard markup languages, including WML, CHTML, XHTML, HTML, DHTML, and Cascading Style Sheets (CSS)

✦ Dynamic content, such as JSP, ASP, PHP, and other languages

✦ JavaScript 1.2

✦ Image and video streaming support

✦ Full form functionality and form queuing

✦ GPS integration via JavaScript

✦ Viewing supported document formats (`.doc`, `.pdf`, and `.ppt`)

With Research In Motion's online form cache, a user can open a Web page and fill out an online form at any time. If the user's device is out of coverage when he or she clicks the Submit button, the information is queued and sent when the device has sufficient data coverage to send the form. The user doesn't need to click Submit a second time.

Flash is not supported by the native BlackBerry Browser. If you want to make your site visible by a BlackBerry, create a non-Flash version and redirect your mobile users to it.

Being able to display a full site on a BlackBerry without making any changes to your content or code may seem like a great idea, but it *does* have its drawbacks, especially if your site contains Flash programming, large pictures, animations, or video content.

The BlackBerry smartphone simply can't display the Flash, and it has to download all the images, video, and animations, as well as rendering them. This can be a time-consuming proposition, especially for users with older devices and no high-speed Internet connection. Additionally, sites that use the following (which is, like, the majority of the Web!), will slow down page loading:

+ JavaScript
+ Foreground and background colors
+ HTML tables
+ Background images
+ Embedded media
+ Style sheets

In Book VII, Chapter 2, we provide you with some guidelines and other information you need to know to develop a mobile site.

When planning a mobile version of a site, don't forget to provide the user an option to switch between mobile and desktop versions. This choice allows users to take advantage of your full site when they have access to a more robust experience.

An example of a site that is formatted for both desktop and mobile is Yahoo! (www.yahoo.com). In Figure 1-2, the full Yahoo! site is on the left. This looks pretty similar to the version on your desktop, except that because your BlackBerry device's screen size is smaller, the text and graphics are very small. The version of the site on the right is the Yahoo! Finance mobile site (http://m.yahoo.com). The mobile version looks much more readable.

**Book VII
Chapter 1**

Getting a Start in
BlackBerry Application
Development

**Figure 1-2:**
The full version of the Yahoo! Finance site (left) and the mobile version of the same site (right).

## Pushing content with message push

If you're developing applications for BlackBerry smartphones connected to a BlackBerry Enterprise Server, you can use all the different Push content methods. Again, *pushed* Web applications are simply Web pages, but to users, they appear like a typical BlackBerry application because they're not accessing them through Browser.

Message pushing is done when an administrator pushes a Web page to a device or group of devices. This message shows up like a message with a globe as an icon on both the Home screen next to the user's message count icon and in the user's message list. It also has a separate message notification icon. Using message push has many benefits:

+ **Instant notifications:** Message pushing is used to communicate to all users on the enterprise at the same time. Once you push the information, the data is immediately sent to all the devices that the administrator specifies, with a message indicator.

Messages can be pushed to some users or groups of users or to all users on the enterprise. The enterprise administrator has the power to decide when messages are pushed and to whom. For more information on the BES, see Book VIII, Chapter 1.

+ **Better battery life:** Because the device isn't used repeatedly to retrieve information updates, the battery life on users' devices is extended. A good example of this scenario is when an enterprise user needs to know the completion of a task via a Web page. Instead of the user checking the page regularly, an update to the page can be pushed to the device.

+ **No learning curve for users:** From an administrator's point of view, message pushing is a great way to convey important data quickly across the board. Users are accustomed to checking their e-mail, so pushed messages are in a familiar place and are much likelier to be read.

Message pushing can be done in two ways:

+ **Link:** In this situation, the message acts like a bookmark. When the message is opened, BlackBerry Browser launches and navigates to the associated link. The device needs to have data coverage to open the Web page. One enormous benefit of using a link is that if changes are made to the URL, the user sees the update right away.

+ **Cache:** In this situation, the link is pushed to the user along with the page content. The Web page is stored locally in the device's cache and can be viewed when the device is offline. The upside of this option is that the content can be viewed at any time. The downside is that this data is static and can't be changed after it is pushed to users.

When content is pushed to users, messages are displayed in the users' BlackBerry Messages. However, users don't see the information when they use their desktop e-mail client.

Even if users have configured their BlackBerry to ring bells, play a special ringtone, or vibrate whenever an e-mail message or SMS text is received, the BlackBerry smartphone doesn't have an audio or vibration alert when a push message is received. The only indication that a message has been pushed is the notification that appears in the BlackBerry mail client.

### Pushing content with browser channel push

A browser channel push happens when a user clicks an icon that goes to a specific Web page. It's like having a bookmark on your computer's desktop. The only difference between channel push and message push is that no message notification appears in the Messages app.

Figure 1-3 shows the icon on the BlackBerry Home screen, as well as the site that appears after the icon is clicked. However, the content owner of this icon has a few more options than just having the icon open a Web page.

In addition to linking users to a URL that opens a Web page, you can also link to documents.

The push can contain two icons that can be used to indicate that new content is posted. After the user selects the icon, the second appears, indicating that the content has been read. For the weather push icon shown in Figure 1-3, the icon changes based on the current weather condition.

**Figure 1-3:**
Browser channel push allows users to click an icon to access an application or content.

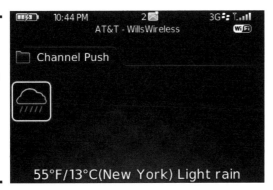

When a channel push icon is added to a user's BlackBerry, a link to the related site is added to the BlackBerry Browser's bookmarks under Browser Channels. If the user deletes this link, the icon disappears until the next time a channel push occurs. Users must unsubscribe to the push service to stop the icon from reappearing.

The developer or administrator can change the text for the icon that appears on the bottom of the BlackBerry screen. You can use this text to show users what's changed since the last update, or you can include other important information and the name of the application. For example, a weather icon might change based on the temperature, location, and current weather conditions.

If the service is no longer being used, the administrator has the ablity to delete the icon from the BlackBerry smartphone by using a Browser Channel Delete Push.

### Pushing content with browser cache push

Similar to channel push, browser cache push allows the developer or administrator to push an icon to the BlackBerry smartphone. Unlike channel push, the icon in browser cache push isn't updated when the content is changed.

Using a content cache push sends more than just an icon that is linked to a Web page. The content cache push also sends data to the device's browser cache. Content that's pushed using browser cache pushing is viewable even when the user isn't connected to the Web.

## Developing BlackBerry apps using BlackBerry Widget SDK

If you want to develop a feature-packed application with some integration to native BlackBerry apps and a rich user interface, a simple Web page just doesn't cut it. To give you the ability to integrate with native BlackBerry applications, you may want to consider using BlackBerry Widget SDK, which we introduced earlier in this chapter. Before you consider diving into BlackBerry Widget, you need to know the pros and cons.

### Understanding the advantages of using BlackBerry Widget

There are good reasons why you should consider using the BlackBerry Widgets SDK:

+ **Leverage Web technologies.** If you understand Web technologies like HTML, cascading style sheets, and JavaScript, you can easily adapt to the BlackBerry Widget SDK.

+ **Test your app in a browser.** Because your code is written in a combination of HTML, JavaScript, and CSS, testing your app on any Web browser is very convenient. Creating user interfaces is also easier than creating user interfaces with Java.

+ **Develop apps quickly.** Because you can write and test apps on a Web browser, you will find yourself making it to the finish line much quicker. Being ahead of schedule *never* hurts.

### Knowing the disadvantages of using BlackBerry Widget

BlackBerry Widget is a fairly new technology, and with any new technology comes challenges, including these:

+ **Widgets run only on BlackBerry smartphones having OS 5.0 and above.** A majority of BlackBerry devices are still running on an older OS. To some, this is enough reason not to use Widgets. If you're writing a consumer app, you want to broaden your user base as much as possible. A major OS limitation is a blow to the goal of having a wider user base.

+ **BlackBerry Widget API is only a small subset of the BlackBerry Java API.** You can't perform complex integration with the built-in BlackBerry apps. For example, you can write JavaScript code to use the native BlackBerry e-mail app to write and send e-mails. However, you don't have the option to program your app to react to an incoming e-mail.

+ **Few developers use BlackBerry Widget SDK.** With fewer developers using the technology, you won't find a good history of reported issues and resolutions. You're more likely to be the first one to report an issue and have to wait for RIM support to reply to your queries.

## *Java application development*

If you want the greatest flexibility developing rich client applications for the BlackBerry, using Java and BlackBerry APIs is your best choice. An API is an application programming interface. APIs allow one software program to interact with other software programs (usually through the magic of programming languages). Like most technology companies, BlackBerry has an open API that you can access to develop applications that enhance users' experiences with their BlackBerry devices.

*Java* is an object-oriented programming language developed by Sun Microsystems in the 1990s. Because of its flexibility and portability, Java's pretty much the standard bearer for relatively easy programming. It is platform independent and secure, meaning that it works just as safely on a PC as it does on a Mac, or a Web page, or an Android phone, or a BlackBerry. It uses classes and objects to organize commands, called *strings*.

Java programmers will be pleased to know that BlackBerry smartphones are full Java-based devices. BlackBerry devices support MIDP 1.0 and CLDC 1.0, and any devices using BlackBerry Device Software 4.0 or higher support both MIDP 2.0 and CLDC 1.1.

Some of the BlackBerry APIs allow you to develop the following options for the BlackBerry:

+ You can create customizable user interfaces.

+ You can make sure that local data is stored on the device.

+ You can listen to smartphone-specific events, such as a low battery or a dropped Bluetooth connection. You can code your app so that it reacts to these events.

## Java-specific file types

When you develop applications for the BlackBerry (or for any platform, for that matter) using Java, you have to deal with some proprietary file types, courtesy of your friends at Sun Microsystems, the creators of Java. The three files that are associated with BlackBerry applications are

✔ .cod: This is the program file and the one that is installed into the device.

✔ .jad: This is a file that is stored on a Web server. Users can direct their browsers to the file for an over-the-air, or OTA, application download.

✔ .alx: This file describes an application and where the COD file is located. This is used to install applications through BlackBerry Desktop Manager or when deploying an app through BlackBerry Enterprise Server.

The significance of these files comes into play when you're about to test an app in a simulator or install an app in a device.

---

✦ You can integrate rich multimedia (graphics, animations, video, and more) into the user experience.

✦ You can use cryptography and security APIs to ensure both individual and corporate security.

✦ You can easily integrate all BlackBerry-specific functions (called PIM, or *platform independent model,* functions) with the applications you develop.

✦ You can write animations using the BlackBerry's implementation of the *Open Graphics Library,* or OpenGL. OpenGL is an API standards designed to manipulate graphics for animations. Most programming languages have support for this standard, and that includes Java. The advantage of using the BlackBerry OpenGL library is the ability to port your code with ease to any other languages that have OpenGL implementations. For example, you can write games using Java OpenGL library for the BlackBerry and later port the code to Objective-C to target the iPhone market.

## Getting the Right Tools

After you decide which platform makes the most sense for your application development, the next step is to download the applications that will help you begin your project.

To find more details and download the tools, go to `http://na.black berry.com/eng/developers/resources/devtools.jsp`.

## An app developer is an artist

Obviously, your app provides functions that are useful to the end users. But the look of your app can also influence adoption. A critical requirement for a great app is a nice and cool user interface, especially for a consumer app. Images can convey information easily, and this is true for a BlackBerry application. You may find yourself needing good images as you start building a smartphone app, and good skills in using graphics and image-editing software is a must.

A lot of image-editing software is out there, and if you are already familiar with one, you can stick with it. If you haven't used one, you can't go wrong with either of these:

✦ **Adobe Photoshop or Adobe Illustrator:** Photoshop is by far the most popular image-editing software, so you have a lot of help online. You can purchase and download a copy from `www.photoshop.com`.

✦ **GIMP:** This is open source software. It's free and includes lots of free add-ins and scripts. You can download GIMP from `www.gimp.org`.

## Getting the right simulator for the job

The first tool you need for any application development project is a simulator, which lets you emulate just about every BlackBerry model's interface (regardless of the carrier) so that you can visualize how your application will appear and function. A simulator is certainly a more logical choice than buying every device on every carrier with every smartphone operating system. Choose as many simulators as you can get your hands on. Here are must-have simulators:

✦ **BlackBerry smartphone simulators:** Each of these simulates a certain BlackBerry model running on a particular OS. Expect to download and use many simulators if your app is targeted to run on multiple models and operating systems. You can download the simulators at `http://na.blackberry.com/eng/developers/resources/simulators.jsp`.

You can run only one BlackBerry smartphone simulator at a time. If you're testing your application across different device models, you need to close a simulator before starting a new one.

✦ **MDS (Mobile Data Service) Simulator:** This simulator is crucial if you're building applications that connect to the Internet. This is also needed if your app is targeted on an enterprise environment. You use this in conjunction with a BlackBerry smartphone simulator, allowing you to simulate a device that is connected to an MDS connection service component of the BlackBerry Enterprise Server. The MDS connection service is the tool that allows the BlackBerry simulator to access the Internet.

✦ **BlackBerry Email Simulator:** This tool allows the BlackBerry smartphone simulator to send and receive messages. You'll find this useful if you're building an app that sends or receives e-mails.

Accessing *every single possible* device configuration, from very old BlackBerry devices with old OS software to brand-spanking new BlackBerry devices with the latest OS upgrade, may not be the most logical testing scenario. Usually, the better way to go is to follow these steps:

1. **Evaluate the market.**

   Knowing whether your "customers" are users in your enterprise or strangers downloading your app from an app store helps you determine how many BlackBerry devices you need to build for.

2. **Determine which devices and configurations are the most prevalent.**

3. **Choose a reasonable number of phone configurations for which you plan to design.**

   You can always modify your development later to reach more customers. (And remember, you'll have to update your app to meet new technology needs down the road, anyway.)

   Decide what you can live with. Sometimes, determining what's most prevalent may mean alienating customers who haven't done a phone or software update in a while. If you can live with alienating a small percentage of your possible users who are using the smartphone equivalent of two soup cans connected by a string, then you're probably going to save yourself a lot of trouble.

4. **Set up a systematic testing experience.**

   Your test should follow a strict protocol. Being rigorous about testing the most common configurations will reap more benefits than lazily "testing" every possible configuration.

---

# Cutting through OS version red tape

Research In Motion doesn't provide one package with all the different devices and operating systems. You need to download a separate simulator for each combination of device model and OS. You're probably wondering now how many simulators you need to download, because BlackBerry has so many OS revisions. Not to worry! Each OS revision usually comes with *release notes,* which indicate changes to the operating system. Usually, these changes are bug fixes that probably don't concern you when developing apps. As a result, RIM usually has just one simulator for the main OS version — the latest and greatest simulator. For example, an OS 5.0 has several minor revisions in the likes of 5.0.0.122, 5.0.0.168, and 5.0.0.442, but you need only one simulator for all the ranges of version 5.0.

### Getting tools for BlackBerry Widget

Luckily, everything you need to build an application using the BlackBerry Widget SDK can be found on RIM's Web site. The main page for the SDK is `http://na.blackberry.com/eng/developers/browserdev/widgetsdk.jsp`.

On the Widget SDK page, you can also find useful video tutorials and sample applications. Watching those videos can shorten your learning curve dramatically.

The Widget SDK page clearly describes and explains what tools you need. Here's the quick rundown. We recommend you also check what's listed on the Widget SDK page, as this could change:

✦ **Sun Java Application Development Kit:** Although you're not writing Java code while building an app using BlackBerry Widget, the Widget SDK converts your code to actual Java code, and you need the Sun Java Application Development Kit to do this conversion. You can find this kit here: `http://java.sun.com/javase/downloads/index.jsp#jdk`. Choose Java JDK 6 Update under the Java Platform Standard Edition section.

✦ **BlackBerry Widget SDK:** This is the actual SDK that you need for writing an app. At this writing, the available version is v1.0 and can be downloaded from `https://www.blackberry.com/Downloads/contactFormPreload.do?code=DC727151E5D55DDE1E950767CF861CA5&dl=82EE71AE5F6ECC5164F2080FB805405D`.

✦ **BlackBerry Widget SDK Installation Guide:** Download and read the guide that corresponds to the version of your Widget SDK. For v1.0, the guide can be downloaded here: `http://docs.blackberry.com/en/developers/subcategories/?userType=21&category=BlackBerry+Widgets&subCategory=BlackBerry+Widget+Development+Guides`.

✦ **BlackBerry Widget SDK Documentation:** This is the guide you need to read to understand the nature and structure of a widget. We highly recommend you read this before you go into a deep dive in writing your first application. You can download the document here: `http://docs.blackberry.com/en/developers/subcategories/?userType=21&category=BlackBerry+Widgets&subCategory=BlackBerry+Widget+Development+Guides`.

### Getting tools for Java

RIM provides the necessary tools for you for building BlackBerry Java applications. Download links on these tools can be found on their developers' Web site at

`http://na.blackberry.com/eng/developers/javaappdev/javasdk5.jsp`

Here are the tools you need:

+ **Java Development Kit (JDK):** The JDK contains all the components to develop Java applications.

  You must install the JDK before you install any other BlackBerry tools.

+ **BlackBerry Java Development Environment (JDE):** The JDE is a standalone application that includes the BlackBerry APIs, smartphone simulators, and the BlackBerry Signing Authority Tool that allows you to access native features in the BlackBerry when developing Java-based applications. To use the BlackBerry Signing Authority Tool, you also need to obtain code signing keys from RIM. RIM uses these keys to track the owner of an application that uses any system APIs that can be used maliciously. For example, if your application uses the mail API to send e-mail, you need a code signing key.

  To obtain a code signing key, you need to fill out a form and pay about 20 bucks. The form (and, the credit card) establishes your identity with RIM. RIM sends you your unique code signing keys. You can order code signing keys at `https://www.blackberry.com/SignedKeys`.

+ **BlackBerry Plug-in for Eclipse:** This plug-in allows you to use BlackBerry-specific functions in the Eclipse application.

Another tool that may be helpful is the BlackBerry Application Web Loader. You can create a Web site using Microsoft ActiveX controls to allow users to install an application from the desktop to the BlackBerry without Desktop Manager. You can find this at

`https://www.blackberry.com/Downloads/contactFormPreload.do?code=996009F2374006606`
`F4C0B0FDA878AF1&dl=E348AB56663CB5BA9450744D74AD9B08&step=4`

## Creating Your Own BlackBerry Theme

BlackBerry themes are great for personalizing the BlackBerry experience. They allow you to change many aspects of your BlackBerry. For example, you can use themes to change the size and location of icons, backgrounds, menu colors, fonts, and more. Here are some of the elements that you can make your own:

+ **Lock screen and background:** You can specify the picture that displays when the device is locked and unlocked.

+ **Fonts:** Anywhere there is text, like e-mail, date and time, and menu options, you can change the font that is used.

+ **Home screen icons:** Design icons in a photo editor and use them as your default BlackBerry icons. You can also change the way you interact with them by having them hide, shake, change color, or do something else when the user rolls his or her mouse over the icons.

✦ **Menu styles:** You can change menu styles so that when users click menu buttons, the color, fonts, and text displays change.

✦ **Active and incoming calls:** You can change the way the screen looks when users make a call, are on a call, or both.

✦ **Animated screen transitions:** You can make interesting animations when the user moves from screen to screen.

✦ **Other notification icons:** Customize the battery indicator, signal strength indicator, the Move option, as well as Caps and Num icons.

Because each BlackBerry model has a different screen resolution, themes must be developed specifically for each device model. A BlackBerry Pearl theme doesn't work on a BlackBerry Bold or on a BlackBerry Storm.

Figure 1-4 shows you a list of categories for themes that are available in App World. You can use the existing themes to give you inspiration in developing your own ideas for themes.

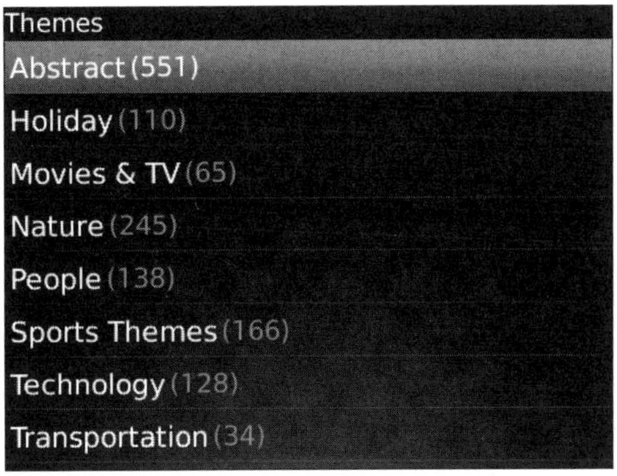

**Figure 1-4:**
Find inspiration for your own theme development by looking at what's already available.

Themes
Abstract (551)
Holiday (110)
Movies & TV (65)
Nature (245)
People (138)
Sports Themes (166)
Technology (128)
Transportation (34)

In this section, we review a few tools that you need in order to begin making your own themes. Best of all, everything you need to make your own themes is free! All you need is imagination and time.

Plazmic, the company that created Brick Breaker (the standard game on all BlackBerry smartphones) initially created the first application, called the Plazmic Content Developer Kit, for creating themes. RIM, however, realized the importance of themes, acquired the software, and built a robust infrastructure for building themes, called BlackBerry Theme Studio. Again, all the tools you need for building BlackBerry themes are all free.

Aside from being graphics-editing savvy, you will also need the following essential tools for creating themes:

✦ **BlackBerry Theme Studio:** A software suite you need to build a BlackBerry theme, as shown in Figure 1-5. Within this suite, you should find the following:

- *Theme Builder:* The program for building a theme, as shown in Figure 1-6.

- *Composer:* You need this tool if you want to add animations in your theme. This tool allows you to easily manage and build graphic sequences for an animation.

- *Documentation:* This is a folder containing documentation, user guides, and help files for Theme Builder and Composer.

- *Samples:* Sample themes give you a template and inspiration to start with.

✦ **Simulators:** Refer to the earlier section, "Getting the right simulator for the job." You need at least one BlackBerry smartphone simulator to test your theme. You also need BlackBerry Email Simulator to see how your theme indicates the appearance of unread e-mails.

✦ **BlackBerry Desktop Manager:** Needed only once, you're ready to install your theme to an actual device. You need to run Application Loader inside BlackBerry Desktop Manager for installing the theme.

**Figure 1-5:**
Tools within
BlackBerry
Theme
Studio.

**Figure 1-6:**
Build Your
BlackBerry
theme here.

# Chapter 2: Brainstorming Your Application

## In This Chapter

✔ Knowing what to code before you start coding

✔ Finding helpful hints to improve your application

✔ Getting more help

*B*efore you write a line of code, taking the time and effort to plan out how you want to structure your application is a really, *really* good idea. Some considerations include defining the scope of the application, deciding what language to support, and determining which platform other than BlackBerry you want to support. Many factors will directly affect how you write your application.

Although this chapter is intended to jump-start you in brainstorming about your application, it certainly isn't a definitive guide. We hope that after reading this chapter you will have the information you need to get started. Feel free to seek out the resources we list at the end of this chapter to find out more about BlackBerry development.

## Defining the Application Requirements

Although books are dedicated to showing developers how to write requirements and specifications for applications, this section is about thinking ahead before you start typing code. Whether you're writing an enterprise or small application, we want you to think about the different features and aspects of your application. In short, ask yourself what you want the application to do.

Depending on what features you require for your application, you can answer the big questions, like (but not limited to):

✦ Will the application run in the background?

✦ Will the application have a lot of graphics?

✦ How will users interact with the application?

✦ Will the application have specific hardware requirements, such as GPS or an accelerometer?

Say you're writing a pedometer application — an application that counts how many steps users take every day. Right away you should know that your application will be limited to BlackBerry models that have an accelerometer. That means the BlackBerry Storm 9530 and the BlackBerry Storm 9550. This info is important to know because you need to factor how much time and money it will take to develop your application and whether there are enough potential users to buy the app and offset your costs.

## Writing a Background Application

On the BlackBerry platform, you have the option to create an application that runs as a background process. There are two types of background applications:

✦ **An application that is started by the user and then continues to run in the background:** A good example of a *user-started* background application is a music player. After the user launches it, the music player keeps playing tunes even if the user is composing an e-mail or browsing the Web.

✦ **An application that starts automatically and continues to run in the background:** An example of a *system-started* application is a message-alert application that is always listening to incoming e-mail messages so that it can notify the user when a new message comes in.

The bottom line is that you need to decide which type of background application you want to create, and this decision should be based on the features of the application and how you expect users to use these features. You should always consider users' needs — and how your application meets those needs — so that you can create the proper code.

As an application developer, you have to very careful of what tasks the application performs while running in the background. Here are some words of caution:

✦ **Using a network:** If you need to retrieve information for your application in the background, consider a push model. You can leverage the BlackBerry Push API so that information is pushed to your application rather than pulled from it. The BlackBerry Push API is available for BlackBerry ISV partners. If you're going with a pulled model, be sure to make the pulled interval long enough so that the application does not drain the battery after a few hours.

✦ **Using location and GPS:** Like using a network, using GPS on the BlackBerry is very battery intensive. Consider using GPS only when the user is interacting with your application.

✦ **Using battery-level information:** If you do need to use the network or GPS constantly to meet the application requirement, then use the battery-level indicator to your advantage. You can structure your application to use GPS or the network with varying levels of intensity based on the battery level. For example, if you've written an RSS reader and have no choice but to use pull technology, then you can write the code so that if the battery level is 100 percent, it can poll the network for new RSS data every 15 minutes. When the battery level reaches 50 percent, you can set it up so that the application polls the network for new RSS data every hour.

If your application requires the user to input something, it's a foreground application. We don't cover how to create a foreground application in this book.

# Understanding Development Best Practices

Although we try to give you helpful hints throughout this chapter, we thought you might appreciate a whole section dedicated to the topic of Best Practices. The bottom line is that we know, and you know, that you can design your application in any number of ways, using any number of tools and based entirely on your level of commitment, creativity, and programming prowess, not to mention your budget, time frame, short- and long-term goals, staffing, and whether you hope to turn a profit from your creation. And don't forget about your personality, patience, and ability to push through to the end. Considering all these factors, your measure of success for a completed application might just be survival.

Assuming that you do survive (and we have faith that you will), when you're done with your project, we wouldn't be surprised if you think of a few things (or a thousand things) you might do differently next time. This section provides a few items that you should consider so that you don't need to rewrite you application because you left them out.

## Keeping one code base

With many different types of BlackBerry models, you want to try your best to keep only one version of your code (*one code base*). One way to make sure you manage one code base is through checking for different capabilities in your code and adjusting the function accordingly.

## Assuming nothing about your user

One of the mistakes you can make is to make assumptions about potential users, as well as the devices they're using. Just because Research In Motion just rolled out a new device doesn't mean that most people have upgraded.

You'd be surprised to know how many people are using older devices from three or four years ago.

For example, if one of the features needs to use the built-in camera, you want to first find out whether a camera is available in most models. For example, you can use this line of code:

```
net.rim.device.api.system.DeviceInfo.hasCamera();
```

Then you can implement the camera function. As a developer, you don't want to assume that the user who downloaded your software has a camera-enabled BlackBerry.

### Using preprocessing directives

*Preprocessing directives* are basically directions that you give to the compiler indicating what should be compiled and what shouldn't. Why do you care about preprocessing? Well, a preprocessing directive is another way to keep one code base. For example, the API call `hasCamera()` is only available from BlackBerry OS 4.2.0 and higher. If you are supporting BlackBerry OS 4.1.0, then your code wouldn't compile. What do you do? This is when preprocessing directives comes in handy.

When you're writing an application, you want to reach as many potential users as possible. (And why wouldn't you want to do that?) You want to reach as many models and as many operating systems as possible. You can use preprocessing to make sure that only certain sections of the code compile for a specific OS or a specific build. The goal is to have one code base — even if you have to layer in code for varieties of models and operating systems.

You *don't* want to have to manage multiple code bases for one application. Consider the implications — you would have to change your code in 20 different places every time you make a minor change. Egads!

For more on preprocessing directives, read the information posted here:

```
http://docs.blackberry.com/en/developers/deliverables/12002/
        Specifying_preprocessor_directives_657636_11.jsp
```

### Picking a minimum-supported OS

The latest and greatest OS from RIM is OS 5.0. However, many people are still running OS 3.7. It is your job to pick a minimum OS version to support so that your application runs on a majority of the BlackBerry smartphones out there. (You don't have to go all the way back to OS 3.7 if you don't want to, by the way.)

After you decide the minimum OS version your application will support, we guarantee that keeping one code base will be a lot easier.

## Keeping user data

People frequently buy new devices. Smartphones are easily lost, stolen, broken, or flushed down a toilet. In addition, some people always want something new or may get a free phone when they renew their service contract.

With all these phones being put out of service and so many new phones being brought into service, valuable stored data needs to make its way to the new device. You should design your device so that information can be easily backed up and restored when needed.

This feature is essential for any application that holds information. If users store any data at all using your application, we're talking to you.

Don't forget to include user configurations and settings in your backups. It's great when users have their data, but this is an expectation, not a perk. If the application looks and acts exactly how it did before restoring the data, users will appreciate the perk.

You can achieve application backup and restore functionality in a few ways:

+ **External memory:** If you develop an application for a device that has external memory capabilities, your application can make a copy of the data to the external memory card. If the user gets a new device, he or she can easily import the data to the application. Obviously, this method doesn't work if the user loses the device and memory card.

+ **Desktop sync:** If you create a desktop client that interacts with the mobile application, consider building a backup and restore option into it. This method will work even if the user drops his or her smartphone over Niagara Falls. (We'd love to hear the story about how that happened.)

+ **Over-the-air sync:** You can develop a back-end tool that synchronizes the data wirelessly OTA from the device to the server. This method requires additional development time and funds to support the server, but depending on the application, these additions may be well worth the time, money, and additional resources.

# Dressing Up Your Application

We entreat you to jazz it up. Customers are begging for it. We're not pointing any fingers at that certain phone that revolutionized the way mobile apps are presented and used, but it's pretty iSafe to iSay that those apps are the new gold standard. If design ain't your thing, *please hire a designer!*

Unless you're writing a background application with minimum interaction with the users, you should make sure the application goes beyond being merely functional. It must be pleasing to the eye. At the very least, instead of a text-based button, add an icon to the button relating to the function.

Don't forget to consider the following options for graphics, icons, and logos:

✦ **Add motion.** A simple rollover image has become the industry standard for all buttons in user interfaces for desktop applications, so why not add them for mobile apps? Or go further — a little movement goes a long way. Make your icon highlight, shake, or change in some other way to indicate that the user has selected it. As long as your rollover or action isn't too complicated or distracting, and your button design reflects your logo and other corporate branding, your app will set a higher standard.

✦ **Make little tweaks.** Sometimes small touches say the most important thing about a business's attention to detail. Building icons is actually pretty easy to do, so if your icon looks like everyone else's — a square with a graphic in the middle — it just isn't appealing. Remove the image's borders and add a transparent background, and you've made something unique and professional.

✦ **Make it pretty.** First impressions are everything. A user will already have a judgment about your application when he opens it and sees how it looks, regardless of whether it does the job. The user interface must be simple and aesthetically pleasing. Consider a unified color scheme for your icon design that echoes your application's functions and your corporate logo. See Figure 2-1 for a simple example.

**Figure 2-1:**
A simple application with a matching color scheme.

Make sure you do your research on what complementary and contrasting color palettes are harmonious and pleasing to the eye.

Or hire a designer (hint, hint, hint).

## Maximizing space

Because the screen on a BlackBerry smartphone is limited, you need to be creative in reducing the number of menu options and choices you display on the screen. Users need to see the most important stuff first and not be distracted by stuff that doesn't help them accomplish their goals.

If you're not a designer, hire a designer. If you're a programmer, you're interested in how things work. Any designer will tell you that how things work is determined by the user's *perception* of how they work.

Tabs are great for giving more screen real estate to an application. They also keep users from having to dig too deep into the application to get to the information or tools that they want to use.

## Giving users what they're used to

BlackBerry smartphones have been successful for their ease of use and quick access to commonly used tools. Deliver the best of both worlds by creating a great application that looks good and that extends users' expectations with solid shortcuts and short menus.

✦ **Shortcuts:** For any actions that users will need to repeat over and over again, such as creating a new entry, searching, or deleting an item, you absolutely need to create a letter or Alt+*key* shortcut so users can quickly get the job done. Of course, making sure you document the existence of the shortcut is crucial so that users know about it. The shortcut isn't an Easter egg that exists only for your amusement.

✦ **Short menu:** For devices with a trackpad, pressing on the trackpad provides a short menu, as shown in Figure 2-2. Using this menu to include common tasks allows users to work without taking their hands off the trackpad. Short menus are usually well received by users who like to multitask. (And who doesn't like to multitask?)

**Book VII Chapter 2**

**Brainstorming Your Application**

**Send Using:** [Default] ▾

To:
Cc:

**Subject:**

|

Save Draft
Full Menu

**Figure 2-2:**
A short
menu.

The key to smart short menus is anticipating what users want to do most often.

## Providing Timely Updates

Updates are essential to keeping your application competitive in the ever-changing applications market. Believe us when we say that as soon as your brilliant idea hits the InterWebs, two more copy cats are going to try to take away from your market share. Survival will mean not just keeping up but staying ahead of your competition with new features.

In order to continue providing new elements to your application for new and existing users, you should provide a mechanism to notify them about updates. Set up the application so that it queries a database to automatically prompt the user to download the latest version. Alternatively, have your users sign up to a mailing list so when new features are available you can send them an e-mail with an over-the-air download link and a list of the changes.

## Getting Help

Okay, so you have a problem, a question, or just want to browse for some ideas. Well, good news. The BlackBerry Developer Resource portal is just the place to start. If you can't find an answer there, try checking out some of the other useful sites that we list here.

Go directly to this link to find the following resources: `www.blackberry.com/developers/resources`.

+ **Discussion forum:** If you are not sure about how to do a certain thing, most likely you are not alone.

  This is *the* place to start. Remember to also contribute to the forum after you become an expert (and we know you will)!

+ **Audio and video library:** BlackBerry has posted many videos and audio presentations to help you get started with your project. These multimedia presentations give you overview information as well as detailed instructional information for each of the development platforms used for BlackBerry applications.

+ **Tutorials:** Get step-by-step instructions about setting up each of the tools you need, creating your application, and getting it ready for deployment.

+ **Documentation:** This resource is a central repository of all user manuals and guidelines for all the tools that you're using or thinking of using.

+ **Tools and downloads:** Download all the tools to get you started.

✦ **Knowledge base:** If you need an answer to a technical question, BlackBerry offers a detailed knowledge base with BlackBerry-posted articles on issues and resolutions.

BlackBerry hosts developer's conferences. Go to `www.blackberry developerconference.com` for more details.

# Chapter 3: Getting Your Applications Out There

## In This Chapter

✔ **Using BlackBerry App World**

✔ **Finding more places to publish your application**

✔ **Publishing an application to your own Web site**

✔ **Promoting an application**

*Y*our goal in creating a BlackBerry application is to get the application used by as many people as possible as soon as possible. Whether you're creating applications for BlackBerry devices connected to a BlackBerry Enterprise Server (BES) or for individuals using a BlackBerry Internet Service account, you have various options from which to choose.

Probably the hardest part of application deployment is getting the word out. People need to be excited enough about your application to go to the trouble of installing it. You need to strategically market the application, generate buzz, and then make the application available from a variety of BlackBerry-related sites and stores. We give you a few marketing tips and tricks to encourage consumers to look, test, and buy your application.

Standing out in the crowd is crucial to your success because you're bound to have two or three copycats trying to steal customers away from you. We also provide some techniques that you can use to help your application stand out from the others.

## Getting an Application Installed on Users' Devices

Research In Motion provides different ways for you to get your application out there and installed on devices.

### Understanding your installation options

These terms are commonly used to describe the publishing and installation process:

✦ **Pull:** When an application is *pulled,* the user initiates the action to install the application file over the air (OTA). The user pulls the application by directing the BlackBerry Browser to an application download link. Pulling is a user-generated process because once you put the application out there and promote it, it's up to the user to act — he or she must initiate the download and install the app.

As the application developer, pulling means that you've made the application available through BlackBerry App World, through your own Web site, or through a third-party Web site. The user clicks a link, pays a fee, and installs the application.

✦ **Push:** As opposed to pulling an application, which is a user-generated installation process, *pushing* is a developer-generated process. A push method can be either over the air (OTA) or tethered through the user's computer.

When an application is pushed, one of three methods is being used to transfer and install the application:

- *E-mail or content pushing:* This is a sort of hybrid between push and pull. E-mail and content pushing are ways of sending an application OTA to users. You simply send a link or a clickable icon to a Web site where the application can be downloaded. This method can be used for both enterprise users and consumers who purchase the application.

  There are pros and cons to this method of deployment. The advantage is that through a marketing e-mail, you can send a link to a free trial of your app. The disadvantage is that your e-mail is static. Consider this potential problem: A user saves an e-mail you sent in May. In July, the user decides to click the link. Unfortunately, if you made changes to the app between May and July, the link won't work, and you've lost a customer.

- *Software configuration:* This push deployment technique is used to install applications OTA to devices that are connected to a BlackBerry Enterprise Server. An administrator can send an application OTA without the users' involvement. We discuss this deployment method more in Book VIII.

- *Desktop:* Whether the application is installed using an executable (.exe) file on the users' desktop computer or with an ALX file through Application Loader via Desktop Manager, in order for this push method to work, the BlackBerry needs to be physically connected to a desktop computer to transfer and install the application. This type of install is commonly known as a *desktop install* or *tethered install.* Therefore, this installation is not OTA. It can be used in a retail or enterprise environment.

You're probably not going to be surprised to know that you don't have to choose one installation method at the expense of all others. You should understand which installation options make the most sense to your target audience, and you should understand which methods are most common. Although, generally, using OTA as a primary installation option is desirable, it doesn't hurt to provide a tethered installation method as a secondary installation option, especially in a retail environment. Some folks don't like downloading apps — even free apps or apps with a free trial — because downloading uses bandwidth and can cost precious download minutes. Using the tethered method, the app is downloaded through a desktop computer, skipping any wireless bandwidth costs. Here are some specific recommendations:

✦ **If your primary end user is part of your enterprise:** We recommend software configuration. This option gives you good control and allows the administrator to know exactly which apps are installed on which users' devices. This kind of knowledge is a must in a corporate environment.

✦ **If your primary end user is a consumer:** We recommend that you use all available installation options possible. Provide a link for users to download your app, send an e-mail or a content push, as well as post the app on a Web page. Using as many options as possible allows potential users to install your app using their desktop computer.

✦ **If you're trying to reach both enterprise users and consumers:** This means your product is targeted to multiple companies as well as retail customers. We recommend you provide two links on your Web site to clearly drive people to the correct Web page. With companies, you will be dealing with BlackBerry Enterprise Server administrators. They're mostly knowledgeable about the BlackBerry platform and will want to know the requirements for your app to run on their users' smartphones. Follow the preceding recommendation for a consumer Web page.

## Getting to know installation files

Research In Motion makes it relatively easy to get an application onto a BlackBerry smartphone. However, you need to know which installation files to create for your chosen deployment method. You may need to create more than one type of installation file. (See the preceding section, "Understanding your installation options," for our recommendations.)

### OTA installation files

When a user installs an application without connecting his or her device to a desktop computer, the application is installed over the air (OTA). OTA installation can occur whether the user pulls the application from a Web site or BlackBerry App World or whether the application is pushed to the user. OTA installations are also used in enterprise environments.

A file with the .jad extension is used for all OTA installations on the BlackBerry. The file extension stands for Java Application Descriptor. When this file is accessed from BlackBerry Browser, a download screen appears displaying the description of the application. (See Figure 3-1.)

**Figure 3-1:** BlackBerry Browser accessing a JAD file for OTA installation.

Keep in mind the size of your application. If the size is very large, OTA installation may take too long. If the user is in a poor coverage area or loses data coverage, the installation may fail. Understanding how long a file takes to download should be a fundamental part of your application testing.

### Tethered installation files

Whenever a user attaches his or her BlackBerry device to a desktop computer in order to install an application, he or she is using a *tethered* installation method. This method is associated with a specific type of installation file called an Application Loader File (.alx).

An ALX file must be used for every tethered installation. This file type is a proprietary file format from RIM that's used to install all kinds of files, from games to themes to enterprise applications.

The application file is downloaded to the computer and installed using the Application Loader via Desktop Manager. The user browses for the ALX file and adds it to the list of applications that will be loaded onto the BlackBerry smartphone. (See Figure 3-2.)

Using ALX files to install applications through the Desktop Manager is useful, especially if the size of the application is large. Large files may take a considerably long time to download OTA. Depending on their service plan, users may be charged for time downloading applications OTA.

## Getting a COD

Cash on delivery? A type of fish used to make fish'n'chips? Hardly. A file with the .cod extension is a compiled code document. What's that? It's the program file for your application — the thing that makes your application function. This file is associated with your application whether you choose to have it installed OTA or in a tethered installation.

**Figure 3-2:**
An ALX file is used for tethered installations.

**Book VII
Chapter 3**

Getting Your
Applications Out
There

This method is also helpful if your application has a sync feature associated with it. If you built an application with a desktop component that the BlackBerry smartphone application synchronizes to, then it's smartest to require a tethered installation using the Application Loader. By associating your application with an ALX file, you allow users to install the application at the same time as the desktop software.

The downside to using a tethered installation is that many people never connect their BlackBerry devices to their PC and never use Desktop Manager. Do your research!

## Getting Your Application from Point A to Point B

You understand about pushing and pulling; you know your .jad from your .alx. Now you need to get the job done — get that application into the hot little hands of your customers, clients, or enterprise users.

As a security feature, BlackBerry smartphones that are connected to a BlackBerry Enterprise Server may be associated with IT policies that block the device from installing applications from third-party vendors. This is because some applications are bad or can cause security breaches. If you're a third-party vendor with a good application that works well and doesn't cause security problems (which we hope), there's not much you can do to allow your application to be installed if it's being blocked. However, if you're a BES administrator and you want to know more about how to use IT policies to control which apps can be installed on your users' BlackBerry devices, check out Book VIII.

## Publishing an application to a Web site

Publishing an app through a Web site requires you to create a JAD file.

Creating a JAD file is described in detail in BlackBerry Development Guides, which you can find on RIM's Web site at

```
http://docs.blackberry.com/en/developers/group.jsp?userType=2
    1&group=Java+Application+Development
```

After you create a JAD file, all you need to do is upload it to a Web server. Then provide users with a link to the file. An example link might be

```
www.mysite/Application.jad
```

The user accesses your Web site using BlackBerry Browser, selects the link, and starts the application download and installation process. Refer to Figure 3-1 to see what the screen looks like.

## Publishing an application using Application Loader

Applications that are installed using Application Loader can be found on the Home screen of the BlackBerry Desktop Manager. This option allows users to update, add, and remove applications. But before the app makes it through to a user's desktop machine, you need to provide a Web page where she can download the installation files. The following short sections describe what you need to do.

### Creating installation files

The files needed from your app for this method of installation are `.alx` and `.cod` files. Refer to BlackBerry Development Guides on how to generate these files. You can find the guides, and all other BlackBerry developer-related documentation, at

```
http://docs.blackberry.com/en/developers/group.jsp?userType=2
    1&group=Java+Application+Development
```

Instead of giving users the ALX and COD files, what developers typically do is zip the files and make the Zip file available for download. Of course, you should also include an accompanying instructions page so that users know what to do with the Zip file.

### Creating installation instructions of your app

The best way to help users through an installation process is to use steps. Numbered instructions are easy to follow and give users a sense of where they should be at every point in the installation process. The main component of your installation instructions page should contain the following steps:

1. How to download the app's installation Zip file.
2. How to extract the Zip file to a local folder.
3. How to run the BlackBerry Application Loader.
4. How to open and run the app for the first time.

One trick to writing good instructions is knowing how much information to give. Most technical types tend to assume that everyone knows as much about technology as they do. The last thing you want is for people to give up installing your app because they can't figure out how to install it. Try reviewing other app developers' installation instructions to get a sense of what's clear and what's not clear before you write your own.

You can see detailed discussions on installing apps using BlackBerry Application Loader in Book VI.

### Making your app's installation available on the Web

After you've written the instructions Web page with a download link to the Zip files, your next step is to make this information available on the Web. Of course, you need to have your own Web site.

You can provide potential customers with a link to download BlackBerry Desktop Manager, which contains Application Loader. Send them to `http://na.blackberry.com/eng/support/downloads/`.

In order to use the Application Loader, users must connect their BlackBerry smartphones to a computer that has the BlackBerry Desktop Manager installed on it.

After selecting Application Loader and then selecting the Start button under Add/Remove Applications, you will then be able to remove Applications. By selecting the Browse button, you will be able to find the application `.alx` file you want to install on the device.

## Publishing an application using BlackBerry Application Web Loader

The Application Web Loader is an alternative way to make your applications available for users to install. Users connect their BlackBerry to a PC and use the PC's Web browser to download your app and install it on their BlackBerry. The difference of this approach to using BlackBerry Application Loader is that users don't need to have the BlackBerry Desktop Manager installed on their computer. In order to use this method, you need to publish a Web page that has JavaScript code for installing your app. This JavaScript code uses the BlackBerry Application Web Loader, which is an ActiveX component. The following sections describe what you need to do.

### Creating app installation files

The files needed from your app for this method of installation are `.jad` and `.cod` files. Refer to BlackBerry Development Guides on how to generate these files. You can find the guidelines on RIM's Web site at

```
http://docs.blackberry.com/en/developers/group.jsp?userType=2
    1&group=Java+Application+Development
```

### Downloading BlackBerry Application Web Loader

The download Web page you need to publish requires the BlackBerry Application Web Loader ActiveX component called `AxLoader.cab`. Your users don't need to download `AxLoader.cab` but you need to place it on your download Web page.

To obtain the `AxLoader.cab` file, follow these steps:

1. **On your desktop computer's Web browser, open the following Web page:**

   ```
   https://www.blackberry.com/Downloads/
       contactFormPreload.do?code=996009F2374006606F4C0B0FD
       A878AF1&dl=E348AB56663CB5BA9450744D74AD9B08&step=4
   ```

   A Web page appears, asking for your business information: name, address.

2. **Enter your information and click Next.**

   A page displaying the software eligibility requirements appears. You must agree to these requirements before you can proceed to the next page.

3. **Click the Agree check box and click Next.**

   A page showing the download button appears.

4. **Click Download and select Save File.**

   This is an executable file. Take note of the location where this file is saved on your hard drive.

5. **Double-click the downloaded file.**

   The file runs, and an installation welcome page appears.

6. **Click Next.**

   Another screen appears, displaying the folder location where the files will be installed. Take note of this location.

7. **Click Next.**

   BlackBerry Application Web Loader is installed. In the folder location indicated in Step 6, you should see the `AxLoader.cab` file and a sample subfolder containing a sample page and sample app.

### Creating a download page

The simplest way to create a download Web page for your app is to mimic the sample page found in the BlackBerry Application Web Loader installation folder. Just replace the application filenames with the name of your app. This page needs your app's JAD and COD files and the ActiveX installer `AxLoader.cab`.

You can test your installation locally by opening your Web page using a PC Web browser. Your users' basic experience will be similar to the following steps:

1. **The user is prompted to visit a URL.**

   By default, the download page you created displays a message that says No BlackBerry Found. (See Figure 3-3.)

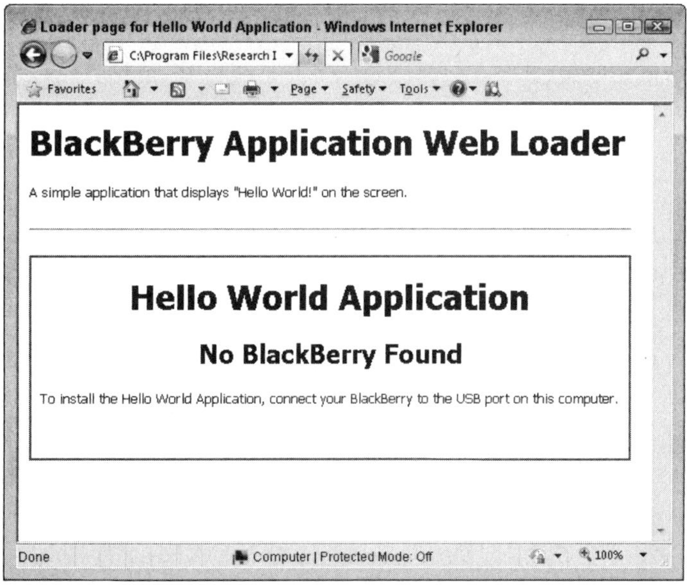

**Figure 3-3:** The page users see when they visit a download site if their device isn't connected to a computer.

2. **The user connects his or her device to a desktop computer that doesn't have the Desktop Manager installed.**

   Once the user connects the BlackBerry smartphone to the computer, the page automatically updates and shows that the BlackBerry is connected. (See Figure 3-4.)

**Figure 3-4:** The page users see when they connect their device to a computer.

3. **If the device requires a password to use or install applications, the user is prompted to enter the password on the Web site.**

   A new button appears, prompting the user to load the software.

4. **The user clicks the button.**

   The application file is installed on the BlackBerry smartphone.

Even though users don't have to have Desktop Manager installed, they must have the BlackBerry USB drivers installed for Application Web Loader to work. That said, most PCs will automatically detect and download the BlackBerry drivers as soon as a BlackBerry is connected to the computer using a USB connection.

### Making your download page available on the Web

Your last step, after you create the download page and instructions for your users, is to make them available to the Web. This, of course, requires you to have your own Web site describing your product. To publish your download page, all you need to do is copy the files to your Web server and provide a link to the HTML page.

### *Publishing an application with BES push*

If you're a BlackBerry Enterprise Server administrator, you can offer every method to users who need to install applications. You can push full Java-based applications to every device connected to the BES.

As the administrator, you set IT policies that make installations fall into one of these categories:

✦ **Mandatory:** A mandatory installation means that the application is pushed down to the device and the user will not be able to remove or uninstall it.

✦ **Optional:** In an optional installation, users can install and uninstall applications at their convenience.

## *Submitting Your App to BlackBerry App Stores*

If you build it, they will come. Maybe this was true in *Field of Dreams,* but it's a little more complicated in real life. First, you have to make it easy for people to find your field of dreams, or no matter how badly they want to get there, they won't. You can publish the files to your own Web site, but users need to know that your Web site exists, what its URL is, and why they should visit it, before they will go to the trouble. If you intend to sell your application, you also need a backend infrastructure to accept different payment types. You also need systems to store and send the activation numbers for the application. All this extra stuff needs to be secure, safe, and functioning well.

Book VII
Chapter 3

Getting Your
Applications Out
There

Even if you plan to upload your app to every single app store in the universe, you should also provide all available popular payment methods to purchase the app on your own Web site. Besides credit cards, several well-established payment methods like PayPal, Google Checkout, or Amazon are familiar to users. Most people who are buying apps online are accomplished online shoppers, and you can benefit from the trust that comes with these payment methods.

Putting your application online at a third-party app store offers several advantages:

✦ **Catch your target audience in its native habitat.** You can market your app to your target audience. Using descriptions that match the needs of your audience, you can catch customers who are shopping for what you're selling.

✦ **Take advantage of your secondary market.** You can catch a secondary market of people who happen to be browsing around looking for something new. They may not go online looking for your app, but your awesome marketing will lure them in.

✦ **Benefit from the brand name.** You get the benefit of the app store's brand name. Customers are inclined to purchase from a site that is well known and trusted.

✦ **Take advantage of the tools and features of a big site.** You can let the app store handle the stickiness of collecting money. Established app stores have the infrastructure to allow multiple payment methods, so you don't need to worry about how you are going to manage accepting payments from around the world.

You may face some challenges when you provide your app through a third-party app store.

✦ **So many app stores, so much red tape:** Getting your app to a store takes time because of all the paperwork and communications you need to do. With so many app stores out there on the InterWebs, the time spent quickly adds up.

✦ **Stiff commission rates:** Each individual app store charges a commission rate for every app that's sold. Commission rates can change from time to time, but they range from 20 percent to 65 percent of the purchase price. Stores like Handango and MobiHand have partners or affiliates, and if your app is sold by one of their affiliates, the charges usually double. For example, MobiHand charges 20 percent, but if your app is purchased through CrackBerry's App Store — a MobiHand partner — the charge is 40 percent.

Don't get down. If you have a high volume of sales, these commission rates won't seem so terrible. This leads us to reiterate our advice — you need marketing, marketing, marketing.

✦ **Your app could be rejected:** An app store has the right to reject your app for any reason. This fact is especially true for carrier app stores. Mobile carriers are very picky about which apps they want in their stores. Sometimes a carrier has its own apps that it wants to sell. You app is more likely to be rejected if it directly competes with the carrier's proprietary app.

✦ **Your app might not fit available app categories:** Users typically browse apps using categories. If your app doesn't fit well into an existing category, users will have a hard time finding it. For example, if your app provides some social networking features but also has business productivity features, you have to choose between the two categories. This leaves your potential customers divided, with a good percentage not finding your app.

✦ **Recovering from a bad rating is tough:** Most people don't bother to write a rating on an app, but if they do, it's because they had an *exceptional* experience. An exceptional experience means that the app performed exceptionally well or it performed exceptionally poorly.

Guess which type of review occurs most frequently! If a customer has a bad experience, you're probably going to be on the receiving end of a terrible review (even if you react quickly to fix problems). One bad rating for an app is all it takes for other users to shy away from it like it's the swine flu. Recovering from a bad review takes time.

Ask beta testers to review the app. Ask testers for their opinions and encourage them to write reviews. The more good reviews your app gets, the better visibility it will have in the app store. More good reviews also downplay the effects of bad ones.

✦ **Issues at the app store may reflect on your app:** Software infrastructure sometimes goes bonkers, and an app store is not immune to this problem. When users downloading or purchasing your app experience problems, they think it's your app that's problematic. Unfortunately, you can do nothing about this problem (except wish it would go away quickly).

BlackBerry applications can be found on hundreds of Web sites. In the following sections, we highlight just a few of our favorites. We describe briefly how to get your app to these stores. Note that the app store business is highly competitive, and each one keeps updating their infrastructure and also their partners' Web sites. We point you to their Web sites, but bear in mind that there's a good likelihood that they'll change by the time this book is published.

Try conducting a quick Web search using the keywords *submit BlackBerry app* to see a list of app stores. This strategy is useful because the app store market is dynamic — new players enter the game every day.

## Uploading applications to BlackBerry App World

Research In Motion provides an application for the BlackBerry smartphone called BlackBerry App World, which provides apps organized by category and type (shown in Figures 3-5 and 3-6).

**Figure 3-5:**
BlackBerry
App World
categories.

> Categories
> Business (261)
> Education (123)
> Entertainment (316)
> Finance (134)
> Games (664)
> Health & Wellness (266)
> IM & Social Networking (129)
> Maps & Navigation (140)

You're probably pretty well acquainted with this application because it's the main source you use as a BlackBerry owner to download and install consumer apps. We talk about BlackBerry App World extensively in Book VI. If there is one app store that you need to have your application on, it's BlackBerry App World. BlackBerry App World commission rate is 30 percent.

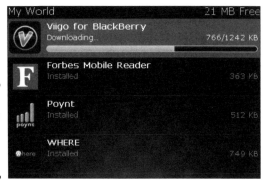

**Figure 3-6:**
Downloading
an app from
BlackBerry
App World.

Getting your apps published to BlackBerry App World is actually quite simple. All the information you need is described at the BlackBerry App World Vendor Support page:

```
http://na.blackberry.com/eng/developers/appworld
```

You need to do the following before you can submit your apps to the store:

*1.* **Create a vendor account.**

You need to agree to the vendor agreement and provide your information with your PayPal account, which is one of the payment methods available to BlackBerry App World at the time of this writing.

You can fill out the vendor registration form here:

```
https://appworld.blackberry.com/isvportal/signup/
    signupterms.seam?pageIndex=1&cid=503196
```

After registration, RIM sends you your site login information. You use this login to submit your apps.

*2.* **Make sure your application(s) adhere to the BlackBerry App World Vendor Guidelines.**

You can find the guidelines here:

```
https://appworld.blackberry.com/isvportal/home/
    guidelines.seam?pageIndex=1&cid=503060
```

You won't find anything unusual in the guidelines. Your app needs to be stable, and you need to inform the user if your app requires airtime or using the wireless bandwidth. Other guidelines require you to follow simple legal terms about truthfulness of the information you provide, and so on.

3. **Sign in to RIM's Vendor Portal.**

   ```
   https://appworld.blackberry.com/isvportal/home/login.
      seam?pageIndex=1&cid=503327
   ```

4. **Submit your app for approval.**

   RIM reviews your application to make sure that you're following the guidelines. After you receive a stamp of approval, you can finally submit your app to the store.

For more detailed information, be sure to read the frequently asked questions on RIM Vendor Portal at

```
http://na.blackberry.com/eng/developers/appworld/faq.jsp
```

## Publishing your app in MobiHand

MobiHand (`www.mobihand.com`) is one of the biggest publishers of BlackBerry apps. Its app store is *white labeled* or *rebranded* by many Web sites selling BlackBerry apps, such as BlackBerryOS.com and CrackBerry. MobiHand takes care of all the hassles related to handling customer purchases. For example, you don't have to worry about shopping carts and billing issues.

**Book VII**
**Chapter 3**

**Getting Your Applications Out There**

Many developers have good experiences dealing with MobiHand, and we recommend you submit your apps to this site. The process to get your app published to the MobiHand store is quite simple. You just need to sign up with the MobiHand App/Content Developer Program, which you can find here:

```
http://corporate.mobihand.com/developers_signup.asp
```

After you fill out the registration form, you are contacted by MobiHand with more information on how to submit your apps.

MobiHand charges 20 percent commission, if your app is sold directly from their app store, and an additional 20 percent if it is sold from one of their partners or affiliates.

## Handing your app to Handango

Handango (`www.handango.com`) is a very clean and organized site for finding applications. And it's also one of the pioneers in the app publishing space. It's a popular site to find apps, and it makes sense that your apps should be published there.

Unlike BlackBerry App World, Handango allows users to download the application files to their PCs so that they can install applications though BlackBerry Desktop Manager's Application Loader. Handango also provides an OTA installation option, as well.

Handango provides a much wider range of categories to select from than App World does. Unfortunately, the site doesn't provide free applications. The lowest prices you can list your app for on this site are in the $10 range.

Handango charges a 50 percent commission on all sales. The site also has other charges on top of the commission, which can go as high as another 15 percent.

Handango has a mobile application that you can download by going to Handango.com from your BlackBerry Browser. (See Figure 3-7.)

In order for your apps to be published by Handango, you need to sign up to the site's Partner Program, which you can find here:

`www.handango.com/info/Partner.jsp?storeId=2218`

This Web page also has other information you need to know as a developer partner.

**Figure 3-7:** The Handango mobile client.

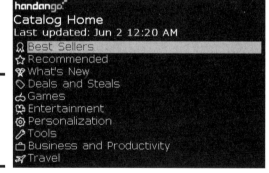

## Getting your app on CrackBerry Superstore

CrackBerry.com is one of if not the biggest BlackBerry enthusiast Web site, with tons of reviews on BlackBerry apps. It has its own app store, which is powered by MobiHand. This means that if you submit your app to MobiHand, your app also appears in CrackBerry Superstore, which is shown in Figure 3-8.

CrackBerry.com is a great resource for finding out the latest rumors and official releases of software, devices, and other information. As a developer, you can benefit from the information the site provides.

After you submit an app to MobiHand, e-mail one of the bloggers at CrackBerry to announce the availability of your apps in its app store. If you're lucky, they'll blog about your app or review it. Publicity from this site surely drives sales, so remember this tip; you'll thank us when you get a boost on your bank account.

MobiHand charges 40 percent commission for apps bought in CrackBerry Superstore. You may think that the commission rate is stiff, but our experience tells us that CrackBerry brings more users to download our apps. And the additional 20 percent is certainly worth the price to pay.

**Figure 3-8:**
The
CrackBerry
Superstore.

**Book VII Chapter 3**

**Getting Your Applications Out There**

## Going directly to mobile carriers

All mobile carriers, including AT&T, Verizon Wireless, Cellular One, Nextel, T-Mobile, and Sprint, also provide their own sites for downloading applications.

One of the benefits for users who download apps directly from their service providers is the fact that charges for apps go directly to the users' monthly mobile bills. Figure 3-9 features the AT&T Media Mall app store.

Unfortunately, each carrier has its own app store and different requirements in order for your app to get published in their app store. Usually, the process is to submit your app for approval. The carrier checks your app to make sure it's not malicious and that it's running what you described it to. And you need to answer the carrier's questions regarding your app during the review process. This could be a time-consuming process, and we recommend you submit your app to only the biggest carriers that carry the BlackBerry models your app runs. Here's a list of the developer Web sites for the top three carriers in the U.S.:

✦ **Verizon Wireless:** `http://developer.verizon.com`

✦ **AT&T:** `http://developer.att.com/devcentral`

✦ **T-Mobile:** `http://developer.t-mobile.com`

**Figure 3-9:**
The AT&T
Media Mall
app store.

# Promoting Your Application

If you're developing your app for an enterprise, your enterprise administrator can push your application to devices without users requesting it. But in the dog-eat-dog world of retail apps, users need to find your application and download it.

So, how do you get someone to select your application? You can use a few marketing tips that will attract customers and increase the number of downloads for your application.

## Writing awesome text

The worst enemy for your app is obscurity. Even if you have the best app in the world, the difference between a bestseller and an obscure one is simply marketing.

Take advantage of all online avenues to promote your apps. Don't hesitate to give a free license out as a promotion. Some bestseller apps always offer discounts and run promotional offers. Reach out to enthusiast Web sites and publish promos on those sites.

Most application stores ask you to provide some information about your application. The information you provide is essential for two reasons:

✦ **Your words should be tied to the terms for which users are likely to search.** Otherwise, your app gets trapped at the bottom of a long list of search results.

✦ **Your words should accurately describe the features and benefits of the app.** If they don't like what they see, users *will* move on.

Here's what information you typically need to include:

✦ **Write an SEO-laden short description.** The short description is usually two or three sentences long. The short description appears next to the icon for the app when it is listed on a Web-based mobile application store or in BlackBerry App World. The short description should quickly summarize the application using as many SEO terms as possible, targeting the value the app adds or the need that it addresses.

SEO stands for *search engine optimization.* Your short description will have the most bang for its buck if it uses the keywords that your target audience is most likely to search for when looking for an app like yours.

✦ **Choose the best keywords.** To increase the possibility that people using the search feature in the app store will find your app, include any words that are key elements of your application. Also include keywords that you know users are interested in.

Google's AdWords tool displays which search words are most commonly used so that you can incorporate these keywords into your description. For example, if you type the term **apps** into the Google AdWords tool, you see the number of times this term has been searched for. In addition, you can see related keywords, such as *software.* You can also see that the term *applications* is widely searched. By maximizing the keywords users are looking for, you can guarantee more hits for your application. Visit `https://adwords.google.com/select/KeywordToolExternal`.

✦ **Keep your descriptions professional.** Don't write in ALL CAPS or include ASCII art such as "(-(-(-Best App-)-)-)".

✦ **Write a superior long description.** First you capture buyers' attention with a great short description and some awesome screen shots. Then, once you've fished them in, you provide the details they need to finish the deal. Here's some basic advice for your long description:

  • *Just because it's called a long description doesn't mean you should make it long.* As it turns out, people don't really like to read all that much — especially when they're shopping. Keep it simple. No massive blocks of text allowed!

- *Provide bullet points that highlight the app's key features.* Incorporate the same keywords and SEO terms that you use in your short description. This strategy helps your app turn up when people search for those keywords.

## Including awesome screen shots

Screen shots are really important when you're trying to make an application stand out against its competition. People like to buy something they can see. When you provide screen shots of a good-looking application that does what they need it to do, potential customers will be more inclined to purchase it. Consider the following when taking screen shots:

✦ **If your application works on several BlackBerry models, make sure to provide pictures of all possible models to show how it looks on those devices.** Some users don't know the name of the models of their BlackBerry and hesitate to try an app, thinking it's not compatible with their model. When they see a picture of the app on their specific model, it will give them the assurance that the app works on their device.

Don't just show the splash screen of your application; also show the screens that people are likely to use most.

✦ **Some app stores allow you to upload only one picture.** To show as much real estate as possible, create an animated GIF file that cycles through a few different pictures. One GIF, five images!

To create an animated GIF, you don't have to go crazy with Photoshop. You can upload your screen shots to an online site. You upload several pictures, and the site creates the animated GIF for you. Try `http://picasion.com`.

## Creating demos, tutorials, and installation info

In combination with keywords, descriptions, and screen shots, you can provide video demos showing off the very powerful tool that is your awesome app. When users can see how others use the application, they will make a direct connection to how the app can fulfill their needs.

YouTube is a great sales tool for posting product marketing and demo videos. It's also a great free advertising tool because users may stumble on your video while searching for videos on YouTube. Your demo should direct visitors to the appropriate app store to download the app.

Demos should include information about installing and uninstalling the app. If your app has special shortcuts, don't forget to include information about them in your demo.

TIP

## Giving good support

The best way to get your app bashed all over the Web is to appear to be unresponsive to customer feedback, complaints, or issues. If you don't set up a meaningful feedback and support protocol and you don't give users realistic expectations of your response times, you will start to see nasty comments on the public forums bashing your apps. To avoid this negative advertising, follow these steps:

✔ Make sure that you allow users to provide feedback and ask questions.

✔ You don't necessarily need a phone support hotline, but it's critical that your users can reach you when they're having trouble with your app. Provide feedback options and a support link in your app and

also on your Web site (and any third-party sites where the app is distributed). If they have a support need, users need to be able to report the issue quickly and expect a timely response.

✔ Make sure that the support or feedback link in your app and on your site offers a reasonable time frame for a response. Most customers don't mind waiting for a reply for about 24 hours, as long as you set the right expectations at the time that they report an issue.

✔ Set the expectations low and over-deliver. Nothing makes customers happier than when you beat a self-imposed deadline!

## *Luring users in with free trials*

Everyone loves things that are free, so by providing a free trial period for your application, you will get more people to download the application. And the more people who download the free version of the application, the more likely they will be to fall in love with it and purchase it when the trial ends. You can provide trial versions a few different ways. Depending on your application, you may want to select one of the following, but always give users the option to skip the trial and pay for the service from the start:

✦ **Limit the number of times the app can be used.** You can limit the number of times a user can launch the app. After the specified number of times, the app disables, and the purchase screen appears. For example, if you create a tip calculator and it expires after three uses, when the user gets to her fourth use, the application prompts her to purchase it before she can continue.

✦ **Set an expiration date.** You can have an expiration set after a specified number of days. This option forces users to purchase the app if they like what they're getting. The disadvantage is that if users don't like what they're getting or if they like "free" more, they will move on.

✦ **Limit functionality.** You can provide the base application and reserve premium tools for users who take the plunge. Make sure that the base tool is useful enough on its own that people don't feel strong-armed. You can also think about limiting the number of actions they can take for a specific task. An example would be to limit users to searching for only ten items a day or using the tool only three times a day. Then ask users to pay a fee for the additional or full features.

# Book VIII

# Enterprise Communications

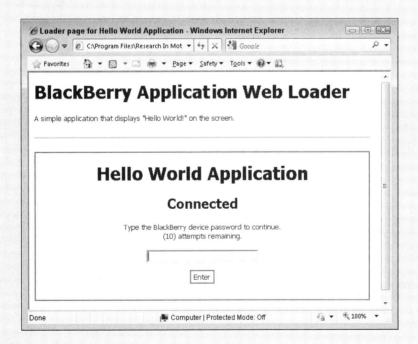

# Contents at a Glance

# Chapter 1: BlackBerry Enterprise Server

## In This Chapter

✔ Understanding the role of BlackBerry Enterprise Server

✔ Knowing why you should use BlackBerry Enterprise Server

✔ Deconstructing BES components

*S*o far, you've mastered the features of your BlackBerry, familiarized yourself with Desktop Manager, and discovered everything there is to know about BlackBerry applications (including how to find and install some great ones!). Now it's time to explore communication across the enterprise. That's a jargony way of saying that you can connect your BlackBerry to a company network so that whether you're on the road or in the office you never miss a thing. The key to enterprise communications is BlackBerry Enterprise Server, which is — not surprisingly — the subject of this chapter.

The term *enterprise* is used to refer to a large corporate system. While an enterprise network may encompass a single office in a local area network, many business enterprises encompass national or international networks. Enterprise applications are applications that work across that massive network, enabling all employees to use the same tools to collaborate with each other.

BlackBerry Enterprise Server, or BES for short, is used to connect your BlackBerry with your company's e-mail servers and enterprise applications. In this chapter, we explore the role of BES in enterprise communications, give an overview of how the BlackBerry enterprise infrastructure works, and explain the various BES components that make the magic happen.

Your company may have a strict policy about which features and enterprise applications can be accessed on your BlackBerry. It may also have a policy declaring which models of BlackBerry smartphones are compatible with the enterprise network. We recommend discussing your options with the IT professional in charge of managing mobile devices before making a phone purchase or making changes to your BlackBerry's software.

## Understanding What BES Does

BES wirelessly synchronizes data between your desktop or server and your BlackBerry smartphone. It handles all your enterprise messaging and collaboration, hooking up with software such as Microsoft Exchange, Lotus Domino, and Novell GroupWise. You can count on BES to link with your company's

+ E-mail

+ Calendar

+ Tasks

+ Contacts

+ Notes

+ Files and documents

+ Some types of instant messages and other information

Basically, BES makes your desktop applications (like Microsoft Outlook or Lotus Notes) accessible from your BlackBerry.

In addition to syncing all your important business-related communications, appointments, and contacts, BES has over-the-air, or OTA, functions that allow BlackBerry applications to be installed, removed, and upgraded on one or more BlackBerry smartphones without connecting to a computer. So if your company decides to upgrade one of its enterprise applications, your network administrator can make the update on your BlackBerry as well as on the BlackBerry smartphones of all your colleagues, all at once, over the air.

## Understanding How BES Does What It Does

Don't worry — we're not going to get too technical here, but it is good to know why BlackBerry Enterprise Server is safe and won't swallow your precious files and messages.

When BES is installed on your company's network, it connects to the company's corporate messaging server, its application server(s), and Research In Motion's Network Operations Center, which is a fancy term for the place that passes data between a BES and your BlackBerry. By maintaining a constant connection between these systems, the BES can wirelessly synchronize access to enterprise resources, including documents and e-mail messages.

BES has built-in security features and works with your company's firewall to ensure that outsiders can't intercept or otherwise access files or messages. It also has systems to prevent viruses and other malware from infecting the BlackBerry and enterprise applications.

We could go into more detail than this, but the one thing you should know is that when you receive messages and open and change attachments on your BlackBerry, you are working with a *copy* of the original attachment or message that BES sent to your BlackBerry. The original message and attachment remain on the corporate messaging server, safe and sound, unless you delete them.

The BlackBerry doesn't sync a message and attachment changes back to the corporate messaging server. You would have to send a reply with the revised document.

Figure 1-1 provides a high-level overview of a BlackBerry Enterprise infrastructure.

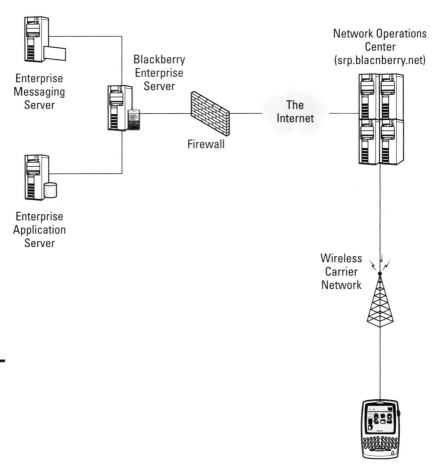

**Figure 1-1:**
Overview
of a
BlackBerry
Enterprise
infra-
structure.

You should note that the process is different for contacts, notes, tasks, calendar entries, and any other applications that your company allows you to access with your BlackBerry. These applications directly synchronize changes between your BlackBerry and the corporate messaging and application servers.

# Convincing Your IT Guy to Get BES

If you've been using your BlackBerry for a while now, you're probably familiar with BlackBerry Internet Service (BIS). It's that thing that allows you to hook up your BlackBerry with your Gmail, Yahoo!, Hotmail, or other e-mail address. Take a look at Book III, Chapter 2 for more information on BIS.

## Introducing BES flavors

The most common messaging and Internet service provided by BlackBerry is BlackBerry Internet Service. This service has some of the same functions as BES, but it's relatively limited. We recommend it if your business is very small — less than 10 people — and doesn't have specific security or compliance concerns that require the features of BES. If you're part of a larger business, or have specific security requirements, consider one of the two flavors of BES:

✦ **BES Express:** This is a good option if your business wants to implement the basic security and features of BES without paying thousands of dollars for the server and client licenses, and the slightly more expensive BlackBerry enterprise data plan needed for the full version. This version of BES is *free,* and allows you to receive your enterprise e-mail, calendar, contacts, memos, tasks, and use most of the other great features that BES provides using any data plan that's on your BlackBerry.

The price of "free" is that this version comes with limitations. BES Express can be used only with Microsoft Exchange, so Lotus Domino and Novell GroupWise users are out in the cold on this one. It also has only 35 IT policies, which provide decent control over device features and functionality but are nowhere near as detailed as the 450 IT policies provided with the full version of BES. BES Express also doesn't support BlackBerry Enterprise add-ons like Mobile Voice System (MVS) — discussed further in Book VIII, Chapter 4 — and doesn't allow over-the-air activations, which means you have to plug your BlackBerry into Desktop Manager to activate it with this version of BES.

✦ **Full version of BES:** This is the best option for a large business that needs the best possible security, control, and administration options. In addition to the features offered by BES Express, BES supports

BlackBerry Enterprise add-ons like MVS, includes advanced features that are important to your IT staff, and can be used with all of the major messaging server software systems, such as Microsoft Exchange, Lotus Domino, and Novell GroupWise. BES also supports OTA activation, which means your IT staff can get you back up and running if you're out of the office and accidentally wipe your device after too many incorrect password attempts. With over 450 IT policies available, your IT staff can secure just about every aspect of your BlackBerry. That might sound like a bad thing, but trust us — it's actually a good thing because it helps to keep your work environment as safe as it can be.

Our focus in this chapter is on the full-featured BES software.

## Understanding the features and benefits of BES

BES gives your IT department a lot more control than BIS. However, you might not have a whole lot of control over whether your company allows you to link your BlackBerry with your enterprise servers. If you *do* happen to have a little bit of sway with the guys and gals in IT, here's what you should tell them:

✦ **Good news for IT:** BES allows for centralized BlackBerry user administration, which means the IT folks can create, change, or apply IT policies (explained later in this chapter), deploy BlackBerry applications, and disable lost or stolen devices, all from one console. BES also allows IT to monitor what model, carrier, and handheld software version and phone number employees are using. This is less about playing Big Brother and more about identifying how many users are affected by a particular issue, who might need a software upgrade, and who's being billed for a device that's not being used.

✦ **Employees will save time and increase productivity:** You have access to so much more with BES, from corporate e-mail to contacts, notes, tasks, and calendar information — not to mention files and documents. Because employees can rest assured that their information is safe and protected, they will be more likely to work remotely using their BlackBerry smartphones. That means they'll be working during lunch, while they're waiting at the doctor's office, while they're at their kids' soccer games. . . . Wait a minute — maybe you shouldn't say anything about employee productivity.

If your IT manager still isn't convinced, tear out Table 1-1 of this book and slip it anonymously on your Chief Geek's keyboard.

**Book VIII**
**Chapter 1**

**BlackBerry**
**Enterprise Server**

| Table 1-1 | Comparison between BES and BIS | | |
|---|---|---|---|
| Feature | BlackBerry Enterprise Server (BES) Full Version | BlackBerry Enterprise Server (BES) Express | BlackBerry Internet Service (BIS) |
| Cost | $2,999 per server license / $99 per client license | Free | Free |
| Push e-mail | Yes | Yes | Yes |
| Internet browsing | Yes | Yes | Yes |
| Centralized user administration | Yes | Yes | No |
| Wireless application deployment | Yes | Yes | No |
| IT pros can see user and device info (model, carrier, pending messages, and so on) | Yes | Yes | No |
| Wireless synchronization of contacts, memos, tasks, and calendar | Yes | Yes | No<br><br>While some third-party e-mail providers such as Google and Yahoo! have synchronization applications, they are limited. |
| Security features | E-mail encryption and 450 IT policies that allow IT Pros to secure almost every aspect of your BlackBerry. | E-mail encryption and 35 IT policies that allow IT Pros to secure important aspects of your BlackBerry. | Basic E-mail encryption offered by BIS servers. |

| Feature | BlackBerry Enterprise Server (BES) Full Version | BlackBerry Enterprise Server (BES) Express | BlackBerry Internet Service (BIS) |
|---------|---------|---------|---------|
| Access to MDS (Mobile Data System) Studio applications for those who are interested in developing apps for the BlackBerry (For more information on MDS Studio, see Book VII, Chapter 1.) | Yes | Yes | No |

# Getting the Low-Down on System Requirements

There are several requirements that a machine must meet before BlackBerry Enterprise Server software can be installed. These requirements can be broken down into three categories:

✦ Hardware

✦ Operating system

✦ Additional software applications

In the following sections, we break down the requirements.

## Knowing your software and hardware requirements

BES is compatible with several Windows Server operating systems as well as several different enterprise messaging platforms. The messaging platforms that BES is available for are:

✦ Microsoft Exchange (version 5.5 or later)

✦ IBM Lotus Domino or IBM Lotus Domino Express (version 7.0 or later)

✦ Novell GroupWise (version 6.5, Service Pack 4 or later)

**Book VIII**
**Chapter 1**

**BlackBerry Enterprise Server**

BES is supported on the following Windows Server operating systems:

✦ Windows Server 2003, Service Pack 1 or later

✦ Windows Server 2003, Release 2

✦ Windows Server 2003 (64-bit)

✦ Windows Server 2003, Release 2, Service Pack 2 (64-bit)

✦ Windows Server 2008

✦ Windows Server 2008 (64-bit)

Your system must meet certain hardware requirements in order to support BES. The minimum hardware requirements vary based on your messaging platform and the number of users you plan to add to each BES. Table 1-2 outlines the minimum requirements for up to 1,000 users, which is a common configuration for a large business or business segment.

**Table 1-2    Hardware Requirements for BES up to 1,000 Users**

| *Messaging Platform* | *Hardware Requirements* |
| --- | --- |
| Microsoft Exchange | **Processor:** Two, 2.0 GHz Intel Xeon processors |
| | **Memory:** 3GB |
| | **Hard disk drives:** Two drives, RAID 1 |
| IBM Lotus Domino | **Processor:** One, 3.0 GHz Intel Xeon 5100 series (dual core) or two, 1.6 GHz Intel Xeon 5100 series (dual core) |
| | **Memory:** 4GB |
| | **Hard disk drives:** 4 drives, RAID 1+0 |
| Novell GroupWise | **Processor:** One, 2.8 GHz Pentium 4 dual processor |
| | **Memory:** 3GB |
| | **Hard disk drive:** 60GB |

## Additional software requirements

The additional software that is required varies based on your messaging platform, but at the very least, the following software applications are required, regardless of whether you're using Microsoft Exchange, Lotus Domino, or Novell GroupWise:

+ Microsoft SQL Server

+ Microsoft Internet Explorer (version 6.0 or later)

If you're using Microsoft Exchange, you also need to have the latest Microsoft CDO library installed on BES. If you're using Novell GroupWise, you need to have Novell ConsoleOne 1.3.6c with Novell GroupWise 6.5.4 Snap-ins.

## Working with BES on Your BlackBerry

If you're one of those lucky people whose BlackBerry is already connected to a BES, then you don't have to worry about convincing anyone of its value. But you might want to know more about maximizing your BES experience.

Your BlackBerry is connected to a BES through a process called enterprise activation. See Book III, Chapter 2 for more information on activating your BlackBerry in an enterprise environment.

Not sure if your BlackBerry is connected to a BES? You can check using these steps:

*1.* **From your BlackBerry Home screen, select the Options icon.**

Depending on your BlackBerry model, you may have to select the Settings icon before you see the Options icon.

*2.* **Select Advanced Options.**

*3.* **Select Enterprise Activation.**

If you see Activated On with a date next to it, as shown in Figure 1-2, your BlackBerry has been activated on a BES. If not, it's time to speak with your IT department to see if BES is an option.

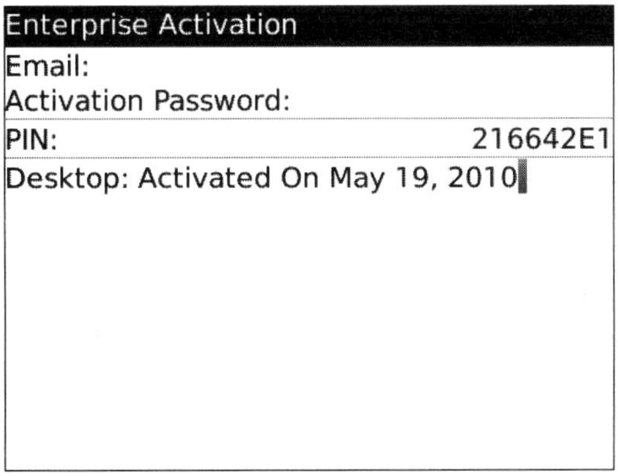

**Figure 1-2:**
Your
BlackBerry
is
connected
to a BES if
Activated
On shows a
date.

```
Enterprise Activation
Email:
Activation Password:
PIN:                          216642E1
Desktop: Activated On May 19, 2010
```

## BES-only e-mail features

If your BlackBerry is on a BES, you have access to several cool e-mail features that may make your personal BlackBerry friends jealous.

### Using BlackBerry Lookup

With BlackBerry Lookup, you can send an e-mail to your coworkers even if you don't know their e-mail addresses! All you need is their name — or part of their name. Then follow these steps:

1. **Compose a new e-mail on your BlackBerry.**

   See Book III, Chapter 2 for more on working with e-mail on your BlackBerry.

2. **Start typing the name of your coworker in the To: field.**

   Try to be as accurate as possible so that you can limit the number of results that are returned.

3. **Select Lookup:** *name* **from the list that appears, as shown in the left side of Figure 1-3.**

   Your BlackBerry sends a request to BES, which searches your messaging server for matches and returns them to your BlackBerry, as shown in the right side of Figure 1-3.

4. **Select the person you want to e-mail.**

   That person's name populates the To field.

5. **Repeat Steps 2–4 for all coworkers you would like to add to your e-mail.**

6. **When you're finished composing, send your e-mail.**

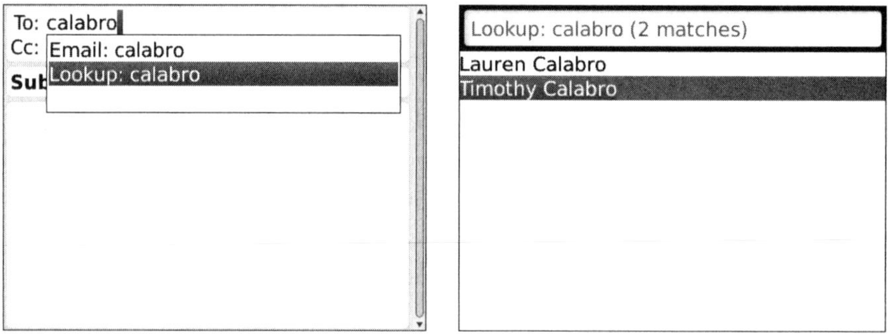

**Figure 1-3:**
Select
Lookup,
then
select the
coworker
that you
want to
e-mail.

To: calabro
Cc: | Email: calabro
Sub| Lookup: calabro

Lookup: calabro (2 matches)
Lauren Calabro
Timothy Calabro

### Adding a BlackBerry e-mail auto signature

If your BlackBerry is on a BES, you can set up a signature that's automatically added to the end of every e-mail that you send from it. Just follow these steps to add your John Hancock to your BlackBerry e-mails:

1. **From your BlackBerry Home screen, select the Messages icon.**
2. **Press the Menu button and select Options.**
3. **Select Email Settings.**
4. **Change Use Auto Signature to Yes.**
5. **Type the information you would like to show in your auto signature, as shown in Figure 1-4.**
6. **Press the Menu button and choose Save.**

   Your signature is added to the end of every message that you send from your BlackBerry.

**Figure 1-4:** Change Use Auto Signature to Yes, then type your auto signature.

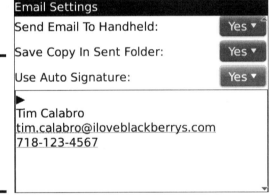

```
Email Settings
Send Email To Handheld:        Yes ▾
Save Copy In Sent Folder:      Yes ▾
Use Auto Signature:            Yes ▾
▶
Tim Calabro
tim.calabro@iloveblackberrys.com
718-123-4567
```

### Setting an Out of Office response from your BlackBerry

For those who aren't familiar with Out Of Office responses, this setting sends responses to incoming mail senders automatically, letting the people who contact you know that you're not around. If your BlackBerry is connected to a BES, you can set up an Out of Office response right from your BlackBerry! To set up an Out of Office response, follow these steps:

1. **From your BlackBerry Home screen, select the Messages icon.**
2. **Press the Menu button and select Options.**

3. **Select Email Settings.**

4. **Change Use Out Of Office Reply to Yes.**

5. **Type the information you want to send in your Out Of Office reply, as shown in Figure 1-5.**

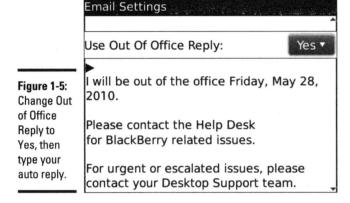

**Figure 1-5:** Change Out of Office Reply to Yes, then type your auto reply.

Email Settings

Use Out Of Office Reply:   Yes ▾

► I will be out of the office Friday, May 28, 2010.

Please contact the Help Desk for BlackBerry related issues.

For urgent or escalated issues, please contact your Desktop Support team.

6. **Press the Menu button and select Save.**

   Anyone who e-mails you automatically receives your Out of Office reply.

To turn off your Out of Office reply, follow these steps:

1. **Repeat Steps 1–3.**

2. **Change Use Out of Office Reply to No.**

3. **Press the Menu button and select Save.**

## BES-only calendar features

BlackBerry users who are on a BES also have access to wireless calendar synchronization as well as a few additional calendar features.

### Checking your coworkers' calendar availability

Have you ever sent a meeting invitation from your BlackBerry only to find that someone you invited wasn't available at that date or time? With a BES-activated BlackBerry, you can check who's available to attend a meeting before you send the invite! Just follow these steps:

1. **From your BlackBerry, open your calendar.**

2. **Create a new meeting and add attendees.**

   Refer to Book II, Chapter 2 for more information on using your BlackBerry Calendar.

3. **Scroll down and select View Availability, as shown in the left side of Figure 1-6.**

   The availability screen opens and shows if everyone that you've invited is available, as shown in the right side of Figure 1-6.

   If some or all of your attendees aren't available at that time, select Next Available Time. Your invite is updated to a time when everyone is available.

4. **Press the Escape button and select Save if prompted.**

5. **When you're finished creating your meeting, send your invitation.**

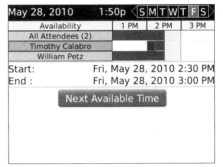

**Figure 1-6:**
Select View Availability to see if everyone you invited is available.

### Resetting and resynchronizing your BlackBerry Calendar

Every now and then, something may go awry with the background magic that synchronizes your BlackBerry Calendar with your corporate PC and messaging server through the BES. When that happens, your BlackBerry may miss some calendar updates, or even some appointments altogether. Luckily, your BlackBerry has a built-in way to reset and resynchronize its calendar with your enterprise messaging server. Follow these steps:

These steps clear all entries from your BlackBerry calendar. You should use them only if you're sure that your BlackBerry is connected to a BES server and you don't have any calendar items that are on your BlackBerry but not on your desktop e-mail client. If you haven't synched your calendar entries, they will be lost.

1. **From your BlackBerry Home screen, select the Calendar icon.**

2. **Press the Menu button and select Options.**

3. **Scroll to and highlight Desktop.**

4. **On your BlackBerry keyboard, type** rset.

Yes, you really don't need to type the first E in reset. Just type **rset**. Whether you type **rset** or **RSET** doesn't matter. They're treated the same.

A pop-up menu appears, as shown in Figure 1-7. Don't panic — your BlackBerry is erasing the calendar named Desktop on your BlackBerry, not the calendar on your Desktop computer.

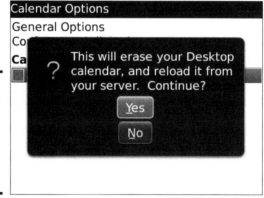

**Figure 1-7:**
Go into your
Calendar
options
and reset/
resync your
Calendar.

5. **Choose Yes.**

   Your BlackBerry Calendar clears itself out and resynchronizes with the BES and your enterprise messaging server within a few minutes.

## Using BlackBerry Manager

BlackBerry Manager is the brains of the whole operation. It's the interface on your administrator's computer that enables a connection to BlackBerry Configuration Database to perform administrative tasks, including the following:

✦ Add or remove BlackBerry users

✦ Set enterprise activation passwords

✦ Move users between BES servers

✦ Create and assign IT policies (*IT policies* are templates that are sent to individual devices remotely to configure them.)

✦ Create and assign software configurations

✦ Reset device passwords

✦ Wipe and disable lost or stolen devices

✦ View device information and usage statistics

✦ View, add, or edit device filters

✦ Resend or resync calendar information

## *For IT professionals only: Accessing BlackBerry Manager*

You can open BlackBerry Manager using the following steps:

*1.* **On the BES, or a remote computer with BlackBerry Manager installed, click the Start button.**

*2.* **Select All Programs.**

*3.* **Select BlackBerry Enterprise Server.**

*4.* **Select BlackBerry Manager.**

To send IT commands to individual BlackBerry phones, follow these steps:

*1.* **Open BlackBerry Manager.**

*2.* **Click the BlackBerry Server name in the upper-left corner.**

*3.* **Click the Name field near the top of the screen and enter the name you are looking for.**

If you'd like to send an IT command to multiple users, you can click the Users tab, located in the upper-middle portion of BlackBerry Manager. Then Ctrl-click each user whose settings you want to change. If you do this, skip to Step 5.

*4.* **Click Search.**

BlackBerry Manager filters the user list so you see only matches for your search.

*5.* **Right-click the username to which you want to send an IT command.**

*6.* **Choose the IT command you would like to send from the list that appears.**

The BES queues the IT command and sends it to the user's BlackBerry. Figure 1-8 gives you a glimpse of BlackBerry Manager.

 BlackBerry Manager can administer only BlackBerry Enterprise Servers that are part of the same configuration database (also known as a domain). If your administrator has a separate pilot or development environment, he or she needs to set up a different computer with BlackBerry Manager pointing to the pilot or development database, or remotely connect into the pilot or development BES and then launch BlackBerry Manager.

**Book VIII
Chapter 1**

**BlackBerry
Enterprise
Server**

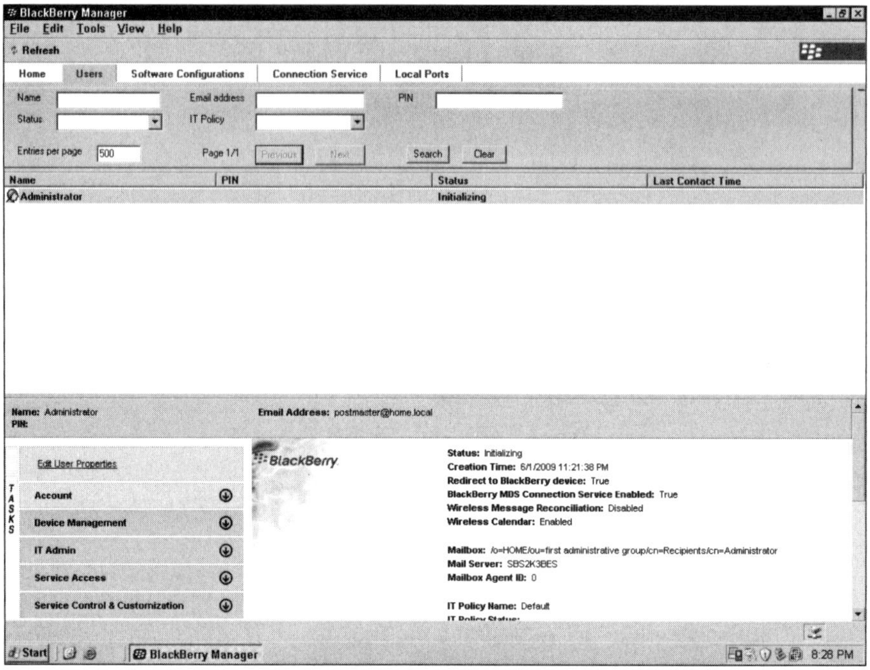

**Figure 1-8:**
BlackBerry
Manager.

# Chapter 2: Security and Compliance

## In This Chapter

✔ Useful IT policy settings for most environments

✔ Methods for archiving e-mail, SMS, and PIN messages sent via BlackBerry

*I*f you've owned your own BlackBerry and are attaching it to the enterprise for the first time, you're in for a rude awakening. Although the BES allows you all the convenience of accessing your work-related e-mails, applications, and documents, all this access comes at a price — security.

The BlackBerry platform is built around security and offers many options to meet the demands of a secure and compliant messaging environment. Using BlackBerry Mana ger, which is part of BES, your IT professional can go hog- wild assigning security policies that limit your use of BlackBerry apps, third-party apps, and other special configurations that you've become accustomed to.

In this chapter, you'll find out what an IT policy is, how your IT manager might be using it, and explore some of the groups that IT policy options are divided into. You also find options for BlackBerry e-mail archiving and SMS and PIN logging. If you're the tech geek in the windowless room who is in charge of maintaining the BES on your company's enterprise, we have some tips for you, too.

IT policies are also categorized into groups, called *policy groups*.

## Protecting Yourself and Your Company

Before we get into the nitty-gritty about security settings, configurations, and whatnot, you should understand a few general principles. Using your BlackBerry in an enterprise environment can leave your company open to security breaches if the BES isn't managed properly.

In the following sections, we explain some of the simple things your IT department will most likely require, and why. If you find that your organization isn't following these guidelines, you can share this chapter with your IT manager.

### Using external identification

It may seem simple, but putting your name and an alternative phone number on the outside of your phone (using the magic of a label maker) can mean the difference between "bye-bye, phone" and "hello, good Samaritan." For reasons we're about to explain in the section "Password protection," your phone should be locked up tight when you're not using it. That means if you lose your phone, no one will be able to access it to find out who the owner is. That's why an external contact is so helpful.

Confidential to IT departments: You may want to include a serial number instead of a name. That way, anyone who finds the number can call your department and let you know that it was found, but you won't have to reveal the owner's name. Also, avoid putting the company name on that identification. It's like announcing that the phone contains state secrets.

### Providing password protection

Your IT department will probably require you to use a password to unlock and use your BlackBerry. You probably won't be able to disable this feature. You may be asked to change your password frequently and you may be required to use a password that is difficult to crack.

If you're not used to using a password with your BlackBerry, you may find this feature annoying. However, passwords are vital. Imagine this scenario: You attend an industry-wide conference and accidentally leave your BlackBerry in the restroom. If your BlackBerry is password protected, the worst-case scenario is that you'll be mildly inconvenienced because you won't have a phone. If your phone isn't password protected, your mortal enemy could have access not just to your contacts, but to your e-mails. With a little extra work, someone can access (and copy) secret files on your corporate network, or worse. We know we sound a little 007, but corporate espionage *does* happen.

### Conducting regular backups

It may seem fairly obvious that you need to frequently back up data on your BlackBerry. But then again, it seems fairly obvious that you need to back up data on your desktop computer. And guess what! People still don't do it. They say they'll do it once a week or once a month, and then they forget, and the world doesn't come to an end, so they get more and more lax until the next thing they know they can't remember the last time they backed up their system, and who knows if they'll ever bother.

Notice we said "they," and not "you." Because that's not you, is it? Never! Seriously, backing up your data makes it less likely that you'll lose the *only* copy of The Most Important Document in the World in the event that your phone is disabled, lost, stolen, flushed down a toilet, dropped in the ocean, or run over by a train.

Backing up your BlackBerry doesn't clear space or improve performance — it doesn't remove anything from the device.

By the way, backups happen automatically on BES. Talk to your network administrator to find out how often data is backed up and what type of data is stored because the BES doesn't back up everything.

Data that's backed up by the BES isn't placed somewhere that users can get to it, nor is it stored in a type of backup file. Instead, it's sent to the BlackBerry configuration database, which automatically restores it when a BlackBerry is activated/reactivated with the user's mailbox. Also, it doesn't back up all types of data. For example, documents, pictures, and e-mails beyond the last 350 sent/received items aren't backed up by the BES.

## Getting periodic software updates

Your IT department will probably schedule a ton of periodic updates — from server patches to application patches. A *patch* is an update created specifically to fix a security breach or problem, to enhance an existing feature, or to introduce a new feature. Just like those patches on the knees of your Toughskins when you were a kid, software patches close holes — big holes that would otherwise be left wide open for hackers and other creeps. They also make the applications you know and love more usable. Make sure you follow any instructions given to you about rebooting your BlackBerry when these updates occur. Patches are vital to the health of your device, as well as the health of the enterprise.

## Handling Bluetooth the right way

Bluetooth technology is great — it allows you to talk hands-free and wirelessly. It also allows you to quickly and efficiently transfer data between two smartphones or between your smartphone and your PC. However, your IT department may lock down some of your Bluetooth capabilities in order to protect your security and the security of the enterprise. For one thing, your settings probably will prevent other Bluetooth devices from discovering your device. Another security setting prevents you from using Bluetooth to exchange information from your Address Book with other devices.

## Being wary of third-party apps

Third-party apps are great — except that they may open your BlackBerry (and, consequently, your enterprise) to security issues. That's because viruses, worms, Trojan horses, spyware, and other malware can attach themselves to Java-enabled apps. Your company may put the kibosh on allowing you to download apps over the air — or it may prevent you from downloading apps all the way around.

**Book VIII
Chapter 2**

**Security and
Compliance**

# Introducing Common IT Policy Settings

BlackBerry *IT policies* are kind of like the security permissions that are used to configure Windows computers — the only difference is that IT policies impact how you can use your BlackBerry. An IT policy is applied to you on the BES, and the configuration is pushed out to your device over the air (OTA).

You wouldn't believe how many IT policy groups there are, and each policy group has a set of options and configurations. It's enough to make your IT manager go bonkers. And it's good for you to know the general settings so that if your BlackBerry goes haywire (or operates less than optimally), you can help your IT department troubleshoot accordingly. If you *are* an IT manager, you can find additional information about these settings and configurations using the BlackBerry documentation.

## Understanding default IT policy settings

BES is centered on its capability to manage passwords and other security issues. The IT policy names may vary depending on the BES version and how your company is set up, but more likely than not, your company is using one of these configurations:

✦ **Default:** Unless it was changed by your BES administrator, this IT policy is the most lax of all the security settings. As the BlackBerry user, you don't need to use a password to access enterprise e-mail and apps. You can set a password on your own, but you can disable it any time. You can also download third-party apps to your heart's content and can change your BlackBerry's default settings. You might find this type of open configuration at a small company, but if you work for a large enterprise, you shouldn't get your hopes up because Default probably indicates you're in the most restrictive IT policy.

✦ **Basic password security:** With this IT policy, you have a password of any length, which you're prompted to update every 60 days. If the BlackBerry is idle for 30 minutes, it locks up tighter than Alcatraz — and only the password will unlock it.

✦ **Medium password security:** With this setting, your password must have at least one letter and one number, and it must be at least six characters long. You must change your password every 30 days and your BlackBerry locks up after being idle for 10 minutes.

✦ **Medium password security with no third-party applications:** This setting is just like Medium password security, except that you're not allowed to download any third-party applications onto your BlackBerry.

✦ **Advanced password security:** The password settings are exactly the same as with Medium password security, but there are additional security features in place restricting your use of Bluetooth technology and USB mass storage. In addition, it requires encryption of file systems.

✦ **Advanced password security with no third-party applications:** The strongest security there is — everything you find with advanced password security, plus you can't download any third-party apps.

It's possible for you to check your BlackBerry's password and security settings. Depending on how your IT policy is set up, you may even be able to change some of them. Just follow these steps:

**1.** **From your BlackBerry Home screen, select the Options icon.**

Depending on your BlackBerry model, you may have to select the Settings icon; then select the Options icon.

**2.** **Select Password.**

The password screen opens, as shown in Figure 2-1.

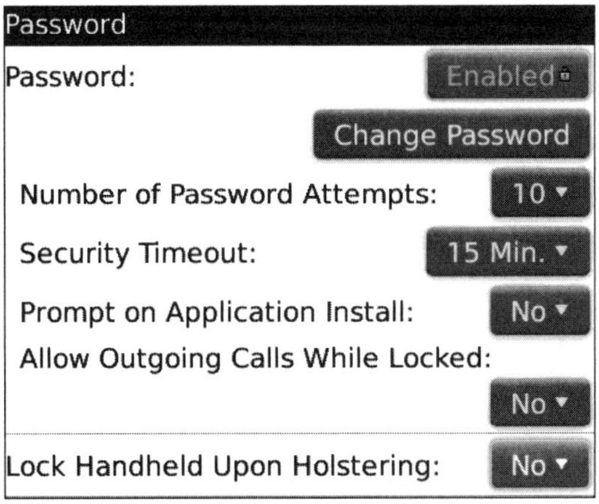

---

**Password**

| | |
|---|---|
| Password: | Enabled 🔒 |
| | Change Password |
| Number of Password Attempts: | 10 ▾ |
| Security Timeout: | 15 Min. ▾ |
| Prompt on Application Install: | No ▾ |
| Allow Outgoing Calls While Locked: | |
| | No ▾ |
| Lock Handheld Upon Holstering: | No ▾ |

**Figure 2-1:**
A password setting can't be changed if it has a lock next to it.

---

If a setting has a lock next to it — like the one shown next to Password Enabled in Figure 2-1 — you can't change it.

If there is no lock next to any of the following settings, you can change the settings to your heart's content by scrolling to the desired setting and selecting the option you would like:

- *Password:* Specifies whether you're required to use a password to unlock your BlackBerry.

- *Change Password:* Allows you to change your BlackBerry unlock password.

- *Number of Password Attempts:* Sets the number of times you can type the wrong password before your BlackBerry wipes all of its data.

- *Security Timeout:* Specifies how long your BlackBerry can go unused before it locks itself and requires a password unlock.

- *Prompt on Application Install:* Specifies whether you're prompted when applications are installed on your BlackBerry.

- *Allow Outgoing Calls While Locked:* Specifies whether your BlackBerry can place phone calls while it is locked.

- *Lock Handheld Upon Holstering:* Specifies whether your BlackBerry locks when you place it in its holster.

## *Knowing what else your IT policies might restrict*

The previous section lists basic security settings. The sticky wicket is that your IT guru can modify a whole slew of settings for everyone on the enterprise and on a case-by-case basis — seemingly on a whim. It helps to know your company's policies so that you can troubleshoot your BlackBerry appropriately.

The good news is that your IT guru would have to be some kind of glutton for punishment to randomly change configurations. Most likely, any deviations from the default security settings have to do with security and compliance regulations for your company, how high up you are in the company, your job title, and what work functions you perform using your BlackBerry. In other words, there *should* be a method to the madness.

Here are some additional settings that you might find have been modified:

✦ **Allow Peer-to-Peer Messages:** This setting determines whether PIN messages can be sent to other BlackBerry users. This setting is a big deal. To find out why, check out the nearby sidebar, "Pin messaging and security."

✦ **Allow SMS:** This setting allows you to send or prevents you from sending text messages, also known as SMS (short message service) messages.

✦ **Do Not Save Sent Messages:** This setting specifies whether a copy of sent e-mails is saved to the Sent messages folder of your desktop PC's e-mail program. Don't worry — we haven't seen many companies use this setting, and your sent messages still pass through your enterprise messaging server. A copy is archived in case you ever need your IT guys to retrieve it.

✦ **Disable MMS:** This setting prohibits you from sending and receiving Multimedia Messaging Service (MMS) messages — that is, text messages that include music, photos, or other multimedia.

## PIN messaging and security

If you've read Book III, Chapter 3, you may know a little about PIN messaging. If not, flip back and check it out. When you communicate PIN to PIN, that means you send messages directly to another BlackBerry user without e-mail or anything else. Rockin', right? Even more rockin' is that PIN messages aren't generally saved — anywhere (except your device, of course). That's not good news to your network administrator because that means that you could inadvertently (or intentionally) share information that isn't supposed to leave the enterprise.

Your system administrator can (and, if she's smart, will) change that setting so that PIN-to-PIN communications are saved to the BES. She may even prevent you from sending and receiving PIN messages all the way around. The moral of this tale is that you should never expect absolute privacy — ever. If you're thinking of using your BlackBerry to send personal messages, company secrets, or love letters, you should probably just whisper them instead.

### *Getting the straight scoop on password settings*

A good password is probably the best security feature when using your BlackBerry in an enterprise environment. In addition to the default password settings that we've already discussed, Table 2-1 shows you some password-related configurations that you might run into.

| Table 2-1 | Password Policy Security Options | |
|---|---|---|
| *Policy Option* | *Description* | *This Is Useful Because. . .?* |
| Set Password Timeout | Your administrator can set the time, in minutes, before your BlackBerry locks and requires a password. | This is a classic anti-theft strategy. |

*(continued)*

**Book VIII
Chapter 2**

**Security and Compliance**

**Table 2-1** *(continued)*

| Policy Option | Description | This Is Useful Because. . .? |
|---|---|---|
| Set Maximum Password Attempts | If this setting is configured, after the maximum number of incorrect passwords attempts is hit, the BlackBerry deletes all of its data. | This is a real-life self-destruct button — crucial if you have access to highly sensitive corporate secrets, but overkill if your access to sensitive data is limited or non-existent. A high number of failed attempts indicates that a thief is using decryption technology to crack the password. Your administrator shouldn't set this setting at two attempts, but if he does, don't try to unlock the phone while eating a jelly donut. |
| Maximum Password History | This configuration allows the administrator to set the number of previous passwords that the BlackBerry checks new passwords against before a password can be recycled. | Prevents users — ahem, that would be you — from recycling easy-to-remember passwords. By the way, reusing passwords is like inviting criminals to have another crack at accessing your information. Don't do it, even if you can. And if your password is coff33, don't cop out and change it to coff333. |
| Duress Notification Address | Types the e-mail address that the BlackBerry device notifies when a user types a password under duress. The duress password is always the user's password with the first character moved to the end. So if the user's password is coff33, the duress password would be off33c. | When the duress password is used, the maximum number of password attempts is cut in half. That's because once the duress password is used, you're prompted to use the correct password. In the meantime, the specified e-mail contact is notified that there's an issue. |
| Forbidden Passwords | The admin can create a list of words that users can't use as passwords. | Our guess? If you work for Starbucks, your password can't be coff33. |

# Getting Some Serious Security (and Maybe a Little Bit of Control)

The basic IT policy settings already enable IT administrators to disallow you from downloading third-party software (that is, software not created by Research In Motion). The following settings can also be added:

✦ **Disable BlackBerry App World:** This setting might be used to keep you from downloading and installing bad software that could damage the whole network or individual smartphones, slow down performance, or — heaven forbid — allow you to have a little fun on the company dime.

If you're an IT manager, this IT policy setting is not included in BES 4.1.6 by default. It must be downloaded and imported.

✦ **Force Lock When Holstered:** This setting specifies whether the BlackBerry device is security locked when placed in a holster.

✦ **Content Protection Strength:** This setting controls the cryptography strength the BlackBerry device uses to encrypt data.

- *Strong:* This setting is adequate for most situations and doesn't sacrifice security for performance.

- *Stronger:* This setting provides higher security, but slows down the performance of the BlackBerry. If you use this setting, RIM recommends that you set the Minimum Password Length IT policy rule to 12 characters.

- *Strongest:* This setting is ironclad, which comes at a heavy cost — performance is molasses slow. If you use this setting, RIM recommends that you request that the user set a password of at least 21 characters. That shouldn't cause any problems. Right? Right.

✦ **Disable Forwarding Between Services:** This setting prevents you from forwarding or replying to a message using a different e-mail address than the one the BlackBerry device received the message on. It also specifies whether you can forward or reply to a PIN message with an e-mail address, or reply to an e-mail message with a PIN message.

✦ **Disable External Memory:** This setting specifies whether to disable the expandable memory (microSD) feature on applicable BlackBerry smartphones.

✦ **Disable USB Mass Storage:** This setting disables the USB Mass Storage feature.

✦ **Disable Media Manager:** This setting disables the Media Manager.

✦ **Firewall Block Incoming Messages:** This setting allows the enterprise firewall to block certain incoming messages, such as SMS, MMS, PIN messages, and BIS. This setting is important because these types of messages would normally *not* be blocked by the enterprise firewall, because they are specific to mobile devices.

---

# Memory cleaner policy group

Although keeping memory clean isn't necessarily a security concern, it's a good idea for your administrator to keep device memory clean with these settings so that you can prevent memory leaks.

A memory leak happens when you close an app on your BlackBerry but the app refuses to let go of the memory it was using before it was closed. Say you downloaded this cool app that lets you find local restaurants. After you find a restaurant and close the app, it's supposed to release the memory it was using to look up restaurants. The app is supposed to reuse this memory the next time you open the app and search. Instead, the app might hold onto the memory it used and gobble up a little more memory each time you search until your

BlackBerry runs out of steam. If you notice that your BlackBerry has slowed down, randomly reboots itself, and sometimes even crashes, you might want to clean out its memory.

Here are the policy settings that apply to memory cleaning:

- ✔ **Force Memory Clean When Idle:** Specifies whether the memory will be cleaned when the device is idle.

- ✔ **Force Memory Clean When Holstered:** Specifies whether the memory will be cleaned when the device is holstered.

- ✔ **Memory Cleaner Maximum Idle:** If the device is idle for a specified length of time, the memory cleaner starts.

---

✦ **Remote Wipe Reset to Factory Defaults:** Talk about having ultimate control! That's right, with the click of a button, your IT administrator can reset your BlackBerry to its factory default settings and erase all your data, third-party apps, contacts, e-mail, and so on. This setting usually isn't used unless the phone is stolen or unless you're upgrading to a newer (usually better) phone.

# Synchronizing Your BlackBerry

You need to know about some rules when it comes to synchronizing the personal information on your BlackBerry. Personal information management (PIM) refers to all of the personal information stored on your BlackBerry, like contacts, memos, tasks, phone call logs, and SMS and PIN messages. The PIM Sync policy group applies to the wireless synchronization settings for your personal data, including specifying whether:

✦ PIN messages (which are usually not saved) are synchronized to the BES along with other data.

✦ SMS and MMS text messages are wirelessly synchronized to the BES. The default setting is to not sync these items.

✦ The call log is synchronized wirelessly to the BES.

These settings are important to keep in mind because you are more likely to receive personal texts and calls than you are to receive personal e-mail messages to your business e-mail account. As we mention in the nearby sidebar about PIN messaging, you might have an unrealistic expectation of privacy with your PIN messages, so it doesn't hurt to find out what your company's compliance requirements are, as well as its policy about using company-issued smartphones for personal communication.

Your BlackBerry is connected to BES using a process called *enterprise activation.* When your BlackBerry is connected to a BES, all of the PIM synchronization happens wirelessly, and automatically, without any additional action on your end. Read Book III, Chapter 2 for more information on enterprise activating your BlackBerry.

# Auditing and Compliance

Many enterprises are required to comply with federal regulations to capture and archive all electronic communications, such as e-mail messages, SMS messages, PIN-to-PIN messages, instant messages, and so on that are sent through company-owned systems such as Microsoft Outlook, Lotus Notes, or a BlackBerry smartphone. Adhering to industry-wide, national, and international regulations is often called *corporate compliance* or, simply, *compliance.*

Which regulations your organization adheres to might depend on your business and its geographic location. Some common regulators include:

+ **Federal governments:** In the U.S., laws such as the Sarbanes-Oxley Act of 2002 have rules that regulate information revealed by publicly traded companies.

+ **Industry regulators:** Industry regulators set the bar for a whole set of professionals, such as those in the healthcare industry, business or legal professions, and so on.

+ **International regulators:** The most well-known international regulator is the International Organization for Standardization (ISO), which sets standards for business, government, and just about everything in between.

Based on these legal and industry-wide rules, as well as internal compliance rules that the company set for itself, a business may be asked to save communications for a period of months or years. The communication may be needed in the event of a legal dispute or compliance audit.

If a company can't reliably record certain communications, such as SMS messages, the company must disable this communication option to avoid penalties. Research In Motion allows sent e-mails, SMS messages, and PIN-to-PIN messages sent through enterprise BlackBerry smartphones to be archived.

**Book VIII**
**Chapter 2**

**Security and Compliance**

## E-mail archiving

When a new message is received or a reply is sent from a BlackBerry smartphone, a copy is automatically sent to the messaging server mailbox — whether it's Microsoft Exchange or Lotus Domino.

RIM has created an option to automatically send a blind carbon copy, or BCC, of all e-mails sent from a user's enterprise BlackBerry to a specified e-mail address. This means that the messages can be received in a compliance mailbox, which has an archival application set up to record them. That means that if your company adheres to regulations with regard to archiving messages, you don't have to do anything special in order to keep those messages.

## SMS and PIN logging

SMS and PIN messages are not sent or received through the BlackBerry Enterprise Server. This provides a unique challenge for recording, archiving, and auditing, which is why RIM added options for SMS and PIN message logging.

If your company has SMS and PIN logging enabled, the BES creates daily log files, where all of the SMS and PIN messages that you and your coworkers send and receive are stored. The BES tells the BlackBerry smartphones that logging is enabled and the smartphones automatically send the BES information on every SMS and PIN message you send.

# Chapter 3: Understanding How BlackBerry Applications Are Deployed across the Enterprise

## In This Chapter

✔ **Understanding deployment options**

✔ **Working around deployment methods**

Chances are that even though your enterprise administrator *can* banish you from downloading third-party applications, he or she probably won't block them completely. (See Book VIII, Chapter 2 to see more about how enterprise admins might prevent you from accessing third-party apps.) That's because apps give employees like you exactly what you need — added functionality that helps you do your job efficiently, saving the company both time and money. However, it's very likely that your IT administrator controls which third party applications you can install. In short, the benefits of each third-party app are weighed against the security risk, as well as the added maintenance. Those are the factors that your administrator is probably weighing when deciding whether to allow an app into the enterprise environment.

Your IT department will probably create a *whitelist* — a list of acceptable third-party apps. It might even create a *blacklist*. You can probably guess which kinds of apps go on that list — the ones that are untested, are a security risk, or present a maintenance nightmare.

This chapter talks about that maintenance. New applications are released every day, and the applications you already have are updated almost as frequently. Your enterprise administrator will be spending a lot of time keeping apps up to date and meeting the demand for new applications.

To make things easier for those tireless admins, BlackBerry has created several options for deploying and updating applications across the enterprise. In this chapter, we explore all of the deployment options. If you're an administrator, we even include some hints for how to set them up. We also describe the pros and cons for each option so that you can choose the best one that works for your environment.

# Understanding Is Half the Battle

If you're a decision maker in the IT department or an app developer who is creating applications for your company, this chapter can help steer you toward the right choice for deploying applications based on the following criteria:

+ How many BlackBerry users are in the enterprise?

+ How many applications does each BlackBerry owner use?

+ Has your company developed proprietary BlackBerry apps for use across the enterprise?

If you're not a decision maker (or you're a decision maker, but not in the IT department), you're pretty much stuck with what your IT department comes up with. This chapter can still give you some hints about what might be going on when you see that e-mail message with the globe next to it, and it definitely explains what's going on when applications are pushed onto your device.

# How Shall I Deploy Thee? Let Me Count the Ways

Your BlackBerry smartphone's apps are maintained using one of four options we outline in the following sections:

+ **BlackBerry Browser Push:** BES is used to place in-house applications or content onto BlackBerry smartphones through BlackBerry Browser. Users have to select an icon to complete the install.

+ **BlackBerry Application Web Loader:** Users connect their BlackBerry to a company PC using a USB cable and visit an intranet site to install applications.

+ **OTA installation:** Users visit a URL on their BlackBerry smartphone to install applications wirelessly.

+ **BES software configuration:** BES wirelessly installs and configures applications on BlackBerry smartphones. It can also mandate which apps must be installed and/or which apps aren't allowed.

Each application deployment method works differently and has its own set of pros and cons.

## BlackBerry Browser Push

BlackBerry Browser Push uses the BES to place a message or icon on your BlackBerry smartphone automatically. You see this new message or icon appear automatically, and when you select or click the icon or message, your BlackBerry Browser directs you to a specific Web page or Web application. The application is installed or updated without any further steps on your part.

### Pros

This method is generally used by *in-house* developers, meaning the application was created by a team of developers that works for your company instead of an outside source. Most of the legwork is handled by the developers, so there's little setup or maintenance needed on the administrator's end. But you should still be aware of this method and how it works.

### Cons

You have to actually click the icon to make it happen. For administrators, the choice to use this option depends on whether you need to roll out in-house BlackBerry applications, rather than how many people in the enterprise will be receiving the deployment. In other words, it's not always the best choice, but you may experience it from time to time.

---

## Getting the lowdown on browser pushing

BlackBerry Browser pushing doesn't require any setup or maintenance on the administrator's end. When in-house developers write applications, they just include code that tells the application how to push it to the users' BlackBerry smartphones through BES, and BES handles the rest. Take a look at Book VII, Chapter 1 for more information about BlackBerry Browser pushing. In case you're wondering, there are three different types of BlackBerry Browser pushing:

✔ **Message Push:** You receive a message that looks similar to an e-mail message. (It has a globe next to it instead of an envelope icon.) This message contains a locally stored Web page that automatically opens when you read the message.

This type of push can only be received by BlackBerry smartphones that are connected to a BES.

✔ **Channel Push:** A new icon appears on your BlackBerry Home screen. Select the icon to direct BlackBerry Browser to a specific Web address. Unlike the message push, you must have a wireless carrier signal to load the page.

✔ **Cache Push:** A new icon appears on your BlackBerry and stores the Web page information on your device, which means you can see it even when the device isn't receiving coverage. This type of push can be received by only BlackBerry smartphones that are connected to a BES.

## BlackBerry Application Web Loader

BlackBerry Application Web Loader places all BlackBerry applications on one central intranet Web server. You then connect your BlackBerry to a PC via a USB connection, visit the application site on your PC, enter your BlackBerry password, and install all the applications or updates.

Figure 3-1 shows you what you may see when using BlackBerry Application Web Loader to install an application.

**Figure 3-1:**
A
BlackBerry
Application
Web Loader
page.

### Pros

The great thing about this option is that if your company has developed its own proprietary applications for the BlackBerry, your company's developers can place newer versions directly onto the Web server, and you can download them immediately, all without any intervention from the enterprise admin — no fuss, no muss.

### Cons

This is a manual deployment method that depends on you (and all the other members of your enterprise) to connect your device to a PC and visit an internal site to install or upgrade the application(s), or for the enterprise administrator to do it for you. There are also several desktop PC requirements that can be a headache to manage. Additionally, the initial setup can be tricky for administrators.

We recommend this method for only small enterprises, or large enterprises with a small number of BlackBerry users. It also makes a great backup plan if you decide to use BES software configurations and they fail for some reason.

## OTA installation

*OTA installation*, short for over-the-air installation, means that you can install applications using your wireless carrier network instead of using a PC and USB cable. You have to click a link from your BlackBerry. This link starts downloading and installing the application from the external Web site or an internal intranet site where it's stored.

OTA activation and OTA installation are two completely different things. *OTA activation* is the process of wirelessly connecting your BlackBerry to a BES so you can receive your corporate e-mail and PIM information, apply IT policy settings, receive browser pushes, and have applications pushed down to your device. *OTA installation* is a separate process that refers to visiting a link from your BlackBerry Browser and choosing Download to install an application.

Take a look at Book VII, Chapter 1 for more about the files needed for OTA installation — .jad and .cod files.

### Pros

OTA installation allows you to install or upgrade applications on the go from anywhere a wireless carrier signal is available, without having to connect to a desktop PC.

### Cons

This method is manual because it depends on you selecting a link. Some users are uncomfortable following the steps, and the administrator may end up having to do it for them.

We recommend this method for an enterprise that supports only a small number of BlackBerry users.

### Installing an app OTA

In order for you to install an app OTA, the app has to be available through a Web page on the Internet, or on a company intranet, before it can be installed over the air. Once your application has been published to an Internet or intranet site, it can be installed on the BlackBerry over the air.

1. **From the BlackBerry Home screen, select the BlackBerry Browser icon.**

2. **Next to the http:// in the Address box, type the address of the Internet or intranet page that has the link to the .jad file.**

3. **After the Web page loads, select the link.**

    You're directed to a screen with information about the application and the options to Download or Cancel. (See Figure 3-2.)

**Figure 3-2:**
The
BlackBerry
Application
Download
screen.

4. **Select Download.**

   The application downloads and installs.

5. **You may be asked to reboot your BlackBerry smartphone.**

   If not, you receive a prompt that the application was installed successfully with the options OK (dismiss the prompt) or Run (open the application).

The first time the application is launched, you may see prompts, as shown in Figure 3-3, asking you to allow internal and/or external connections, or to trust the application. You should always choose Allow or Trust to ensure the application works properly.

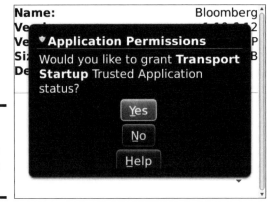

**Figure 3-3:**
An
application
connection
request.

If your BES administrator has blocked you from downloading an application OTA, you see the message "The IT Policy does not allow downloading this application" in the download screen, as shown in Figure 3-4.

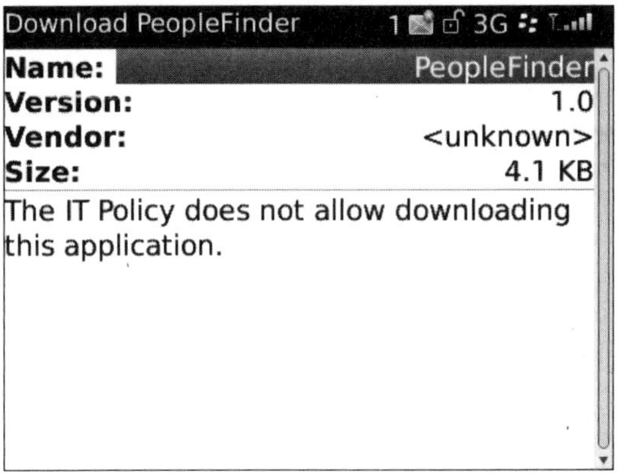

**Figure 3-4:**
If you
see this
message,
your BES
admin is
blocking
the app.

```
Download PeopleFinder      1 ▨ ⛁ 3G ⁑ ▮ııl
Name:                          PeopleFinder
Version:                                1.0
Vendor:                           <unknown>
Size:                               4.1 KB
The IT Policy does not allow downloading
this application.
```

## BES software configuration

BlackBerry software configurations are used to install or remove BlackBerry device applications remotely through the BES. Once your admin sets up the software configuration, it's assigned to users or groups on the BES. The BES then sends the application files out to your BlackBerry, along with everyone else in your group or enterprise. The devices automatically install the application and prompt you to reboot if needed.

### Pros

This application deployment method gives administrators the most control. Admins can specify which applications, and which version of those applications, you can install on your BlackBerry. The admin can also perform the installation remotely instead of relying on you to visit an internal site, click a link, or visiting you in person to take care of it.

### Cons

The only downside to this method is that the more applications you have in your environment, the more complex your admin's job becomes. He or she will need a vacation after creating and maintaining so many software configurations. Be kind.

**Book VIII
Chapter 3**

**Understanding How
Applications Are
Deployed**

# Finding out what you need to know about BES software configurations

As we mention earlier, BES software configurations are by far the best way of managing applications in a large enterprise. The following information is required reading if you're an admin and you will be using the software configurations to make updates. It falls in the interesting but not necessary category if you're an employee whose company uses BES software configurations to get stuff done. So, admins, if you're considering using software configurations, you should be aware of a few items:

✔ Each user or group can have only one software configuration assigned at a time.

✔ One software configuration can install more than one application.

✔ You can deploy only one version of an application at a time. This means all users with that application have to be upgraded at the same time.

✔ BES pushes out applications once every four hours. This is known as a *push interval.*

✔ A BES administrator can manually deploy applications through BlackBerry Manager. When this option is used, BlackBerry Manager ignores the four-hour push interval and attempts to push applications to the selected users or groups immediately.

The rules and functionality of software configurations are greatly improved in BES 5.0. Check out Book VIII, Chapter 6 for more information on BES 5.0, including the enhancements to application deployment via software configuration.

# Chapter 4: Using Mobile Voice System

**Y**ou bought a BlackBerry so that you could stay connected. Your company uses BlackBerry across the enterprise to help you stay connected. Now, if only you could link up your BlackBerry with your office phone so that anyone who calls you can always contact you, even if you're on the road. But wait — you can hook up your two phone numbers so you'll never miss an important call because you're away from your desk. And you'll never miss a call because both lines are ringing at once! And you'll never have to dial more than one voice mailbox to get your messages. What is this magic that allows you to have complete control of all your incoming calls? BlackBerry Mobile Voice System (MVS), that's what.

MVS lets you carry the functionality of your desk phone in your BlackBerry. This chapter explains what BlackBerry Mobile Voice System is, what it does, how it does it, and why you might want it.

This chapter is for enterprise users. However, your enterprise must already have decided to implement MVS in order to use the info in this chapter. Perhaps you can tear this chapter out of the book and present it to a decision maker in your organization, along with a bar graph that shows how much time and money MVS saves.

## Understanding MVS Basics

BlackBerry MVS extends office phone features to BlackBerry smartphones by mobilizing office PBX systems. (PBX stands for *private branch exchange*.) All a PBX really is, is a private telephone network, usually for a business. From the PBX, users can dial extensions within the enterprise, make external calls, and use internal voice functions like call transfer and call conferencing.

The acronym PABX (*private automatic branch exchange*) is sometimes used instead of PBX.

---

# The mother of all acronyms: TDM PBX

TDM PBX (Time Division Multiplexing PBX) systems are legacy PBX systems that are still used by some enterprises today. Many enterprises with a TDM PBX system are in the process of switching to a VoIP PBX system. As a result, some companies are operating both TDM PBX and VoIP PBX systems in the same environment.

---

There are two types of PBXs: TDM (Time Division Multiplexing) and VoIP (Voice over Internet Protocol). Companies usually use one or the other — not both — and most companies that have older TDM PBX systems are in the process of converting to the newer VoIP PBX systems.

## TDM (Time Division Multiplexing)

Time Division Multiplexing (TDM) allows multiple data streams to be combined in a single signal. This telecommunications miracle is accomplished by breaking the signal into segments, each with a very short duration, and then reassembling each data stream at the receiving end of the call.

## VoIP (Voice over Internet Protocol)

Voice over IP, or VoIP, enables people to make phone calls over computer networks. VoIP technology converts analog voice signals into digital data packets and supports real-time, two-way transmission of conversations using Internet Protocol (IP).

## Knowing what MVS does

BlackBerry MVS helps you avoid missed calls by allowing you to use one business number for multiple phone lines. It also provides the advanced security features, wireless administration, and centralized management that you expect from a solution provided by RIM, combining the flexibility you need with the controls you trust.

There are three main benefits for an enterprise to use BlackBerry MVS:

+ You can receive incoming desk phone calls on your BlackBerry smartphone.

+ You can switch from your desktop phone to your BlackBerry smartphone, or vice-versa, during a call.

+ You can make calls from your BlackBerry smartphone that look like they're coming from your desk phone. (We can think of a few ways you might abuse this option!)

As shown in Figure 4-1, MVS combines the power of PBX, BES, and IP (Internet Protocol) to miraculously link your BlackBerry smartphone and desktop phone communications.

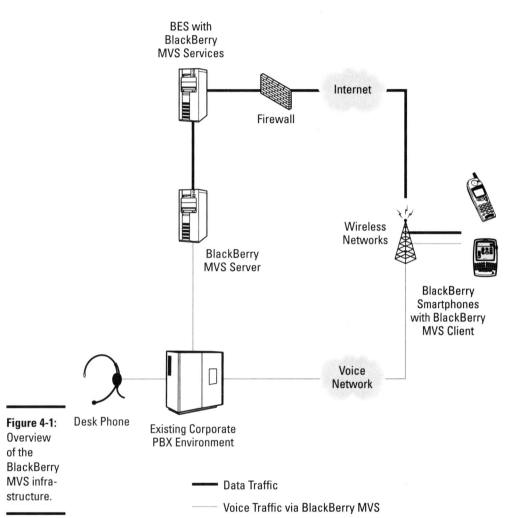

**Figure 4-1:** Overview of the BlackBerry MVS infrastructure.

Data Traffic
Voice Traffic via BlackBerry MVS

Several components need to be in place for MVS to work, as shown in Figure 4-1. You need a working BES, a VoIP-capable PBX system, an MVS server to connect the BES and PBX together, and MVS client software installed on each BlackBerry that you want to use MVS.

# Understanding MVS Advantages

BlackBerry MVS makes it easier for you to do more in less time, especially if you spend a lot of time out of the office and on the phone. By combining desktop and mobile voice communications, you can no longer use the excuse that you were out of the office and didn't get the message. We're not sure if you love that idea, but we're pretty sure your boss does. We find that MVS is especially useful in a fluid enterprise, where a large portion of an enterprise is "out in the field" a good part of the time. The seamless communications strategy makes customers feel like they can always get in touch, and they don't need to memorize a bunch of phone numbers.

MVS extends your current office phone features. Built-in features guarantee security, control, and call-logging features to help make your job easier. The main benefits for you and your users are outlined in the following sections. They include:

+ **You can have a consistent presence.** You have to provide only one corporate phone number and have a consistent caller ID. This helps you manage your calls better, and it also gives you a consistent identity with customers.

+ **Your customers and coworkers have 24/7/365 access to you.** Calls ring your desk phone and BlackBerry smartphone at the same time, so you miss fewer calls.

+ **You can go from talking on the phone to talking in your office without hanging up.** You can transfer calls and even switch between the desk phone and the BlackBerry smartphone during a call.

+ **You have to punch only one code, one time.** You have a single voice mailbox for office calls and mobile calls, which helps you respond quickly.

+ **The system is user-friendly.** The MVS menus are similar to the ones you're used to using, so the system is very user-friendly.

+ **The system is secure.** The BES authenticates all calls from enterprise BlackBerry smartphones. Your network administrator can regulate incoming and outgoing calls, implement security measures, and audit activities.

+ **Your company can stay in compliance.** All inbound and outbound calls are routed through the PBX system, so private information is protected and calls are logged. This fact should be a relief to those in charge of auditing your organization compliance to legislative and industry requirements.

---

## Confidential to decision makers: Understanding MVS disadvantages

We know it's sacrilegious to suggest that there could be any disadvantages to using MVS, but all the same, we do think it's important for decision makers to carefully consider how you want to use MVS technology in your workplace.

Although it's true that your employees will never be able to say that they "didn't get the message" or "didn't get the call," they may start saying that they didn't get a chance to do their work because they were always on the phone.

Here's the bottom line: If an employee's main job function includes talking on the phone (schmoozing clients, providing customer service, and so on), then MVS is a great asset. If you have employees who are sitting behind a desk all day churning out reports, then maybe MVS is overkill. We think a percentage of your enterprise will appreciate the opportunity to extend their image to clients. The rest may come to resent the intrusion.

---

## Knowing What to Expect with MVS

If your company has just rolled out an MVS system, you should know what to expect when you make and receive calls. First, we give you the view from very, very far away. Then we show you what to expect when you're holding your BlackBerry in your hand and taking a call.

### Getting the view from far away: Making a call

If you were flying really high overhead, and you wanted to watch what happens when someone uses MVS to make a call, this is what you might see. Okay, okay, we know that would mean you would also have the power to see wireless signals, but work with us here.

1. A BlackBerry user places a call with the MVS client.

2. A very short encrypted data message is sent over the wireless carrier network to the BlackBerry Enterprise Server (BES).

3. The BES authenticates the call, and then passes it to the BlackBerry MVS Server.

4. The MVS server sends the call to the enterprise PBX system.

**Book VIII
Chapter 4**

**Using Mobile Voice System**

5. The PBX calls the destination party over its interface, which starts off using plain old telephone lines and then changes to cellular signals.

6. The system then makes an authenticated call to your BlackBerry smartphone and bridges in the destination party. (This makes it appear that the call came from the BlackBerry user's desktop phone.)

### Getting the view from far away: Receiving a call

Okay, so imagine you're capable of seeing wireless signals and you're really high up watching a call come in. First, here's an assumption: The BlackBerry MVS server already has stored a profile for you (and any other MVS users in the enterprise). The profile includes the telephone number for your desk phone, the telephone number for your BlackBerry smartphone, and, if you so desire, the telephone number for your home office (or any other number you want to associate with MVS).

Not everyone in the enterprise has to be an MVS user. The enterprise administrator can assign MVS to users who really need it and restrict it for users who don't.

1. The MVS server establishes a connection with the enterprise PBX.

2. When a call comes in to the PBX, it looks to the BlackBerry MVS Server for guidance and rings all of the user's listed numbers simultaneously.

3. When the call is answered by one phone (the BlackBerry, for example), the other numbers stop ringing.

The BlackBerry Mobile Voice System Server does not need to be in the same physical location as the BlackBerry Enterprise Server. Typically, it is installed in the same room as the PBX, which keeps the telecom infrastructure in the same physical location.

## Using the BlackBerry MVS Client

The BlackBerry MVS client adds desk phone features to BlackBerry smartphones. It integrates directly with the native BlackBerry phone application and allows you to place or receive calls from your mobile line or your desk line (shown in Figure 4-2) as well as transfer calls between your BlackBerry smartphone and your desk phone line. This software can be installed wirelessly using a BES software configuration.

The BlackBerry MVS client requires BlackBerry handheld software version 4.2.1 or later. If you install it on a BlackBerry smartphone with older handheld software, you may crash the device. For more information on installing applications on the BES, see Book VIII, Chapter 3.

**Figure 4-2:**
You can receive incoming desk phone calls (indicated by your company's short-dial number) on your BlackBerry with MVS.

## Enabling and disabling MVS

There may be times when you don't want to receive your desk phone calls on your BlackBerry, such as when you're on vacation or in a meeting. For those occasions, you can stop MVS from ringing your BlackBerry using these steps:

1. **From the BlackBerry Home screen, select the Phone (Call Log) icon.**
2. **Press the Menu button and select Options.**
3. **Select BlackBerry MVS client.**
4. **Select General Settings.**
5. **Select your BlackBerry phone number.**
6. **Change Device Status to Disabled.**
7. **Press the Menu button and choose Save.**

To enable MVS calls to your BlackBerry again, follow these steps:

1. **Follow Steps 1-5 for disabling MVS.**
2. **Change Device Status to Enabled.**
3. **Press the Menu button and choose Save.**

## Transferring MVS calls from your BlackBerry to your desktop phone

You can transfer an MVS call from your BlackBerry to your desktop phone using these steps. Remember, you should be on the phone while you perform these steps:

1. **While on a call, press the Menu button.**

2. **Select Move Call to Desk.**

   Your desk phone should start to ring.

3. **Answer the call on your desk phone.**

Transferring the call back to your BlackBerry is just as easy. Follow these steps:

1. **From the BlackBerry Home screen, select the Phone (Call Log) icon.**

2. **Press the Menu button.**

3. **Select Move Call to Smartphone.**

The phone call automatically transfers from your desk to your BlackBerry smartphone.

## Transferring an MVS call to another person

You can transfer an MVS call to another person using these steps:

1. **While on a call, press the Menu button.**

2. **Select Transfer, as shown in Figure 4-3.**

**Figure 4-3:** MVS enables you to move calls between your mobile line and your desk line, as well as providing other useful desk phone options.

3. **Select the contact to whom you would like to transfer the call.**

   - To transfer the call to a number in your call log, select the phone number from the list that appears.

   - To transfer the call to one of your contacts, press the Menu button, select Call from Contacts, and select the contact.

   - To transfer the call to a new number, type the number, and then press the Send button.

   Your BlackBerry places a call to the person to whom you're transferring. Wait for them to answer.

4. **When the person you're transferring the call to answers, press the Menu button and select Complete Transfer.**

## Creating a conference call

You can change an MVS call into a conference call using these steps:

1. **While on a call, press the Menu button.**

2. **Select Add Participant.**

3. **Select a contact to add to the call.**

   - *Add a number in your call log:* Select the number from the list that appears.

   - *Add one of your Contacts:* Press the Menu button, select Call from Contacts, and select the contact you want to conference in.

   - *Add a new number to the conference:* Type the number, and then press the Send button.

   Your BlackBerry places a call to the person you're adding to the conference. Wait for them to answer.

4. **When the person that you're conferencing in answers, press the Menu button and choose Join Conference.**

   You and the person you added are now on the conference call.

5. **Repeat Steps 1–4 to add another person to the conference call.**

# Introducing BlackBerry MVS Services and MVS Server

BlackBerry MVS Services is a combination of IT policies and highly secure authentication stuff that interacts between the BlackBerry smartphones in the enterprise and the corporate phone environment. It helps the BES and the BlackBerry MVS Server communicate, manage MVS data, and makes sure that the system stays secure.

The BlackBerry MVS server goes between the BlackBerry Enterprise Server and the corporate PBX system to help mobilize the PBX environment. The BlackBerry MVS server includes configuration options for TDM, IP, PBX, and mixed environments.

# Chapter 5: Tools for Maintaining a Healthy Enterprise Environment

## In This Chapter

✔ Standardizing the BlackBerry Enterprise Server environment

✔ Introducing the BlackBerry Server Resource Kit

✔ Finding third-party enterprise tools to manage your environment

*E*ach company has its own way of managing the BlackBerry infrastructure, depending on the size of the enterprise, the type of business it conducts, and the needs of employees. Your company can do a few things, but at some point, will need to develop (or hire someone to develop) some tools for you. You can guess that a company that makes its money by doing application development might hire some of its own programmers to develop proprietary BlackBerry apps for monitoring and optimizing the enterprise, but perhaps an organization that is later to the technology game will use third-party systems.

Whether your enterprise falls in the first or second category — or somewhere in between — if you're a BlackBerry Enterprise Server (BES) administrator and you're looking to develop or purchase enterprise-level applications to manage your enterprise, this chapter is for you.

If you're an employee who was given a standard-issue BlackBerry by your BES administrator, don't bore yourself with this chapter. Skip back to the beginning of Book VIII for information that can help you understand the do's and don'ts of using your BlackBerry in an enterprise environment.

## Getting Your Environment in Line

Ever asked the following questions? Why is my e-mail slow? Why can't I send e-mails? Why can't I receive e-mails? Perhaps you've asked the more esoteric (and desperate) question: What is going on with my BlackBerry?

BlackBerry Enterprise Server administrators and help desk agents supporting BlackBerry smartphones know these questions all too well. You can reduce the number of times these questions are asked — we may even eliminate these questions from being asked ever again. It's also important to standardize your environment before implementing the tools we discuss in this chapter.

Before you unleash a major BlackBerry deployment, it's important to standardize your environment. *Standardization* means making sure that all the information that you maintain about users' BlackBerry devices on the BlackBerry server, such as the handheld code, device type, carrier, and software settings exists. (*Hint:* You need to create this information if you don't have it.) Beyond creating these fields of information, you need to make sure that all the fields are in a standard format. For example, if you have two devices, the data fields might look like Table 5-1. In this table, BlackBerry 1 is assigned to a sales manager, and BlackBerry 2 is assigned to an administrative assistant in operations.

| Table 5-1 | Standardization Data |
|---|---|
| *BlackBerry 1* | *BlackBerry 2* |
| Handheld Code: 9981 | Handheld Code: 9982 |
| Username/Role: JHarris/Sales | Username/Role: FHatten/Admin |
| Device Type: Curve 8520 | Device Type: Curve 8310 |
| Carrier: AT&T | Carrier: AT&T |
| Date Received: October 2010 | Date Received: April 2010 |
| Tune-up: October 2011 | Tune-up: April 2011 |
| Rights/Permissions: Sales database share /Manager | Rights/Permissions: Operations share/no management rights |
| Special Software: Twitter | Special Software: N/A |

As you can see in this table, spelling out information about users in a consistent way can make it easier to track and troubleshoot issues. This standard information allows you to quickly adapt to any changes needed when you deploy a new version of the BES, update handheld software, and push out new versions of applications. It also can be used to identify rights and permissions. In the end, the users benefit from being able to utilize the newest tools.

Standardization can be challenging especially if you already have an enterprise with many wireless carriers and a wide range of BlackBerry smartphone types and handheld code versions. You may not be able to deploy the same BlackBerry smartphone to everyone, but you can try some things that may help out. Having the same software on your BlackBerry Enterprise Server, the same device type, device software, and carrier all provide certain advantages.

## Testing and reducing issues

Updates to BlackBerry Enterprise Servers, handheld code, and BlackBerry applications are released every few months. Every device responds differently to updates, depending on the type of device, the handheld code, or the server that the device is installed on.

By standardizing one or all of these items, you can get a good sense of whether a change will have a negative effect on the device or the whole environment.

If a software application has multiple versions, then the test needs to be generated for each scenario. Cutting down on this reduces the time for certifying and reduces unforeseen issues. As a result, it lowers the total cost to support the environment.

## Standardizing help desk support

In any enterprise, users are bound to ask any number of unexpected questions. You can anticipate *some* questions, though. Say the user sees a big red button that says "push here now." You're sure to receive a call to the help desk with a question along the lines of "What happens when I push the red button?" It's a reasonable question, don't you think?

To assist your help desk staff in appropriately answering this type of question, you can give the staff information about each type of keyboard, input device, and operating system that users might have. See Book I, Chapter 5 for a list of all the differences between devices.

## Improving the application development process

The hardest part of developing BlackBerry smartphone applications is ensuring that an application will run on all the devices that you have in your company. By reducing the number of types and BlackBerry device software versions that you have deployed, you can reduce your development and testing time. In addition, you can ensure that all the features you build will work for everyone in your company.

## Improving corporate communications

Many companies do their best to send BlackBerry-friendly e-mail messages. A BlackBerry-friendly e-mail is one that's formatted so that you can easily read it on a handheld device. Knowing the size of the screen for each of the models you're working with, as well as whether HTML is enabled on e-mail messages can help you provide consistent and clear communications to all users, whether they're using a traditional computer or a smaller BlackBerry.

## Getting your inventory and carriers in shape

If you have the power to do so, you should try to deploy BlackBerry devices that all have the same screen size. Further, if you can set up a rule that all users either turn on or turn off HTML, you can definitely standardize communications. We realize that this isn't always possible, but it's always nice to dream.

Here are some benefits: Managing mobile inventory is difficult. Devices are sometimes dropped, flushed down toilets, or lost. By supplying only one type of device, you can reduce support overhead and provide quick turnaround for device replacements and new deployments. You can also save money because your inventory of devices and parts will be smaller.

You can limit the types of BlackBerry devices you use only if you limit the number of carriers that you allow. Each carrier takes the BlackBerry smartphone handheld code from Research In Motion and changes it slightly so that it's unique to that carrier. This means that if you use the BlackBerry Curve model, the device code for the Curve will be slightly different depending on whether the phone is associated with AT&T, Verizon, T-Mobile, or another carrier.

We definitely recommend that you standardize which carriers you allow. You can reduce overhead and support issues, as well as reduce the amount of documentation and testing that your crew has to do. However, we *don't* recommend that you use just one carrier — using one carrier for your BlackBerry devices is never a good idea. Standardization doesn't mean limitation. Here are the advantages to having more than one carrier:

✦ **Better pricing:** Having two or more carriers allows you get better pricing plans for your monthly service charge. The carriers will want to be competitive in pricing and will provide special pricing.

✦ **Redundancy:** Carriers sometimes have outages. If all your devices are connected to one carrier, the whole network will go down. Having more than one carrier reduces the risk of having all of your users without BlackBerry service at the same time.

✦ **International access:** If your enterprise has international locations or if users must travel internationally, it's important to use at least two carriers and models in order to ensure the best international access.

Try to standardize your setup with one model that all carriers offer. We recommend a 3G device that supports UMTS so that your international users who fly to Japan will have devices that work.

---

## Getting the lowdown on the GSM Association

The GSM Association is a global consortium of mobile phone operators and companies. The association's Web site, `www.gsmworld.com`, is a great resource to find information about which countries have 3G capabilities for voice and data. It also helps you find out which carriers operate in a particular country. For more information, go to

```
http://gsmworld.com/
roaming/gsminfo
```
or visit the mobile version of the site at

```
http://gsmworld.mobi
```

---

# Fine-Tuning Devices Before They Go onto the Network

Carriers produce hundreds of BlackBerry smartphones that sit in inventory before someone purchases them. When you receive a BlackBerry smartphone, make it a standard operating procedure to update the smartphone to the latest handheld software version. Refer to Book IV, Chapter 6 for more information on upgrading BlackBerry handheld software.

Just as you need to tune up a car every so often, it doesn't hurt to make sure that users periodically bring their phones into your office to make sure that everything is still working in tip-top shape. Add a date to the document you have about each user's device (see the preceding section, "Getting Your Environment in Line" for more information) so that you can prompt a service call.

# Getting Your Hands on the BlackBerry Enterprise Server Resource Kit

The best applications for a BlackBerry Enterprise administrator are those that automate everyday tasks or make the whole enterprise a little less daunting. For example, apps can give more access to end users so that they can administer parts of their account.

Here are some of the areas you might consider creating or purchasing solutions for:

✦ Monitoring

✦ Device management

✦ Reporting

✦ Auditing

✦ Compliance

✦ Other enterprise-level maintenance

There are really three options for managing the BES with software. You can make one of your developers dazzle you with an in-house solution. You can buy an out-of-the box solution from a third-party vendor. Or you can hire a third-party vendor to create a customized solution. Each option has its pros and cons. We believe that most companies provide some combination of all three solutions depending on their budgets, how much time they have before they must roll out a solution, and the kind of talent they have inside the company. We definitely recommend that you tinker around, especially if you've got some programming mojo, before you purchase a solution.

The BlackBerry Enterprise Server Resource Kit (BRK) is a free resource kit that contains tools and components that you can use to develop your own custom applications to manage your company's BlackBerry infrastructure.

The BlackBerry Enterprise Server Resource Kit is updated frequently. When a new BES update crops up, you should check to see if the BlackBerry Enterprise Server Resource Kit has an update as well. You can download the latest BRK here:

```
http://na.blackberry.com/eng/support/server_resourcekit.jsp
```

By using the BlackBerry Server Resource Kit, you can automate the manual end-to-end process of adding and setting up a user or group of users on the BES. By using the right tools, you can ensure that each user is set up the same way every time in a batch process. Bottom line? Customized processes created with the BRK will save you and your support team time. You can even develop BRK-based tools that empower users or your help desk to reset device passwords, set enterprise activation passwords, and perform other low-level BlackBerry administration tasks. Of course, saving time also means saving money. The BRK is free, and the little time you spend customizing it will pay off for years to come.

## Grasping the power of the BlackBerry BRK

You can develop and create an endless number of applications with the BlackBerry Server Resource Kit. Almost every action that you can do in the BlackBerry Manager Console you can do using one of the components of the resource kit.

These actions are available via a command line or though an application programming interface (API). An API is a software interface that allows you to develop the new application. You can think of the API as a kind of guide to the look and feel of your BlackBerry app. Just as this book has certain conventions that organize the appearance and delivery of information, the API specifies the rules you need to follow.

If you're not a developer, you might ask why RIM would just lift up the hood and let you see everything that's inside. RIM, like Google, Apple, Microsoft, and just about every other techie organization, wants to encourage new and experienced developers to create great apps and services.

The API allows you to build tools for administrators and support staff, and even provide simple tools that allow the BlackBerry smartphone user to administer his account without having access to the BlackBerry Manager Console.

### Getting to know some of BRK's configuration tools

The BlackBerry Enterprise Server Resource Kit consists of the following components for managing configuration settings, which you can use as they are or incorporate into custom applications that fit your needs. In this list, we highlight the tools you're most likely to really need:

✦ **BlackBerry Application Reporting Tool:** This is a great tool for finding unapproved applications that are installed on BlackBerry devices in your enterprise. It lists the applications that are installed in the BlackBerry Domain. You can use it to audit the BlackBerry Domain and find applications.

✦ **BlackBerry Domain Administration History Reporting Tool:** This tool helps you determine a possible cause for an issue. You can see who and what was changed in the environment. You receive a report formatted in the .csv format.

✦ **BlackBerry System Log Monitoring and Reporting Tool:** This tool allows you to create a set of virtual real-time monitoring tools — for example, build tools that will provide you with early warnings if your BlackBerry server has a problem.

**Book VIII
Chapter 5**

**Tools for Maintaining a Healthy Enterprise Environment**

✦ **BlackBerry IT Policy Import and Export Tool:** It's common to tinker with various IT policies to get the right mix. With this tool, when you export your IT policies from the pilot (or testing) server to the production server, you can make sure that you don't forget to change all the IT policies.

### Getting to know some of BRK's log analysis tools

The following log analysis tools can help you track the flow of data on the BES. In this list, we tell you about the log tools you're most likely to need to keep the system running in tip-top shape:

✦ **BlackBerry Message Flow Reporting Tool:** This tool tracks the flow of e-mail messages from the mail server from end to end.

✦ **BlackBerry Thread Analyzer Tool:** This tool analyzes the logs for reports of nonresponsive and slow threads. *Threads* are processes that run on the BES and send data between the various BlackBerry Enterprise Server processes. If a thread *hangs*, or is running slowly, BlackBerry users may experience delays in receiving messages or other data, such as contacts, memos, or calendar appointments.

✦ **BlackBerry Delayed Notifications Monitoring Tool:** If the BlackBerry server isn't receiving messages in a timely manner, usually less than a minute, this tool will alert you.

✦ **BlackBerry Usage Monitoring Tool:** This tool identifies the users who have not sent or received messages for a specified period of time. This tool is great for identifying whether a user's BlackBerry is just sitting in her desk. If she's not using it, you may be able to disconnect it and save your company some money on the monthly fees.

✦ **BlackBerry Message Pending Delivery Tool:** This tool allows you to track all messages that are pending delivery on a BES. A high pending-message count may indicate a server issue. Even if the server is working fine, a high pending-message count for users whose BlackBerry smart-phones are shut off can impact server performance. That's because the BES has to rescan all of those messages periodically.

✦ **BlackBerry MAPI and CDO Error Monitoring Tool:** Use this tool to scan for common Messaging Application Programming Interface (MAPI) and Collaboration Data Object (CDO) errors and events. This is a very useful tool for tracking calendar synchronization issues and message delivery issues.

✦ **BlackBerry Enterprise Server Log Monitoring Tool:** This tool monitors the event log `.txt` file. You can customize the criteria and alerts. The tool lets you know if the criteria have been met. It's a great tool for quickly finding out about a specific issue as soon as it occurs, rather than waiting to hear about it from your users.

✦ **BlackBerry Message Receipt Confirmation Tool:** This tool provides you with virtual real-time verification that the BES is sending messages to BlackBerry smartphones. It's great for getting a quick determination about whether messages are being queued and delivered by your BES when you think there's an outage.

✦ **BlackBerry Enterprise Server User Administration Tool:** This tool allows you to perform user and BlackBerry smartphone administration on the command-line level. You can use it to automate existing tools and create new tools to administer user accounts on the BES.

# Finding BRK Inspiration: Example Applications

Creating your own tools using the BRK does require some basic programming skills, and we don't recommend going into the API and tinkering unless you feel pretty comfortable with software development.

That said, you can do some pretty simple things to configure and use the existing tools we list in the previous sections. Luckily, changing BRK tool configurations is as simple as pointing and clicking.

## Customizing installation and configuration

Say you've got three servers. You can use the BlackBerry Enterprise Server User Administration Tool to build a new application that will dynamically rebalance those servers based on your criteria. You can include any business rules you want. For example, you can set a rule that a group can be added to only one of the three servers. That way, members of the group would never be split among the servers.

If you were to build an application like this, you could also create a page where you could enter a user's name. By clicking the Add button, the user could quickly be added to the BES. The choice of server where the user is added would be based on the logic that you have built in to the application. An example of such logic would be that the user is added to the lowest-population server that was physically located in a specific region.

To add a user account to the BlackBerry Enterprise Server, follow these steps:

*1.* **At the command prompt, go to the folder that stores `BESUserAdmin Client.exe`.**

*2.* **Say you've got two servers you want to rebalance. Type** `besuser adminclient -add` **with the following parameters:**

```
-p <client password>
-u <username>
-PIN <PIN>
```

*Note:* Use the -PIN <PIN> parameter only in conjunction with the -u parameter for the BlackBerry Enterprise Server for MDS applications.

```
-b <BlackBerry Enterprise Server service name>
-sqluser <database username>
```

*Note:* Use the following parameter when you use database authentication.

```
-sqlpass <database password>
```

*Note:* Use the following parameter when you use database authentication.

```
-n <network address of BlackBerry Enterprise Server
    User Administration Tool service>
```

For more command lines for administration of the BlackBerry Enterprise Server, go to:

```
http://na.blackberry.com/eng/deliverables/3359/user_
    administration_guide.pdf
```

After you add a user to the appropriate server based on your criteria, the next step is to assign an IT policy. (For more information on IT policies, see Book VIII, Chapter 2.) At the same time that you add a user, you can set his or her IT policy, allocate a software configuration, and set an enterprise activation password. This customization significantly reduces the amount of work you have to do to manage and maintain the BES.

Depending on the process you follow to set up user accounts, you may have to perform a few additional tasks to complete the process. Here are some additional customizations you can add into the application:

✦ If you work for an organization that generates a help ticket for every new user account, you can automate this step into your setup process. A member of the IT support staff can automatically receive notification to follow up with the user after initial installation to make sure everything is working appropriately.

✦ If you want to be able to send users BlackBerry training material via e-mail, you can customize your application to do this. Say members of a specific group of users have just had new software pushed to their BlackBerry smartphones. You can send the appropriate group a ticket or an e-mail with instructions on configuring the new software.

Countless options are available that you can use to streamline installing and configuring user accounts and software for users.

## Personalizing support

Typically, only the BlackBerry administrator has access to the BlackBerry Manager Console. Your support staff, such as an in-house help desk or other on-site support may not have the tools needed to troubleshoot users' issues.

Who will have these tools? You. And only you. Which means that when the vice president of customer service at your company has a problem with her BlackBerry and calls your help desk, the support tech will have to say, "Sorry. I can't help you." Ay caramba! What's an administrator to do? Change this situation. Quickly.

Obviously, having a good response to a BlackBerry crisis is necessary even if the users experiencing problems aren't VIPs in your organization. However, nothing motivates change like the prospect of inconveniencing a VIP. Be prepared for the emergency call by using the BlackBerry Enterprise Server Resource Kit to give your help team the tools to respond.

You can easily provide support staff with a Web-based tool that can be used to see everything about a device. Some of the items that could be displayed include the type of BlackBerry device, the mobile number, its carrier, available memory, applications installed, the last time the BlackBerry was connected to the BES, and any pending messages that are being sent or received, as well as specific IT policies that apply to the device.

The help desk could then quickly access the documents needed to troubleshoot the problem and resolve it. This process can help reduce the number of questions to the user and the amount of time on the phone with the user.

A help desk Web tool usually has a Web page front end that is based on *single sign-on,* which is a programming method that checks the PC credentials of any user that visits the site. Single sign-on logs users in with those credentials instead of prompting them to do so, and can be set up to allow only specific users (such as your help desk) to access it. On this site, you should add a search field where the support staff can enter a user's display name, e-mail address, PC login name, and any other criteria you want to be able to search. When the staff enters criteria and clicks the Search button, the site queries the BlackBerry database and returns matching users. When a user is selected, it then loads stats that you specify for the user, such as BlackBerry model, last contact time, and pending messages.

You can also include links such as Reset Password, Add to BES, Set Enterprise Activation Password, or Resend Service Books. When the staff selects one of these links, background scripting submits the required information to the BES User Administration Tool, which then performs the job on the BES.

It's a good idea to send a confirmation e-mail or a pop-up message that lets the staff know whether an action was successful or failed. The link to this site can be provided to your help desk team, in its support documentation, and can also be added to your intranet BlackBerry site if you have one.

## Improving end users' experience

No matter how long people have had a device, they can always find a way to forget their password. These users either call the help desk to reset the password or enter their password incorrectly so many times that the BlackBerry locks up. Multiply that experience by a few hundred users, and you can imagine a scenario where your days are filled with password resets. Or perhaps you have already had the pleasure of living this dream of an experience?

If you want to avoid this life of misery, with the BlackBerry Enterprise Server Resource Kit you can eliminate the password problem. You can create a Web site which users can visit to reset their device passwords or generate a random enterprise activation password.

This type of site is usually based on single sign-on, a programming method that verifies a user's credentials and uses those credentials to log the user in. It's set up to allow a user to perform actions for his or her own BlackBerry only. This solution significantly reduces calls to the help desk and ultimately saves time for both you and the users.

# Buying Versus Building Tools

If you've been reading this chapter, you understand that you can use the BlackBerry Enterprise Server Resource Kit in a variety of ways to save time and money, as well as to reduce the number of support calls your department receives. However, to create these applications, you need at least one person who has the knowledge and skills to develop them, the resources to host these solutions, and the money to build them. Otherwise, you need to hire someone to do this work for you.

Your decision to build a solution from scratch or purchase a solution will largely be determined by your interest in creating a proprietary solution in-house.

Several companies provide solutions right out of the box. You might consider creating or purchasing solutions for one or more of the following areas:

+ Monitoring

+ Device management

+ Reporting

+ Auditing

+ Compliance

+ Other enterprise-level maintenance

## Monitoring, managing, and reporting

Nobody pays attention to the BlackBerry administrator as long as everything is running smoothly. But the second things go wrong, believe us, you'll be notified. Everyone will know your name, and you will feel the pressure to fix every problem *immediately.* This is why it is important that you have monitoring tools in place to be able to catch these problems early, before the users do.

When (not if) there's a problem, it's important to quickly generate a report so that you can quickly troubleshoot the cause and find out how widespread the problem is. From there, you can send targeted communications without alarming all the BlackBerry users in the enterprise. Several companies specialize in creating customized monitoring and reporting tools for every type of enterprise.

✦ **Microsoft System Center Operations Manager:** (www.microsoft.com/ systemcenter/en/us/operations-manager.aspx) Also called SCOM, this software solution can monitor, audit, report on, and provide troubleshooting steps for your BlackBerry environment, starting at the cost-effective price of $1,800 for an enterprise monitoring server and license.

- *Pro:* The benefit of this solution is that many enterprises already have it in-house to monitor other types of enterprise servers and services.

- *Con:* The downside is that no BlackBerry-specific monitoring exists out of the box. You either have to purchase a third-party BlackBerry add-on package or develop BlackBerry-specific monitoring yourself.

✦ **HP Operations Manager for BlackBerry Enterprise Server:** (https:// h10078.www1.hp.com/cda/hpms/display/main/hpms_content. jsp?zn=bto&cp=1-11-15-28^43278_4000_100__) Also called HPOM for BES, this solution provides monitoring, alerting, and diagnostics for your BlackBerry environment. It can automatically discover all aspects of your BlackBerry Enterprise environment, and it has a lot of BlackBerry-oriented monitoring and alerting options right out of the box.

- *Pro:* Because HPOM monitors a variety of server and service types, your enterprise may already have it in-house.

- *Con:* The downside is that pricing can vary depending on your current HPOM environment, and if you don't have HPOM already, you need to install it to use this solution. Also, although BlackBerry functionality is included, it requires some fine-tuning to fit the needs of your environment.

✦ **BoxTone Mobile Service Management:** (www.boxtone.com) BoxTone provides a personal portal for users to manage their own device issues, as well as great reporting options for showing trends, monitoring the health of the environment, and troubleshooting history. It also includes some amazing predefined alerts for common issues that cause administrators headaches, such as SRP disconnects, hung threads, high- pending messages per server, and high- pending messages per carrier.

- *Pro:* These alerts allow you to proactively resolve BlackBerry enterprise issues (in most cases before your users even realize there's an issue). BoxTone also includes the ability to monitor your enterprise messaging servers, because issues that affect them can also affect BlackBerry service.

- *Con:* The drawback to this solution is that it relies on BlackBerry Enterprise Server logs for most of its BlackBerry-oriented alerting. Although this tool helps you resolve issues more quickly, it means that issues are already occurring when you're alerted. The alerts also tend to be high-level, meaning that they give you an overview of the problem. You still have to do all the investigative work to find the details. The cost of this solution varies depending on the size and needs of your environment.

✦ **Zenprise MobileManager:** (`http://zenprise.com/products`) This tool proactively monitors your entire BlackBerry environment and, whenever possible, alerts you to potential problems before they occur. Zenprise is a very granular solution, monitoring Windows event logs and overall server and environment trends in addition to the BES log monitoring performed by most other solutions.

- *Pro:* This functionality helps you to proactively identify and resolve warnings before they become full-blown issues. The optional ability to remotely view and even control a BlackBerry device is very useful for troubleshooting. Zenprise MobileManager can even monitor your messaging server environment to identify issues outside of the BES that may cause problems with the BlackBerry service. It also includes user self-help options, as well as a help desk portal, which includes troubleshooting steps and links to RIM's own knowledge base articles.

- *Con:* The downside to this solution is that its predefined e-mail alerts can be difficult to decipher without checking them against the console to see what's really going on. The pricing of this solution varies depending on the size and needs of your environment.

✦ **StealthAUDIT for BlackBerry:** (`www.stealthbits.com/exchange/blackberry.html`) This solution can monitor and report on just about every aspect of your BlackBerry environment, from how long it takes a message to reach your users' BlackBerry smartphone to how much room is available on your BES database server to the load on your BES servers themselves. It also includes a ton of predefined reports, including reports that track how many BlackBerry devices you have on the BES, which carriers they're using, and which applications they're using. StealthAUDIT can also integrate with your Active Directory and Exchange environments to proactively report on external issues that may impact your BES environment.

- *Pro:* The great thing about StealthAUDIT is the fact that it's completely customizable, so if there's not a report for something you need, you can create your own custom queries, reports, and alerts.

- *Con:* Unfortunately, this solution doesn't include any proactive BlackBerry alerts out of the box, so you have to invest some time creating them. Also, the BlackBerry applet that pulls the BlackBerry data places a decent load on your BES servers and can be a challenge to set up, unless you work with STEALTHbits to tweak it to your needs.

This solution works only with Microsoft Exchange environments, and the cost varies depending on whether you use the Exchange and Active Directory modules and the number of BlackBerry smartphones in your environment.

## Auditing and compliance

Each industry has its own set of rules and regulations that it must comply with. These rules can be set internally or they may be enforced by a regulatory agency. In some cases, companies must comply with several sets of internal *and* external regulations. And not adhering to regulations isn't an option, because noncompliance results in some serious fines (*ka-ching!*) and other penalties. Lack of compliance can also really damage a company's reputation.

If your company needs to be in compliance, you probably already know that you need to bring in the big guns to help you out. Here are some companies that specialize in creating auditing and compliance solutions:

- ✦ **Quest Policy Authority:** (`www.quest.com/policy-authority-for-uc/preserve.aspx`) This software collects and parses SMS text, PIN-to-PIN, BlackBerry Messenger, and call logs. It then converts these messages to whatever format your company is using to archive e-mails and sends the messages to your archiving system.

  - *Pro:* The benefit of this solution is that it integrates with Quest Archive Manager, which many companies already use for their archiving needs. It also supports the formats required by many other popular archiving systems if your company doesn't use Quest.

  - *Con:* The drawback is that the archived logs can't be viewed without using your corporate archival software, which could be a problem if your company changes solutions down the road. Pricing varies depending on the size and needs of your organization, as well as whether you're using other Quest solutions.

✦ **StealthAUDIT:** (`www.stealthbits.com/exchange/blackberry.html`) This solution archives, parses, and reports on SMS text, call logs, PIN-to-PIN, and BlackBerry Messenger logs captured by your BES. See Figure 5-1.

- *Pro:* It displays the results in an easily readable format and can even be configured to automatically copy daily reports to a specified location. Or it can send e-mail reports to your compliance department. StealthAUDIT also can audit and report on your Exchange Active Directory environments.

- *Con:* The only major drawback to this solution is that it relies on the native BES SMS, PIN, and BlackBerry Messenger logs, so you have to make sure your BES is properly configured to capture these types of messages. It also takes a little bit of configuration to ensure StealthAUDIT properly captures and sends the data.

The cost of this solution varies depending on which other modules you use and the number of BlackBerry smartphones in your environment.

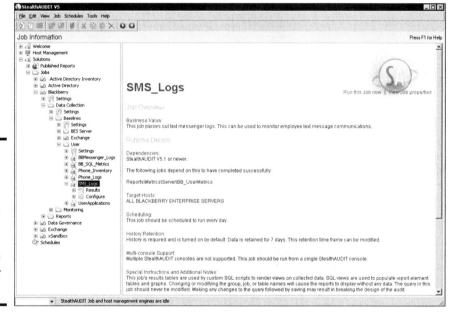

**Figure 5-1:** Stealth AUDIT captures and parses SMSs, PINs, and BlackBerry Messenger logs.

When you have a stable BlackBerry environment, you can consider extending your enterprise tools. Here are a few companies that provide a wide range of tools that can be very helpful in support, training, and other company-related items:

✦ **Chalk Pushcast Software:** (`www.chalk.com/home.aspx`) Allows you to create and send interactive training presentations and important communications that include audio, video, and text to your BlackBerry users. See Figure 5-2.

- *Pro:* You can set reminders for content that has to be completed or read and even track who has or hasn't viewed the content that was sent.

- *Con:* The drawback to this solution is that it can be very data intensive, which may increase BlackBerry costs if you don't have an unlimited data package.

The cost varies depending on the size of your environment.

**Figure 5-2:** Chalk lists items for users to read, watch, listen and take.

✦ **GPSX Mobile Service Manager:** (`www.lifecyclemobility.com/msm.html`) Automates the assignment of BlackBerry IT policies and the deployment of BlackBerry applications.

- *Pro:* If you're a BlackBerry administrator, you're aware that the headache of managing IT policies and applications is almost as bad as reporting on who has what permissions, rights, and software installed. With GPSX, you define rules that determine who gets what policy and/or what applications, and then you can sit back and let it handle the rest.

- *Con:* The only downside is the time and effort required for the initial setup.

Like most solutions, the pricing varies based on your needs and the size of your BlackBerry environment.

✦ **Rove Mobile Admin:** (`www.roveit.com`) Allows you to manage your enterprise environment right from your BlackBerry. See Figure 5-3.

- *Pro:* You can connect to Active Directory, printers, and even servers like the BES to perform administrative tasks — for example, password resets — all while away from your computer.

- *Con:* The drawback to this product is that it requires a back-end server setup in order to work. It can be difficult to use, given the difference between your BlackBerry screen size and a PC's screen size, and can sometimes lag or time out due to your BlackBerry's data connection.

**Figure 5-3:**
Mobile
Admin
Server
Home
screen.

✦ **WIC Messenger and WIC Responder:** (`www.wallacewireless.com/ Solutionsnbspnbsp/BCPCOOP/tabid/121/Default.aspx`) These two options extend business continuity functionality by constantly updating emergency contact information and procedures.

- *Pro:* Mass e-mail, PIN and/or SMS messages can be sent and confirmed delivered during network or system outages. Emergency plans, checklists, floor plans, and maps can be automatically pushed to and stored on BlackBerry devices so they're always available in case of an emergency. WIC Responder compresses documents to significantly smaller files sizes so that documents can be stored on the device without impacting day-to-day functions.

- *Con:* The only drawback to these solutions is the time and effort required to set them up and maintain the accuracy of the data they provide.

Pricing of these solutions is based on the size and needs of your environment.

✦ **Pyxis Mobile Application Studio:** (`http://pyxismobile.com`) Allows you to build a mobile application once and deploy it across multiple mobile device platforms (BlackBerry, iPhone, Windows Mobile, and Android) at the same time. You don't have to build a separate version for each platform! It uses a drag-and-drop interface for basic programs, or uses back-end programming for more advanced developers.

- *Pro:* One of the great things about this solution is the fact that you don't have to push a new version of a Pyxis application via the BES each time it's updated. That's because Pyxis handles all that for you, automatically.

- *Con:* The downside is that there's a Pyxis Mobile client that you have to deploy and maintain on all BlackBerry devices to which you want to send Pyxis applications.

The pricing also scales in relation to the number of devices in your environment and your application needs.

TIP

---

# Other BlackBerry consulting options

The following companies also provide consulting options to help you set up, maintain, monitor, and troubleshoot BlackBerry on the enterprise:

✔ **Research In Motion's Professional Services:** (`http://na.blackberry.com/eng/solutions/industry/professional`) RIM has trained consultants who work with either your company's management team or its IT staff (or both — your choice). RIM consultants stress security and efficiency, and their solutions are based on the assessments they make of your corporate setup and needs, as well as its existing infrastructure. RIM consultants can train on best practices and compliance, as well as maintenance, monitoring, and rights. You can also hire RIM consultants as needs arise. For example, if you're considering an upgrade, you can hire RIM Professional Services to do an upgrade assessment. If you're interested in BlackBerry Application Services, a RIM consultant can help you identify the existing BlackBerry applications that can resolve your business challenges.

✔ **Miles Consulting Corp:** (`www.milesconsultingcorp.com`) Consultants at Miles Consulting Corp are trained to assist companies with BES software installation and configuration. They develop applications, provide technical support, and help you diagnose problems. If you don't currently have BlackBerry devices deployed across your enterprise, these consultants can help you decide who needs a BlackBerry in your company, what access rights and applications they need, and even walk you through finding the right models and carriers for your enterprise needs. They also provide a detailed threat analysis to help you evaluate and remedy security risks.

---

# Chapter 6: Looking at Features in BES 5.0

## In This Chapter

✓ New device management and server availability features

✓ Enhancements to existing device usability and application deployment features

**W**hen you make cutting-edge technology, you always have to remain cutting edge — otherwise, your technology will be obsolete in a snap. Remaining on the cutting edge means improving and enhancing existing technology as well as continuing to introduce innovative technology. Never was this axiom more true than in the mobile phone industry, which has experienced nothing short of a revolution in the last ten years. Research In Motion has made significant advancements with each major BlackBerry Enterprise Server release. Witness the revolution:

✦ Wireless PIM sync in BES

✦ Wireless device activation

✦ Wireless application pushing

✦ Push Now application deployment

RIM continues its tradition of improvement and enhancement with the release of BlackBerry Enterprise Server 5.0. In this chapter, we tell you about a host of new features that BES 5.0 promises, including Web-based BlackBerry Administration Service, wireless handheld software upgrades, and *high availability*. (That's tech-speak for making sure the system is always available.) BES 5.0 has included enhancements to existing features, such as improved application deployment and more flexibility in assigning roles and permissions.

This chapter is written mostly for BlackBerry administrators. However, any techies who want to stay in- the- know may find the info here useful.

## Finding Out What's So New and Improved

BES 5.0 includes a host of new features that give BlackBerry administrators more stability, flexibility, and control than are in previous BES versions. The following sections offer a sneak peek at what you get.

## Getting anywhere, anytime access to the BAS

*BAS*, short for BlackBerry Administration Service, is Web-based. It really is as simple as that explanation. You don't need to connect remotely to a BES or install a desktop client on your PC to administrate BlackBerry users. This innovation in BES 5.0 gives you the freedom to perform BlackBerry-related tasks from *any* computer, not just one that has the BES console installed. Just log in to your BlackBerry account and you have all the BES features at your fingertips. Figure 6-1 gives you a glimpse of the BAS console.

 Are you a developer? You can find an API to write your own plugins for the BlackBerry Administration Service. Bottom line? If you want to create your own BES administrator tool, you should be able to do so pretty easily.

**Figure 6-1:** The Web-based BlackBerry Administration Service.

## Getting ahead of the game on wireless software updates

If you've been doing this BlackBerry administration thing for a while, you'll be happy to know that BES 5.0 finally gives you the long-awaited ability to search for updates for your users' BlackBerry handheld software directly through the BES. In previous installations of the BES, you had to keep checking carrier sites or RIM's BlackBerry downloads page manually to see if new software is available. Now you can just do it through the BES. Even better, you can download and wirelessly push these handheld upgrades to individual users or groups. You can also schedule upgrades in advance to avoid impacting user productivity.

 Over-the-air (OTA) handheld software upgrades are no small matter because they are traditionally one of the most time-consuming and resource-intensive responsibilities for a BlackBerry administrator. In the past, you probably had to schedule physical upgrade appointments with each user, which required you both to stop the other stuff you were doing (including *using* the

BlackBerry) in order to get the upgrade done. With BES 5.0, you can have a user's BlackBerry update on its own at any time you specify — like the middle of the night or on a weekend, when the user is unlikely to be using it.

## Reclaiming your nights and weekends back with advanced scheduling of IT tasks

With BES 5.0, you can schedule additional IT-related tasks ahead of time. Many of the BlackBerry activities and troubleshooting tasks that used to require your assistance can be scheduled and performed remotely, which reduces or eliminates the need for you to perform maintenance or updates during off-hours.

Does this mean you might get your life back? Perhaps. You can schedule to push out a new application or service, or upgrade an existing application, on a Sunday afternoon so that it's available to users on Monday morning. This is extremely helpful for environments that support devices globally and need to adjust for local time zones. Your spouse will be thrilled!

## Getting high availability

High availability, or near-constant access to the functions, tools, and software you need, has been offered by various third-party vendors for some time. With BES 5.0, high availability is enabled out of the box.

High availability (HA) is designed so that BlackBerry service has minimum downtime if any part of the environment fails. It works by installing multiple BES, BAS, and SQL instances, all on different computers. If a primary instance becomes unavailable, the secondary takes over immediately. It's basically a built-in system redundancy so that if a failure occurs, the whole system doesn't crash.

We highly recommend reading the BES Planning Guide for a thorough look at how the technology works. In the meantime, the following are some of the more common high availability options.

You can go here to look at the BlackBerry Enterprise Server Planning Guide:

```
http://docs.blackberry.com/en/admin/deliverables/12057/
    BlackBerry_Enterprise_Server-Planning_Guide--820441-
    1029090714-001-5.0.1-US.pdf
```

### BlackBerry Enterprise Server

If you set up high availability in your environment, each BES instance is installed on two systems. The secondary BES periodically connects to the primary BES to perform customizable health checks. *Health checks* are maintenance checks that look for signs that the system is functioning properly.

(See the BES Planning Guide for more information about how they work and what they look for.) If the primary BES fails one of these health checks, the secondary BES tries to raise its status to be the primary server, demote the failing server to be the secondary server, and take over messaging responsibilities. It can also be configured to alert the BlackBerry administrator if it takes over.

It's fine to install all of the BES components on two separate servers for smaller enterprise environments. However, if you're managing a larger environment, you can customize the installation even more by installing multiple separate services on each server machine.

### BlackBerry Administration Service

You have these two options for high availability for the BlackBerry Administration Service:

✦ **Using a hardware network load balancer:** A *load balancer* is hardware attached to your network that points one URL to all of your BAS servers. When a login request comes in, the load balancer checks the availability of each BAS server, as well as how much load is placed on it by the administrative tasks it's already processing. From there, the load balancer automatically directs the request to the server with the lightest load. If one BAS server or service is down, the load balancer automatically redirects requests to any other server that is available.

✦ **Using DNS round robin:** *Round robin* is a DNS setup process that points one URL to a list of BAS servers. The first login request that comes in is directed to the first BAS server in the list, the second request goes to the second BAS server, and so on — in a round-robin fashion. When the end of the BAS server list is reached, the next request goes to the first server in the list again. If the next BAS server in the list isn't responding, the request is directed to the next BAS server automatically. This method isn't as efficient for load balancing but works well for high availability.

You're prompted to enter the HA DNS pool name for the BlackBerry Administration Service during setup. This flexibility allows you to use a different DNS pool for each installation. From there, you can give your support staff two links to use. If the first link doesn't work, they know the primary BAS service has a problem and they can alert you to then use the second link.

### BlackBerry Attachment Service

High availability for BlackBerry Attachment Service is a little different than it is with other components of the BES because BlackBerry Attachment Service has built-in load balancing, as well.

Instead of one attachment server doing all the work with another standing by just in case it fails, you can split the attachment server load by creating an attachment server *pool*. For each BES instance, you can set up one or more attachment server pools, each with a primary group of two or more attachment servers and an optional secondary group of two or more attachment server instances.

You can also set each attachment server instance to process specific types of attachments. If the primary attachment server group can't process an attachment, it passes the job to the secondary group for handling.

### BlackBerry Configuration Database

With BES 5.0, you can set up high availability for the BlackBerry Configuration Database by using database mirroring with Microsoft SQL Server 2005 SP2 or 2008. In this situation, the primary configuration database mirrors all its information to a secondary location. If the primary database goes down, the BES automatically tries to connect to the mirrored configuration database.

You can also set up HA for these BES services:

✦ BlackBerry Collaboration Service

✦ BlackBerry MDS Connection Service

✦ BlackBerry MDS Integration Service

✦ BlackBerry Monitoring (manual only)

✦ BlackBerry Router

The BES Planning Guide is a great resource for information on high availability for these services.

## Getting blink-and-you-miss-it migration tools

The BlackBerry Enterprise Transporter Tool makes it easy for you to migrate to BES 5.0. The migration is also completely transparent for users on the BES. You can transfer multiple users from one BES domain to another without requiring any action from end users or affecting their BlackBerry smartphones. That's right — they can keep on talking, e-mailing, and using their applications.

The Blackberry Enterprise Transporter Tool makes it possible to thoroughly test your BES 5.0 migration before fully committing to it. You can run parallel BES domains and perform testing in the new BES 5.0 environment. For example, you could assign some of your support staff, or a group of test users, to use BES 5.0 — and work out all the kinks — *before* performing a full migration. You don't need to wipe devices before transferring to BES 5.0, and you can move back to the pre-existing environment with no problems.

## Getting a sense of your network threshold before it's too late

The Threshold Assistance Tool recommends how many users should be on one BES. How does it do it? Well, just as a fortune teller uses information and cues that you provide to tell you what you want to hear, the Threshold Assistance Tool uses information stored in your monitoring database to suggest *operating thresholds* (that is, limits) to your configuration. These recommendations are based on your server and network environment.

Of course, the big difference between the Threshold Assistance Tool and a fortune teller is that sometimes the Threshold Assistance Tool tells you things you don't want to hear. And that's exactly why it's so useful, particularly if your enterprise is likely to grow or is growing already. As the enterprise gets bigger, you have to manage mobile resources accordingly. You can run the tool again to determine new thresholds based on the growing number of users and how they use the mobile technology and applications.

## Giving your users the Windows network access they need

Users of BES 5.0 who also have device software 5.0 can browse and download both local and remote files (shown in Figure 6-2). See Book I, Chapter 6 for more on BlackBerry Device Software 5.0.

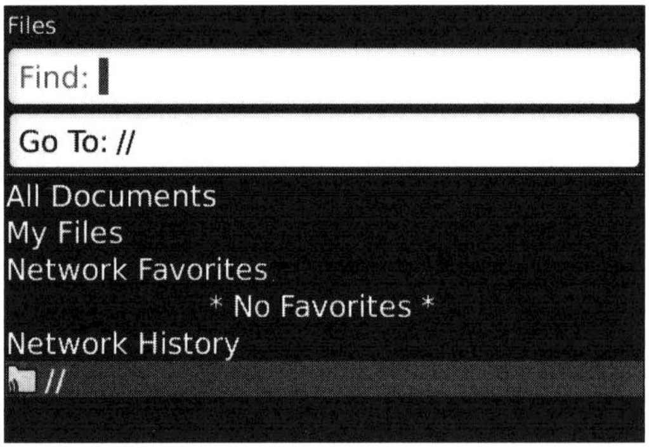

**Figure 6-2:** With BES 5.0 and Device Software 5.0, users can access network files.

> Files
> Find: |
> Go To: //
> All Documents
> My Files
> Network Favorites
>               * No Favorites *
> Network History
> 📁 //

With remote file access, users can view, save, edit, and e-mail files stored on the corporate intranet. Unfortunately, BES 5.0 doesn't allow users to upload files to the corporate network from the BlackBerry smartphone. (Here's hoping that's coming soon!)

Because this is a new feature, your users may not be aware of it. We recommend that you make all users aware of this useful capability. They need to understand that they can't upload changes to the files they access via the network, and that they can access only shares that they have permission to access. Then you should make sure they know the three easy steps they need to follow to access network files:

*1.* **From the BlackBerry Home screen, select Applications.**

*2.* **Select the Files icon.**

*3.* **Next to GoTo://, type the location of the network share.**

If you're the BES administrator, your network administrator needs to customize users' access to network share folders. Make sure the network share rights match each user's job grade.

If users have already accessed the network share folder, they can scroll down to Network Favorites or Network History and select the network share.

Users can access only the same network shares they have access to on their PC. These share permissions are set up by network administrators, not BES administrators.

## *Getting more customized features for BlackBerry Mail*

BES 5.0 made small, smart changes to BlackBerry Mail. With BES 5.0, users with device software 5.0 can create follow-up flags and flag e-mails for follow up. These follow-up flags even synchronize with their desktop PC's e-mail program and work across all corporate e-mail platforms. Previous versions of the BES and handheld software allow users only to mark messages as read or unread.

Inform your users of the steps for flagging messages:

*1.* **From the BlackBerry Home screen, select the Messages icon.**

*2.* **Highlight the message you would like to flag.**

*3.* **Press the Menu button.**

*4.* **Select the Flag for Follow Up option.**

Once a message has been flagged, users can follow a few more steps to define behavior for the flagged message by editing the flag properties. *Flag properties* are settings that define how a message reacts once it has been flagged.

*1.* **Highlight the flagged message and press the Menu button.**

*2.* **Select the Flag Properties option from the menu that appears.**

*3.* **Choose the desired flag properties.**

- *Request:* Specifies what type of flag a user is setting. Users can choose from a list of predefined suggestions or select Custom to create their own. These flags don't force an action; they're just written reminders of what you wanted to do with a message.

- *Color:* Specifies what color flag shows on the message.

- *Status:* Specifies whether a specified flag action has been completed. Changing the status to Completed replaces the flag with a Due check mark and time stamp of when the action was completed.

- *Due:* Specifies a date and time that the flag action should be completed. A reminder appears when the due date and time are approaching.

4. **Press the Menu button and choose Save to save your preferences.**

Users can mark a flagged message complete using these steps:

1. **From the BlackBerry Home screen, select the Messages icon.**

2. **Highlight the flagged message and press the Menu button.**

3. **Select the Flag Properties option from the menu that appears.**

4. **Change the status from Not Completed to Completed.**

5. **Press the Menu button and select Save.**

Figure 6-3 shows an example of the Flag Properties screen.

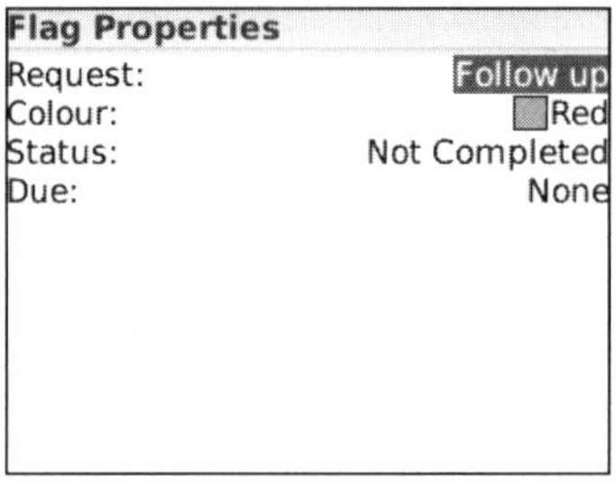

**Figure 6-3:**
Users
can flag
e-mails for
follow up.

```
Flag Properties
Request:              Follow up
Colour:                    ▇Red
Status:            Not Completed
Due:                        None
```

# Introducing Old Features That Have Learned New Tricks

In addition to the new features introduced in BES 5.0, RIM has made some major improvements to some of the features that administrators have grown to love in previous BES versions. In the following sections, we tell you what you get.

## Getting improved enterprise application management

The release of BlackBerry App World inspired RIM to enhance its BES application management capabilities. BES 5.0 allows administrators to assign a variety of application permissions to individuals or groups of users. Here are some examples of how these application management tools can be used:

+ An administrator can restrict GPS location information for all users — except, for instance, your chief information officer.

+ An administrator can apply stricter policies to some users' devices to block application access to features like phone information.

+ Travelling employees can be given full permissions to TeleNav Navigator, but an administrator can disable it for office employees who rarely travel.

BES 5.0 allows administrators to create a white list of applications. (In our opinion, this development is long overdue.) Admins can set default permissions for unknown applications.

In addition, the BlackBerry Administration Service downloads the latest `device.xml` file straight from RIM on a nightly basis, so you don't have to worry about updating to the latest version to ensure that your BlackBerry smartphone users receive their applications.

 One more improvement to application deployment is that you no longer need to run the `loader.exe /index` command after adding third-party application files to the application repository. Instead, this process runs automatically when the BlackBerry Administration Service notices changes in the application repository folders.

## Getting fully customizable roles and permissions

BES 5.0 gives administrators greatly improved control over administrative and user roles and permissions. Here's a little taste of what you can do:

+ **With broad brushstrokes:** You can create roles that restrict access to a specific group of BlackBerry users. For example, you can create a role that lets a support technician administrate the sales team, but prevents that same technician from making changes to the device of the CEO.

+ **With a fine point:** You can create IT policies that restrict access to certain resources, and then apply these policies to individuals or groups. For example, if your company is in the middle of a merger, you may want to restrict access to some shared resources to one side of the company, while allowing employees on both sides to access others.

### Getting improved wireless contact synchronization

Finally! BES 5.0 allows users with Device Software 5.0 to wirelessly synchronize with multiple contact folders, personal distribution lists, and contacts stored in public/shared folders on your enterprise network.

In previous BES and handheld software versions, users could sync wirelessly with only the contacts located in their own mailbox. Users were forced to use the Desktop Manager to sync with other types of contacts. Those days no longer exist!

For you BES admins, personal distribution list contacts automatically synchronize for users with BES 5.0 and BlackBerry Device Software 5.0, without any additional configuration on your part. Unfortunately, public folder contact synchronization does require some setup on your end. Luckily, RIM has a *KB article,* which is RIM's proprietary jargon for a support document, which describes exactly how to set up public folder contact synchronization:

```
www.blackberry.com/btsc/microsites/search.do?cmd=displayKC&
    docType=kc&externalID=KB19111
```

# Index

# p

## ...ple & Macs

...ad For Dummies
...8-0-470-58027-1

...hone For Dummies,
...h Edition
...8-0-470-87870-5

...acBook For Dummies, 3rd
...ition
...8-0-470-76918-8

...ac OS X Snow Leopard For
...ummies
...8-0-470-43543-4

## ...siness

...ookkeeping For Dummies
...8-0-7645-9848-7

...b Interviews
...r Dummies,
...d Edition
...8-0-470-17748-8

...esumes For Dummies,
...h Edition
...8-0-470-08037-5

...arting an
...nline Business
...r Dummies,
...h Edition
...8-0-470-60210-2

...ock Investing
...r Dummies,
...d Edition
...8-0-470-40114-9

...uccessful
...me Management
...or Dummies
...8-0-470-29034-7

## Computer Hardware

BlackBerry
For Dummies,
4th Edition
978-0-470-60700-8

Computers For Seniors
For Dummies,
2nd Edition
978-0-470-53483-0

PCs For Dummies,
Windows
7 Edition
978-0-470-46542-4

Laptops For Dummies,
4th Edition
978-0-470-57829-2

## Cooking & Entertaining

Cooking Basics
For Dummies,
3rd Edition
978-0-7645-7206-7

Wine For Dummies,
4th Edition
978-0-470-04579-4

## Diet & Nutrition

Dieting For Dummies,
2nd Edition
978-0-7645-4149-0

Nutrition For Dummies,
4th Edition
978-0-471-79868-2

Weight Training
For Dummies,
3rd Edition
978-0-471-76845-6

## Digital Photography

Digital SLR Cameras &
Photography For Dummies,
3rd Edition
978-0-470-46606-3

Photoshop Elements 8
For Dummies
978-0-470-52967-6

## Gardening

Gardening Basics
For Dummies
978-0-470-03749-2

Organic Gardening
For Dummies,
2nd Edition
978-0-470-43067-5

## Green/Sustainable

Raising Chickens
For Dummies
978-0-470-46544-8

Green Cleaning
For Dummies
978-0-470-39106-8

## Health

Diabetes For Dummies,
3rd Edition
978-0-470-27086-8

Food Allergies
For Dummies
978-0-470-09584-3

Living Gluten-Free
For Dummies,
2nd Edition
978-0-470-58589-4

## Hobbies/General

Chess For Dummies,
2nd Edition
978-0-7645-8404-6

Drawing
Cartoons & Comics
For Dummies
978-0-470-42683-8

Knitting For Dummies,
2nd Edition
978-0-470-28747-7

Organizing
For Dummies
978-0-7645-5300-4

Su Doku For Dummies
978-0-470-01892-7

## Home Improvement

Home Maintenance
For Dummies,
2nd Edition
978-0-470-43063-7

Home Theater
For Dummies,
3rd Edition
978-0-470-41189-6

Living the
Country Lifestyle
All-in-One
For Dummies
978-0-470-43061-3

Solar Power Your Home
For Dummies,
2nd Edition
978-0-470-59678-4

## Internet

Blogging For Dummies,
3rd Edition
978-0-470-61996-4

eBay For Dummies,
6th Edition
978-0-470-49741-8

Facebook For Dummies,
3rd Edition
978-0-470-87804-0

Web Marketing
For Dummies,
2nd Edition
978-0-470-37181-7

WordPress
For Dummies,
3rd Edition
978-0-470-59274-8

## Language & Foreign Language

French For Dummies
978-0-7645-5193-2

Italian Phrases
For Dummies
978-0-7645-7203-6

Spanish For Dummies,
2nd Edition
978-0-470-87855-2

Spanish
For Dummies,
Audio Set
978-0-470-09585-0

## Math & Science

Algebra I
For Dummies,
2nd Edition
978-0-470-55964-2

Biology For Dummies,
2nd Edition
978-0-470-59875-7

Calculus For Dummies
978-0-7645-2498-1

Chemistry For Dummies
978-0-7645-5430-8

## Microsoft Office

Excel 2010 For Dummies
978-0-470-48953-6

Office 2010 All-in-One
For Dummies
978-0-470-49748-7

Office 2010 For Dummies,
Book + DVD Bundle
978-0-470-62698-6

Word 2010 For Dummies
978-0-470-48772-3

## Music

Guitar For Dummies,
2nd Edition
978-0-7645-9904-0

iPod & iTunes For
Dummies, 8th Edition
978-0-470-87871-2

Piano Exercises
For Dummies
978-0-470-38765-8

## Parenting & Education

Parenting For Dummies,
2nd Edition
978-0-7645-5418-6

Type 1 Diabetes
For Dummies
978-0-470-17811-9

## Pets

Cats For Dummies,
2nd Edition
978-0-7645-5275-5

Dog Training For Dummies,
3rd Edition
978-0-470-60029-0

Puppies For Dummies,
2nd Edition
978-0-470-03717-1

## Religion & Inspiration

The Bible For Dummies
978-0-7645-5296-0

Catholicism For Dummies
978-0-7645-5391-2

Women in the Bible
For Dummies
978-0-7645-8475-6

## Self-Help & Relationship

Anger Management
For Dummies
978-0-470-03715-7

Overcoming Anxiety
For Dummies,
2nd Edition
978-0-470-57441-6

## Sports

Baseball
For Dummies,
3rd Edition
978-0-7645-7537-2

Basketball
For Dummies,
2nd Edition
978-0-7645-5248-9

Golf For Dummies,
3rd Edition
978-0-471-76871-5

## Web Development

Web Design
All-in-One
For Dummies
978-0-470-41796-6

Web Sites
Do-It-Yourself
For Dummies,
2nd Edition
978-0-470-56520-9

## Windows 7

Windows 7
For Dummies
978-0-470-49743-2

Windows 7
For Dummies,
Book + DVD Bundle
978-0-470-52398-8

Windows 7 All-in-One
For Dummies
978-0-470-48763-1

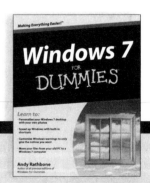

Available wherever books are sold. For more information or to order direct: U.S. customers visit www.dummies.com or call 1-877-762-29
U.K. customers visit www.wileyeurope.com or call (0) 1243 843291. Canadian customers visit www.wiley.ca or call 1-800-567-4797.

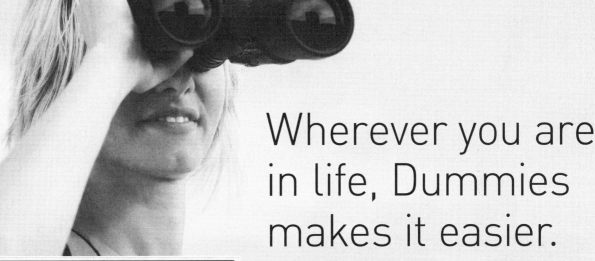

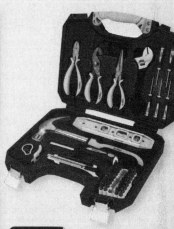